Nanomaterials for Sustainable Energy Applications

This book provides a detailed overview of different devices and nanomaterials for energy storage applications. The application of each nanomaterial is discussed for fuel cells, metal–air batteries, supercapacitors, solar cells, regenerative fuel cells, hydrogen energy, batteries, and redox flow batteries to understand the reaction process and material performance improvement for energy storage devices. In addition, major challenges, case studies, historical, and future perspective are summarized.

Features:

- Summarizes state-of-the-art nanomaterials for energy storage and conversion applications
- Comprehensive coverage of a wide range of nanomaterials, including synthesis and characterization
- Details different energy storage devices, construction, working principles, and major challenges
- Covers specific reactions, nanomaterials, and nanocomposites via audio–video slides/short films
- Includes case studies pertaining to development of energy storage devices and major challenges

This book is aimed at researchers and graduate students in chemical engineering, chemical sciences, nanomaterials, and energy engineering/conversion.

Emerging Materials and Technologies
Series Editor: Boris I. Kharissov

The *Emerging Materials and Technologies* series is devoted to highlighting publications centered on emerging advanced materials and novel technologies. Attention is paid to those newly discovered or applied materials with potential to solve pressing societal problems and improve quality of life, corresponding to environmental protection, medicine, communications, energy, transportation, advanced manufacturing, and related areas.

The series takes into account that, under present strong demands for energy, material, and cost savings, as well as heavy contamination problems and worldwide pandemic conditions, the area of emerging materials and related scalable technologies is a highly interdisciplinary field, with the need for researchers, professionals, and academics across the spectrum of engineering and technological disciplines. The main objective of this book series is to attract more attention to these materials and technologies and invite conversation among the international R&D community.

Nanocosmetics
Drug Delivery Approaches, Applications and Regulatory Aspects
Edited by: Prashant Kesharwani and Sunil Kumar Dubey

Sustainability of Green and Eco-friendly Composites
Edited by Sumit Gupta, Vijay Chaudhary and Pallav Gupta

Assessment of Polymeric Materials for Biomedical Applications
Edited by Vijay Chaudhary, Sumit Gupta, Pallav Gupta and Partha Pratim Das

Nanomaterials for Sustainable Energy Applications
Edited by Piyush Kumar Sonkar and Vellaichamy Ganesan

Materials Science to Combat COVID-19
Edited by Neeraj Dwivedi and Avanish Kumar Srivastava

Two-Dimensional Nanomaterials for Fire-Safe Polymers
Yuan Hu and Xin Wang

For more information about this series, please visit: www.routledge.com/ Emerging-Materials-and-Technologies/book-series/CRCEMT

Nanomaterials for Sustainable Energy Applications

Edited by Piyush Kumar Sonkar and
Vellaichamy Ganesan

CRC Press
Taylor & Francis Group
Boca Raton London New York

CRC Press is an imprint of the
Taylor & Francis Group, an **informa** business

Designed cover image: © Piyush Kumar Sonkar and Vellaichamy Ganesan

First edition published 2024
by CRC Press
6000 Broken Sound Parkway NW, Suite 300, Boca Raton, FL 33487–2742

and by CRC Press
4 Park Square, Milton Park, Abingdon, Oxon, OX14 4RN

CRC Press is an imprint of Taylor & Francis Group, LLC

© 2024 selection and editorial matter, Piyush Kumar Sonkar and Vellaichamy
Ganesan; individual chapters, the contributors

ISBN: 978-1-032-07766-6 (hbk)
ISBN: 978-1-032-07768-0 (pbk)
ISBN: 978-1-003-20870-9 (ebk)

DOI: 10.1201/9781003208709

Typeset in Times
by Apex CoVantage, LLC

Dedicated to our
Authors, Collaborators and the Scientific team
& A heartfelt tribute to
Late Dr. M. Anbu Kulandainathan
(One of the authors of this book)
**Principal Scientist, Central Electrochemical
Research Institute Karaikudi, India**

Contents

About the Editors

Dr. Piyush Kumar Sonkar has received his BSc (Chemistry), MSc (Chemistry) and PhD (Chemistry) degree from Banaras Hindu University, India. He is presently working as an assistant professor in Department of Chemistry, MMV, Banaras Hindu University, Varanasi, India. His research interests include nanomaterials, nanocomposites, fuel cells, electrochemical devices, supercapacitors, biosensors, chemical sensors and new materials. He has published more than 45 international and national research papers in various reputed peer reviewed journals. He has published six book chapters. He presented his research work in various international/national seminars/workshops and conferences.

Dr. Vellaichamy Ganesan received his BSc and MSc in Chemistry from Madurai Kamaraj University and Bharathidasan University, India, respectively. Later, he received his PhD (2000) from Madurai Kamaraj University. After completing postdoctoral research in the United States and France, he is now continuing his research in the fields of nanomaterials, electroanalysis, electrocatalysis, development of fuel cell materials, and photocatalysis in the Department of Chemistry, Banaras Hindu University, India as a professor. He is a recipient of IUSSTF fellowship (India–USA), Commonwealth Fellowship (UK), and DAAD Fellowship (Germany). He has published more than 110 research papers in international/national journals.

Contributors

Uday Pratap Azad
Department of Chemistry, Guru
Ghasidas Vishwavidyalaya, Bilaspur
495009, India

Bhaskar Bhattacharya
Department of Physics, Mahila
Mahavidyalaya, Banaras Hindu
University, Varanasi—221005, U.P.,
India

rer. nat. Somenath Garai
Department of Chemistry, Institute of
Science, Banaras Hindu University,
Varanasi 221 005, Uttar Pradesh,
India

Arindam Indra
Department of Chemistry, IIT (BHU),
Varanasi, UP-221005, India

Murugavel Kathiresan
Electro Organic and Materials
Electrochemistry Division, CSIR-
Central Electrochemical Research
Institute, Karaikudi, TamilNadu,
INDIA

Kyuwon Kim
Electrochemistry Laboratory for
Sensors & Energy (ELSE),
Department of Chemistry, Incheon
National University, Incheon 22012,
Republic of Korea

Biplab Kumar Kuila
Department of Chemistry, Institute of
Science, Banaras Hindu University,
Varanasi-221005, Uttar Pradesh, India

M. Anbu Kulandainathan
Electro Organic and Materials
Electrochemistry Division, CSIR-
Central Electrochemical Research
Institute, Karaikudi, TamilNadu,
INDIA

Dasnur Nanjappa Madhusudan
Department of Chemistry, Institute of
Science, Banaras Hindu University,
Varanasi 221 005, Uttar Pradesh,
India

Shanmugam Manivannan
Department of Chemistry, Institute of
Science, Banaras Hindu University,
Varanasi-221005, Uttar Pradesh,
INDIA

Electrochemistry Laboratory for
Sensors & Energy (ELSE),
Department of Chemistry,
Incheon National University,
Incheon 22012, Republic
of Korea

Ashish Kumar Mishra
School of Materials Science and
Technology Indian Institute
of Technology (BHU),
Varanasi-221005,
INDIA

Shanu Mishra
School of Materials Science and
Technology Indian Institute
of Technology (BHU),
Varanasi-221005,
INDIA

Shankab J. Phukan
Department of Chemistry, Institute of
    Science, Banaras Hindu University,
    Varanasi 221 005, Uttar Pradesh,
    India

Raju Praveen
School of Chemistry, Madurai Kamaraj
    University, Madurai-625021, Tamil
    Nadu, INDIA

Ramasamy Ramaraj
School of Chemistry, Madurai Kamaraj
    University, Madurai-625021, Tamil
    Nadu, INDIA

Manas Roy
Department of Chemistry, National
    Institute of Technology Agartala,
    Agartala 799 046, Tripura, India

Neeraj K. Sah
Department of Chemistry, Institute of
    Science, Banaras Hindu University,
    Varanasi 221 005, Uttar Pradesh,
    India

Kamatchi Sankaranarayanan
Physical Sciences Division, Institute
    of Advanced Study in Science and
    Technology, Guwahati—781035,
    Assam, India

Ashish Kumar Singh
Department of Chemistry, Guru
    Ghasidas Vishwavidyalaya, Bilaspur
    495009, India

Baghendra Singh
Department of Chemistry, IIT (BHU),
    Varanasi, UP-221005, India

Chandni Singh
Department of Chemistry, Guru
    Ghasidas Vishwavidyalaya, Bilaspur
    495009, India

Sunil Kumar Singh
Department of Chemistry, Guru
    Ghasidas Vishwavidyalaya, Bilaspur
    495009, India

Karan Surana
Department of Chemistry,
    Sardar Patel University, Vallabh
    Vidyanagar, Anand – 388120,
    Gujarat, India
    Email- kssurana9@gmail.com

Neus Vilà
Laboratoire de Chimie Physique
    et Microbiologie pour les
    Matériaux et l'Environnement
    (LCPME), UMR 7564
    CNRS-Université de Lorraine,
    405 rue de Vandoeuvre, 54600
    Villers-lès-Nancy, France

Alain Walcarius
Laboratoire de Chimie Physique et
    Microbiologie pour les Matériaux
    et l'Environnement (LCPME),
    UMR 7564 CNRS-Université de
    Lorraine, 405 rue de
    Vandoeuvre, 54600
    Villers-lès-Nancy, France

Jianren Wang
Laboratoire de Chimie Physique et
    Microbiologie pour les Matériaux
    et l'Environnement (LCPME),
    UMR 7564 CNRS-Université de
    Lorraine, 405 rue de Vandoeuvre,
    54600 Villers-lès-Nancy,
    France

Hebei Key Laboratory of
    Applied Chemistry, School of
    Environmental and Chemical
    Engineering, Yanshan
    University, Qinhuangdao,
    P.R. China

Preface

The global energy demand has been increasing exponentially in the last few decades. This energy demand will be expected to increase much higher in the upcoming years. At present, approximately 80% of the global energy requirements are based on fossil fuels. The extensive use of fossil fuels causes global warming, energy crisis, and environmental pollution. Hence, it is necessary to improve renewable sources of energy for sustainable development. Sustainable energy storage devices such as fuel cells, biofuel cells, metal–air batteries, supercapacitors, solar cells, and redox flow batteries may be efficient and suitable alternatives to decrease the excessive use of fossil fuels. Here, nanomaterials play an important role in the development/improvement of the performance of these energy storage devices in a sustainable way. Generally, low efficiency, high cost, and the low abundance of conventional materials mitigate the bulk use of these energy storage devices. Therefore, it is necessary to develop highly efficient, durable, and low-cost materials for energy storage applications. There are the variety of nanomaterials documented in the literature for such energy storage applications. For example, carbon-based nanomaterials, polymer-based nanocomposites, mesoporous carbon nitrides, mesoporous silica, metal nanoparticles, metal-organic frameworks, metal phosphides, and more are reported for this purpose. However, the details about these nanomaterials are scattered in different research papers and review articles. It is laborious for the readers to collect all information regarding these materials. These difficulties motivate us to write this book *Nanomaterials for Sustainable Energy Applications* that can provide a unique platform for the readers.

This book is intended as a critical review of nanomaterials for sustainable energy applications. However, it is clearly impossible to refer to all the work which has been reported in this field. Here, we are providing broad and reasonably complete coverage of this field. Although there is no strict boundary for nanomaterials, some of the composites of nanomaterials have potential energy storage applications. Compact information about different energy storage devices such as fuel cells and batteries, biofuel cells, and redox flow batteries is provided in this book. In addition, specific nanomaterials and their composites for sustainable energy storage applications are also provided in this book. The first part of the book is devoted to a brief introduction to nanomaterials and their application in energy storage devices (Chapter 1). Chapters 2 to 4 provide a brief discussion about the different energy storage devices. Here, the working principle, construction, design, and challenges of significant energy storage devices are discussed. Chapter 5 is devoted to the designing of hollow-structured nanomaterials and their sustainable energy applications. It provides special emphasis on structure–function relationships. Structural features would be correlated with the desired property for a particular application. The detailed applications of individual class of nanomaterials, such as carbon quantum dots and polymer-based nanocomposites, are discussed in Chapters 6 to 9. A brief overview of the synthesis, characterization, and energy storage application of the nanomaterials is discussed in these chapters. Thus, this book is a unique one for beginners and researchers working in the field of nanomaterials.

Acknowledgments

- First of all, we are thankful to our contributing authors for their valuable contribution to this book.
- We are thankful to Prof. R. Ramraj for their continuous help, support, and suggestions.
- We are thankful to Dr. Arindam Indra for their sincere effort and dedication to contribute to this book.
- Our heartfelt tribute to the late Dr. M. Anbu Kulandainathan, principal scientist, Central Electrochemical Research Institute, Karaikudi, India.
- We are thankful to our research group members Mr. Narvadeshwar, Amit Kumar Verma, Mamta Yadav, Devesh Kumar Singh, Smita Singh, Ananya Tiwari, Varsha Singh, and Vikram Rathour for their help and support.
- We are thankful to Mr. Angesh Kumar Maurya for his help in cover page design of the book.
- We are thankful to our colleagues, collaborators, friends, and well-wishers for providing moral support for this book.
- We are thankful to our family members and friends for providing continuous moral support in this book.

1 Introduction
Nanomaterials for Sustainable Energy Applications

*Shanmugam Manivannan, Raju Praveen,
Kyuwon Kim and Ramasamy Ramaraj*

1.1 INTRODUCTION

1.1.1 THE FUTURE IS SUSTAINABLE

Soil, energy, and water are among our extremely treasured reserves, but the way we humans exploit these resources contributes to climate change. Recently, worldwide concern about energy-related climate change has been tied to the spiraling rate of fossil fuels (Turner, 2015). For the moment, Mother Nature, offering all these resources, is itself extremely defenseless to changes in climate. Hence, for improving and modifying functions, effective resource administration is in great need. More than ever, interest in renewable energy are rising due to energy demands and climate change. Because less awareness in the amalgamation of resource evaluations and rule-making precedes contradictory policies and unproductive use of reserves (Yip, Brogioli, Hamelers, & Nijmeijer, 2016). The generalized view of how to procure climate, soil use, energy, and water utilization determines the cure for the forthcoming crisis. In this regard, the imminent energy crisis taken on by the expiration of restricted and non-homogenously scattered fossil fuel resources and the global boost in energy demand has triggered immense research in the development of sustainable energy technologies in the past few decades. In this regard, individuals, as well as communities, have long diagnosed the destruction that can be instigated to our ecosystem and that of the plants and animals we share our planet with; very recently, this has been conceded worldwide. A global energy transition is urgently needed to meet the objectives, and in the meantime, it is essential to protect the environment from pollution as well (Z.-F. Yan, Hao, & Lu, 2010). These problems have been encountered due to industrialization with urbanization from previous decades. Besides, the demand for low-cost, efficient, and sustainable energy production is ever increasing globally. The imminent growth in materials science for the past few decades is a promising innovation in science and technology to meet sustainable energy production. In particular, functional materials for sustainable energy applications offer a vital direction for the advancement and use of these materials in sustainable energy production.

DOI: 10.1201/9781003208709-1

However, the efficiency of most of the proposed functional materials through various technologies to harvest sustainable energy is comparatively insignificant, and consequently, it needs to be increased to finally replace traditional technologies centered on fossil fuels. It is the efficiency increase of the functional materials-based technology that is key to envisioning the suppression of fossil fuel usage. The objective of this book is to give an integrated and inclusive demonstration of the fundamentals, utilization, and strategy of functional nanomaterials for efficient sustainable energy harvesting through various electrochemical energy harvesting and energy storage applications. The book presents general coverage of the use and design of functional nanomaterials for sustainable energy applications. Consequently, the book delivers all the significant aspects, such as nanomaterials' suitability for use in solar cells, fuel cells, and batteries, supercapacitors, biofuel cells, redox flow batteries; potent nanomaterials designed to overcome sustainability issues; and recent advancements in nanomaterials for sustainable energy applications.

1.1.2 NANOTECHNOLOGY: THE HOPE FOR SUSTAINABLE DEVELOPMENT

The production and consumption of energy are basic elements of life. The industrial revolution of the past decades has brought us the imminent energy crisis, which is widely supplied by resources such as fossil fuels. Besides, environmental issues, such as climate change and atmospheric pollution, have mainly risen due to the vast dependency and intermittent supply of fossil fuels. Hence, groundbreaking pieces of knowledge skilled at adapting and preserving energy on a substantial scale are crucial to elevating the sustainable energy harvesting and advancement of our society and economy. Recently, extensive interdisciplinary research on nanomaterials is preceding to give rise to the generation of a pollution-free environment rich in renewable energy resources (Liu, Burghaus, Besenbacher, & Wang, 2010). In this regard, to proficiently utilize the inadequate resources, both transmaterialization and dematerialization should be obeyed (Huang & Jiang, **2019**). The process of moving toward reusable materials is referred to as transmaterialization, and the process of restricted use of available resources while upholding affluence is referred to as dematerialization. (Turner, 2015) This energy transition must occur without leaving any consequences on environmental and human health bound to the resources and their consumption. In this regard, nanotechnologies are not associated entirely with sustainable energy harvesting technologies. But researchers are finding possible pathways in which nanotechnology could help us develop potent nanomaterials and technologies to uplift energy sources. Nanotechnology has already contributed quite useful technologies, such as techniques to retrieve and utilize fossil fuels in a competent way. Nanoemulsions are widely used as corrosion-resistant coatings, and nanocatalysts, as well as nanostructured membranes, have found suitability in extracting fossil fuels and in nuclear power stations. With the aid of nanotechnology, there is huge hope for the advancement of dematerialization of technology. Advancements in the field of catalysis and energy production have already marched in the right direction regarding nanotechnology. Sustainable energy harvesting technologies, which are operational with functional nanomaterials such as fuel cells, supercapacitors, solar cells, rechargeable batteries, and energy conversion and storage devices, are attracting wide attention from researchers. Nanotechnology

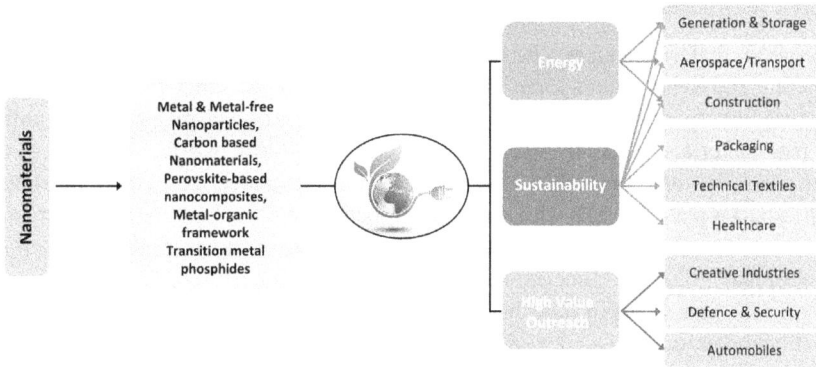

FIGURE 1.1 The strategy of nanomaterials for sustainable energy key challenge areas.

mainly focuses on offering feasible solutions for electrical power generation as well as catalysis, which are majorly contributing to sustainable energy. In addition, functional nanomaterials are being used in fabricating solar films, green coatings, high-efficiency fuels, and sensor devices. Furthermore, advancements in technologies by employing functional nanomaterials will make an immediate and improved influence on economic growth, the environment, and better quality of life across improved processes and products. Figure 1.1 represents the strategy of functional nanomaterials that may be used for key challenging areas.

1.2 FUNCTIONAL NANOMATERIALS

The behavior of materials at atomic, molecular, and macromolecular scales significantly differ from their corresponding larger scale is referred to as the distinguishing property of nanomaterials. Nanotechnology is a multidisciplinary area that involves the synthesis, characterization, and modulation of the properties from either the atomic or molecular level. Nanotechnology is principally the exploitation of functional materials with at least one dimension measuring less than 100 nanometers (nm). By tuning the size, shape, and constituent particles/molecules, the resultant functional nanomaterials do display distinctive physical, as well as chemical, properties. Metal, metal oxides, carbon-based, and quantum dots are the main studied categories of functional nanomaterials. Moreover, it is the high surface-to-volume ratio aroused from the nanosized particles that are responsible for the active surface area and, subsequently, their superior surface reactivity. Such a property is highly resourceful concerning the dematerialization process associated with the energy transition toward renewable energy sources. Due to their nanoscale size, they exhibit significantly enhanced electronic, physical, and chemical properties, and nanotechnology is particularly focusing on these aspects to harvest these exceptional properties of nanomaterials in terms of applications in various fields such as medicine, catalysis, sustainable energy, and electronics. By understanding all these unique properties, researchers could be able to focus on controlling the nanostructures and devices at the atomic, as well as molecular, level. Thereby, novel technological outputs are coming and fulfilling the various needs.

1.2.1 CLASSIFICATION BASED ON DIMENSIONALITY

In general, materials can be classified into natural and artificial materials, organic-, mineral-, and living matter come under natural materials, and materials that are manufactured by synthetic procedures are referred to as artificial materials. Regarding the nanomaterials, most advancements arose through the size effect which is in the form of bulk to the nanoscale. Prevalent methods are widely accessible to produce multidimensional materials with disparate properties to boost boundary performance. Classification of nanomaterials can be made according to their morphology as zero-dimensional (0D), one-dimensional (1D), two-dimensional (2D), and three-dimensional (3D). Our views in this chapter can steer the essential progress in developing functional nanomaterials with various properties for sustainable energy applications. "Functional materials" can be defined in several ways; the utmost suitable statement for functional nanomaterials is the development and study of novel nanomaterials and their assembly into multifunctional structures and devices for their application in key technological areas (Poh et al., 2018). Figure 1.2 depicts the schematic illustration of the different nanostructured materials based on their structural complexity.

Zero-dimensional: In 0D nanomaterials, all the dimensions exist within the nanoscale range. Often, 0D nanomaterials are spherical nanoparticles (NPs) such as gold, silver, and platinum NPs in size with a diameter of 1–50 nm.

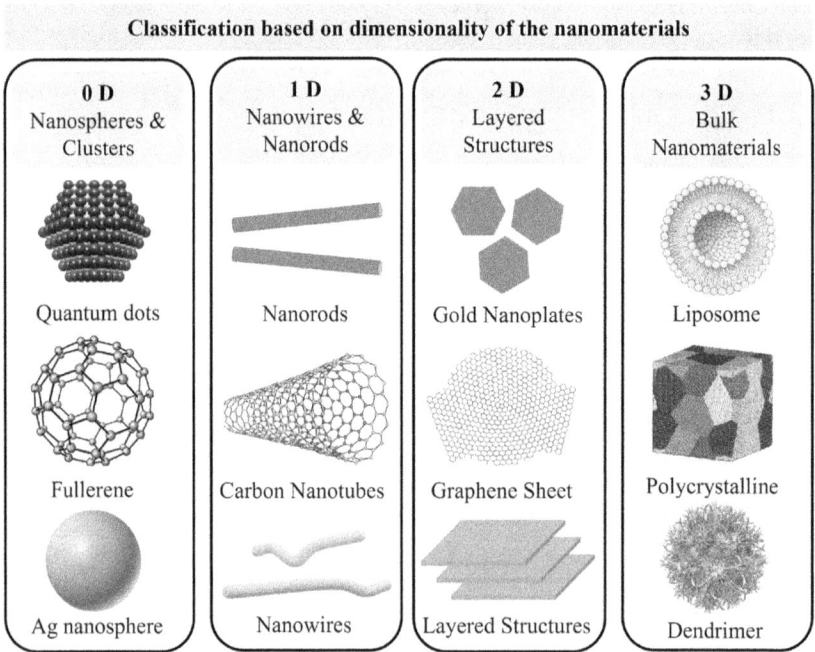

Classification based on dimensionality of the nanomaterials

0 D Nanospheres & Clusters	1 D Nanowires & Nanorods	2 D Layered Structures	3 D Bulk Nanomaterials
Quantum dots	Nanorods	Gold Nanoplates	Liposome
Fullerene	Carbon Nanotubes	Graphene Sheet	Polycrystalline
Ag nanosphere	Nanowires	Layered Structures	Dendrimer

FIGURE 1.2 Schematic illustration of the different nanostructured materials based on their structural complexity with examples of each category: zero- (0D), one- (1D), two- (2D), and three-dimensional (3D).

One-dimensional: In 1D nanomaterials, one dimension is in the nanoscale range, meaning one dimension is in the range of 1–100 nm and the other two dimensions can be macroscale. Nanotubes, nanorods, and nanowires are examples of 1D materials.

Two-dimensional: In 2D nanomaterials (2D), two dimensions are in the nanoscale range and one dimension is in the macroscale range. This class exhibits nanosheets, thin-film multilayers, and nano-thin-film-like shapes. In other words, by maintaining a thickness in a nanoscale 2D material's area can be several square meters. Some examples are graphene, nanolayers, and nano-coatings.

Three-dimensional: Three-dimensional nanomaterials are materials that do not have dimensions in the nanoscale, and all the dimensions are confined to the macroscale only. In this category, bulk materials, nanocolloids, and bunches of nanowires, as well as multilayers can be included. It is observed that these 3D materials have three arbitrary dimensions outside 100 nm. Besides, in 3D materials, electrons move freely within all dimensions since they are fully localized.

1.2.2 CLASSIFICATION BASED ON ELECTRON CONFINEMENT

The term *quantum confinement* is the spatial confinement of electron-hole pairs in one or more dimensions within a material. The energy levels of electrons will be not uninterrupted, similar to that of bulk materials, whereas the energy levels of the electrons will be discrete. Because of the detention of the electronic wave function, they become a discrete set of energy levels to the physical dimensions of the particles as shown in Figure 1.3. The main cause for quantum confinement is when the physical dimension of the potentials is close to the *de Broglie wavelength* and leads to discrete energy levels. This quantum confinement effect can be categorized into three types, 1D confinement (motion is not confined in two dimensions), or quantum wells; 2D confinement (motion is not confined in one dimension), or quantum wire; and 3D confinement (motion is confined in all dimensions (Edvinsson, 2018).

Based on the quantum confinement effect, an electron–hole pair is generated in the bulk lattice during the transfer of an electron from valence bands to conduction bands. This process brought the physical partition in between the electron and hole pair, and it is described as the exciton Bohr radius (rb). The Bohr radius is so sensitive to the configuration of the nanomaterials, especially semiconductors. The density of electron states in 0D, 1D, 2D, and 3D materials is shown in Figure 1.4, where 0D nanomaterials have well-defined and quantized energy levels. The quantum confinement effect for 0D nanomaterials (quantum dots) can be assessed by using a simple effective-mass approximation model. By using the Schrödinger equation, it is feasible to envisage and solve the confined energy level for nanomaterials. Besides, in 0D materials, the diameter of nanocrystal (L) is in the same order of magnitude as the Bohr radius, which is responsible for the quantum confinement of the exciton. This means, if the Bohr radius is very small, discrete energy levels will arise from the quantum confinement of the exciton. Furthermore, the inclusion/exclusion of a single atom will lead to a substantial transformation over the dimension of the

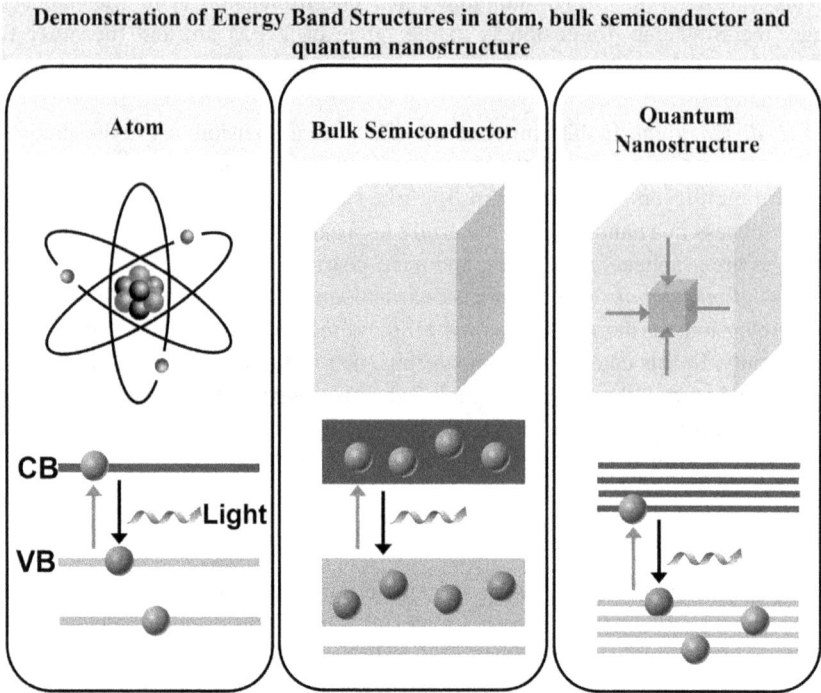

FIGURE 1.3 Schematic representation of energy band structures in the atom, bulk semiconductor, and quantum nanostructure.

nanocrystal as well as the bandgap. The discrete energy (E_n) for different nanomaterials can be defined by Equation 1.1.

$$for\ 0D : E_n = \left[\frac{\pi^2 h^2}{2mL^2} \right] \left(n_x^2 + n_y^2 + n_z^2 \right) \qquad (1.1),$$

where h is Planck's constant, m is electron mass, L is the diameter of the nanocrystal, and n is dimensional coordinates. The 1D materials do exhibit in the nanoscale in two dimensions. According to the quantum confinement effect, the discrete energy for 1D materials can be defined from the following equation:

$$for\ 1D : E_n = \left[\frac{\pi^2 h^2}{2mL^2} \right] \left(n_x^2 + n_y^2 \right) \qquad (1.2)$$

The 2D materials do exhibit in nanoscale in only one dimension. This means that electrons are allowed to move in only one direction, and their free movement to the associated dimensions is restricted. Hence, it is the characteristic of 2D materials

Classification of nanomaterials based on their confinement

No Confinement **Bulk**: Motion is not confined at all *(3D Materials)*	1D Confinement **Q-Well**: Motion is not confined in 2 dimensions *(2D Materials)*	2D Confinement **Q-Wire**: Motion is not confined in 1 dimension *(1D Materials)*	3D Confinement **Q-Dot**: Motion is confined in all dimensions *(0D Materials)*

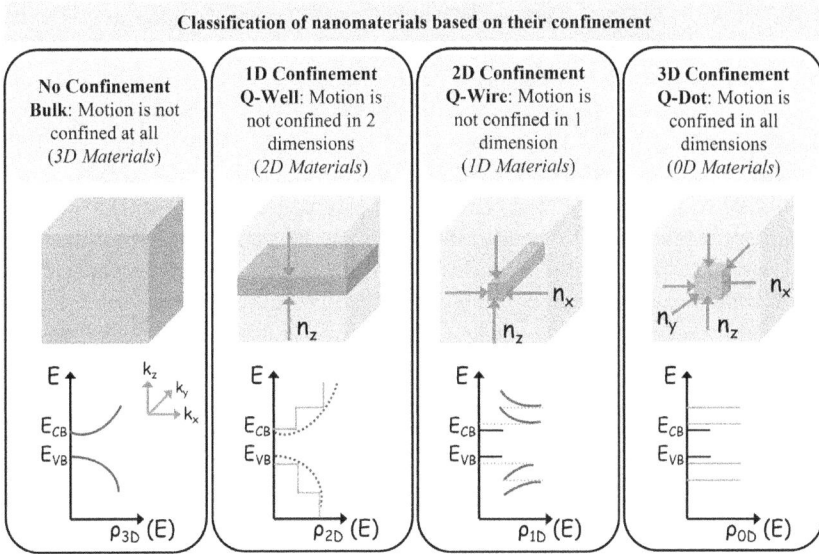

FIGURE 1.4 Schematic representation of the broken symmetry and functional form of the density of states in 1D, 2D, and 3D confined materials.

that usually the length is higher than the width. In 2D materials, electrons will also undergo confinement as well as delocalization; hence, based on the quantum confinement effect, the discrete energy for 2D nanomaterials can be defined from the Equation 1.3:

$$for\ 2D : E_n = \left[\frac{\pi^2 h^2}{2mL^2} \right] \left(n_x^2 \right) \tag{1.3}$$

1.2.3 SIZE AND SHAPE CONTROL OF NANOMATERIALS

The preparation of functional nanomaterials has fascinated rigorous consideration not only for underlying scientific interest but also for their technological applications. In this regard, producing mono-dispersed nanomaterials with controlled size and/or shape are key factors, since their physical, as well as chemical, properties depend firmly on the particle size and shape (Asaumi, Rey, Vogel, Nakamura, & Fujii, 2020). In general, nanomaterials' physical, chemical, and biological properties will be different in individual atoms or molecules or the corresponding bulk materials. The physical shape is the determinant factor for constraining the passage of electrons, holes, excitons, phonons, and plasmons, which have great control over the physical properties of functional materials. For instance, the change in color and shape of the nanomaterials aroused from the fluctuations in energy levels and its corresponding quantum confinement effect. Generally, metal-based, as well as metal oxide-based, nanomaterials are acquired mostly via bottom-up preparation routes in which researchers have the control to tune the size and shape of the resultant materials. Moreover, the synthesis of well-defined nanomaterials has constantly engrossed wide research attention in various fields of basic science and technology aspects, particularly for utilizing them in potential applications, especially towards sustainable energy applications. Besides, to harvest better efficiency in terms of application point of view, precise control over nanomaterials' size and shape is required to boost their assets. In addition to the size and shape, the following two factors will be more useful in judging the efficiency: (1) the inherent nature of active sites, such as facets of the nanomaterials, and (2) the reachable active sites, represented from the specific surface area of the nanomaterials. Nanomaterials of different shapes have different crystallographic facets and have a different fraction of surface atoms on their corners and edges, which makes it interesting to study the effect of nanomaterials' shape on the catalytic activity of various organic and inorganic reactions. Therefore, while designing and synthesizing the functional nanomaterials, the previously mentioned two aspects get the priority in determining the final morphology. Besides, exploring the well-defined nanomaterial system would reveal the fundamental questions to be answered to improve efficiency. Enough contribution has been already made by researchers for improving the properties of the nanomaterials in terms of tuning their size and shape over the past decades. Nanomaterials with several morphologies, such as spheres, cubes, octahedrons, triangles, tetrahedrons, rods/wires, sheets, and so forth, have been successfully prepared at various stages.

1.3   METAL AND METAL-FREE NANOMATERIALS

Metal nanomaterials which are derived by reducing the corresponding metallic ions have been in contemporary research for recent decades, principally due to their exceptional plasmonic properties and promising technological proposals. Figure 1.5 depicts the application fields of nanomaterials. Several industrial reactions do involve either metal or metal oxides as catalysts for various advantageous points of view. However, their use is limited on the industrial scale due to their high cost, predisposition to gas poisoning, and harmful consequences for the ecosystem. In this regard, the need for metal-free nanomaterials comes into the picture, and they must be widely available, environmentally adequate, and corrosion-resistant and have a high active surface area (Lu et al., 2021). Recently, various carbon nanomaterials evolving as metal-free alternatives have been actively engaged in cost-effective industrial catalytic reactions. In general, the final size and morphology of the metal nanostructures rely on their synthetic methods as well as stabilizer matrices. In this context, metal nanomaterials such as gold (Au), silver (Ag), platinum (Pt), and palladium (Pd) are of specific importance owing to their finite band gap between the conduction and valence band, where electrons can freely move between them. The freely moving electrons/oscillation of the electrons is responsible for the unique behavior of metal nanomaterials, referred to as surface plasmon resonance, and it is highly sensitive to the metal nanomaterials' size and shape. Another fascinating property of metal nanomaterials is their color, and it is determined by a combination of the parameters, such as the nanocrystals' size and shape and refractive index of the medium.

When metal/metal-free nanomaterials are being used as catalysts, they should be enriched with surface active sites that are essential for the analyte's adsorption,

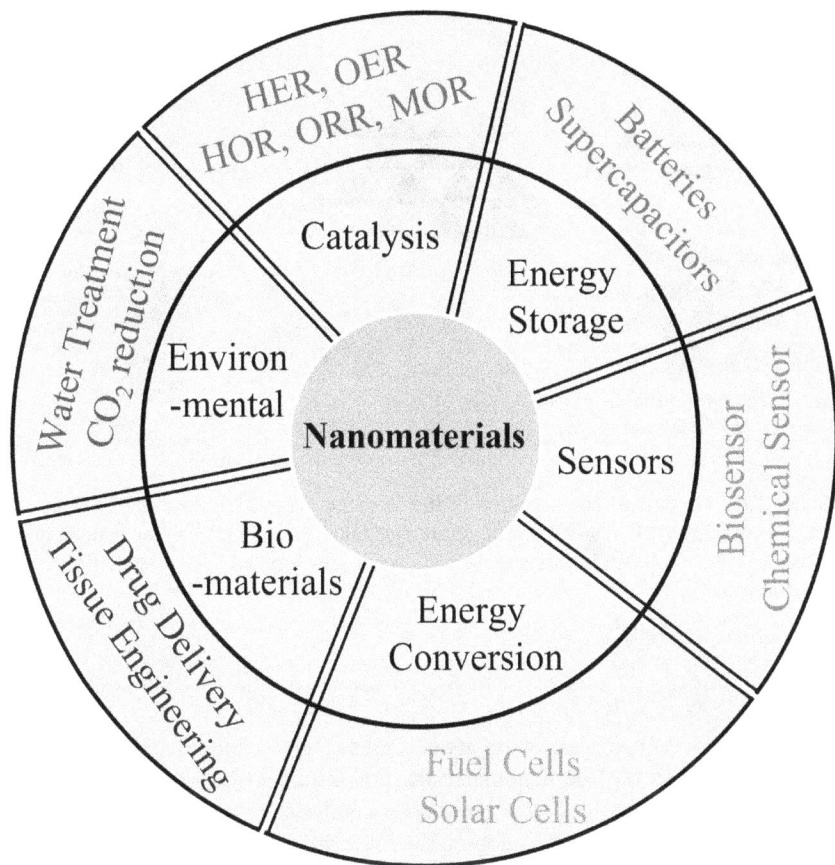

FIGURE 1.5 Application fields of nanomaterials.

bond-formation, bond breaking, and finally desorption of the product from the catalyst's surface. Apart from this, exceptional operational steadiness should be there throughout the catalytic process to enable efficient completion of the reaction. For instance, Pt, Au, and ruthenium (Ru) nanomaterials are used as cathode materials in the fuel cell application to speed up oxygen reduction reaction (ORR). Meanwhile, metal oxides are extensively studied for the hydrogenation and dehydrogenation processes. However, as mentioned earlier, the metal-based catalysts (both metal and metal-oxide nanomaterials) habitually suffer from various limitations, such as being expensive, exercising poor judgment, having low sturdiness, and having adverse effects on the ecosystem through the discharging effluents. To overcome these limitations, researchers are exploring nanomaterials owing to their abundance and their eco-friendly, unique physiochemical properties and have found that carbon-based nanomaterials are a more suitable alternative to use as metal-free catalysts.

The eminent metal-free catalysts are carbon-based nanomaterials such as fullerenes, carbon nanotubes (CNTs), graphene nanosheets, graphitic carbon nitride, and

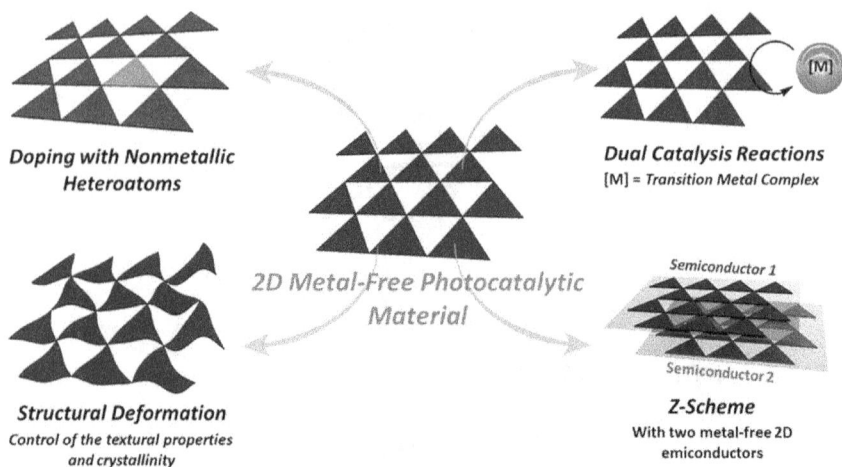

FIGURE 1.6 Graphical representation of the foreseen major avenues to organic photocatalysis by metal-free 2D materials. The figure was taken from the following article: https://pubs.acs.org/doi/abs/10.1021/acsnano.1c00627 with the granted permission from ACS Publications.

Source: Rosso et al. (2021)

other graphene-like materials, there are several strategies demonstrated to tailor the efficiency of the metal-free nanomaterials. For instance, Rosso et al. (2021) have proposed the major avenues to organic photocatalysis by metal-free 2D materials as shown in Figure 1.6. Besides, doping the heteroatoms such as nitrogen (N), fluorine (F), and sulfur (S) will be highly useful in tuning the electronic transport and chemical reactivity of the metal-free materials. Furthermore, metal-free catalysts are widely being used in energy and environmental applications. However, metal-free catalysts are less explored in chemical synthesis and sustainable energy applications.

1.4 CARBON-BASED NANOMATERIALS

Carbon has been the most valuable material in human civilizations, science, and technological developments (Hu et al., 2019). So far, a diverse range of carbon materials like fullerenes, CNTs, carbon nanofibers (CNFs), graphene, graphene oxides, nanodiamonds, mesoporous carbon, carbon nanohorns, and onion-like carbon are availed and offer new opportunities for the development of highly active carbon-based catalyst (Bazaka, Jacob, & Ostrikov, 2016; Gong, Liu, Li, Yu, & Teoh, 2016). Because the carbon materials tendered the highly active specific surface area, high conductivity, tunable wettability, and well-balanced pores, the distribution properties offer improved catalytic properties. The structure of carbon nanomaterials is shown in Figure 1.7. Herewith, we highlight recent achievements and progress of carbon material–based metal-free electrocatalysts material for sustainable energy applications like fuel cells, metal–air batteries, supercapacitors, solar cells, and the like.

Accordingly, substantial modification of carbon materials has recently been directed toward the development of metal-free carbon nanomaterials for various catalytic processes, involving either oxidation or reduction reactions. High electroactive catalyst materials are encompassed in precious metals, and metal-oxide nanomaterials are most utilized. However, the carbon nanomaterials are replaced by the precious metal-based electrode process like their high price, scarcity, and limited commercial applications. The advantage of carbon materials was major earth-abundant, eco-friendly, biocompatible, catalytically active, and durable. The conventional metal catalysts were combined with carbon, and they showed better catalytic activities compared to pristine metal catalysts. The carbon–metal (C-M) composition displays a wide range of accessibility and tunability due to their rich surface chemistries and lack of metal dissolution and poisoning properties. The combination of carbon nanofibers with electroactive metal nanoparticles has demonstrated profound electrocatalytic activities due to the large surface area and high electrical conductivity.

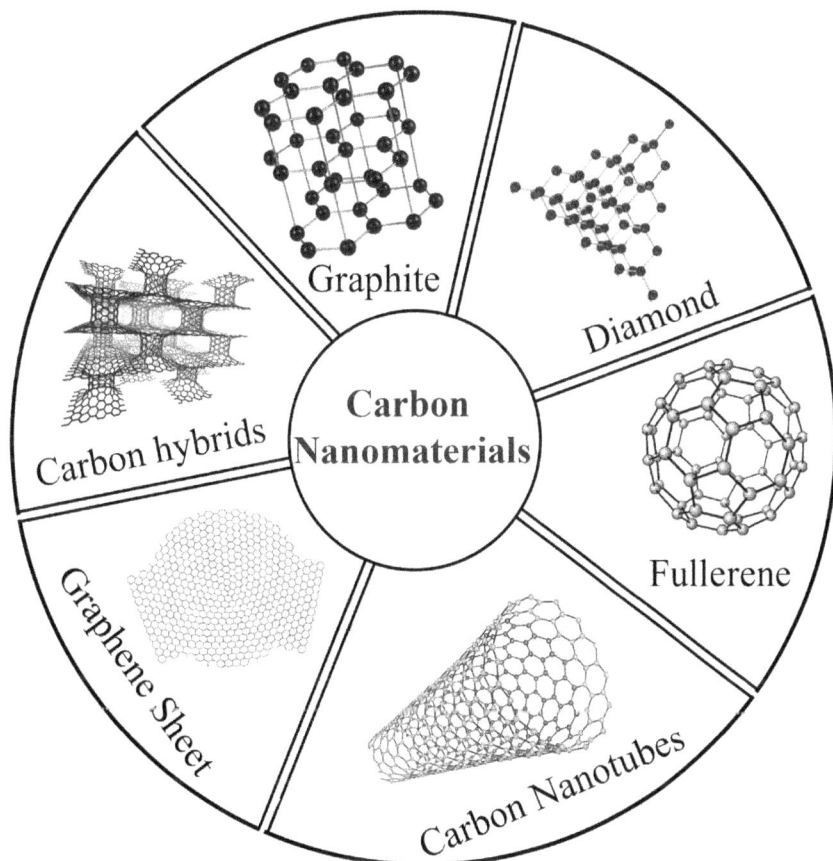

FIGURE 1.7 Structure of carbon nanomaterials: Graphite, diamond, fullerene, carbon nanotubes, graphene sheet, carbon hybrid materials.

The ORR in fuel cell (FC) applications is the most hopeful way for the energy conversion method. The electrocatalyst efficiency can be controlled by the porosity, chemical, and electronic environments of the catalysts. And their important electrochemical parameters like onset potential, durability, and tolerance toward anode fuels properties indicate their capability in commercial purposes. In particular, the carbon-based heteroatom-doped electrocatalysts exhibit improved ORR activity in an alkaline medium, paving the way for the development of alkaline fuel cells, as well as providing long-time durability. The characteristic ORR occurs at two steps either through a four-electron or through two-electron transfer. Among, the 4-e- and 2-e- have proceeded highly efficient and water as the final product and less efficient and hydroperoxide ions as intermediates, respectively. Further, the nitrogen-doped different dimensional carbon (1D and 2D) nanomaterials have emerged as the most promising catalysts, normally the N-doped carbon (1D and 2D) nanomaterials facilitate the surface O_2 chemisorption/activation which improves their profound ORR catalytic performance. In addition, N-doped single or multiwalled carbon nanotubes (N-CNTs) were shown their enhanced electrocatalytic ORR activity. The intrinsic electrocatalytic ORR activity we observed at different N-containing-group-doped CNTs. The real nature of the active sites in N-doped carbon nanomaterials remains unclear, it is generally accepted that pyridine and pyrrole nitrogen atoms contribute differently to the ORR.

1.5   NANOCOMPOSITES

The word *Nanocomposites* is used for materials consisting of a minimum of two nanoscale materials in single phases in order of 100 nm or less. The chemical and physical properties of matter can change significantly when the size changes from the macro- to the microscale and then from the micro- to the nanoscale level. An important feature of a nanocomposite is the large ratio of surface area to volume higher than the pristine component phases. The advantage of nanoscience and technology we can create composites with well-controlled structures on the nanometer scale, these nanostructures have large specific surface areas that could offer a feasible platform for catalysis, sensing, and FC applications. A diverse synthetic methodology has been reported for nanocomposite making with various sizes, shapes, and compositions. Preceding reports confirm that the inorganic nanomaterials are combined with organic materials (polymer or carbon) to prepare a diverse range of nanocomposite materials so for. These nanocomposites have received great attention and solid support in the field of energy conversion, catalysis, and sensors due to their intriguing optical and catalytic properties. The advantage of these nanocomposites is that the organic nanomaterial acts as a host material for inorganic nanoparticles with high specific surface areas and leads to the enhanced durability properties of the nanocomposite. Another advantage of nanocomposite materials is it is mandatory to develop facile, rapid, and low-cost synthetic methods for the nanocomposites (inorganic–organic) from the technological point of view. An important task in this high-performance nanocomposite's synthesis is the uniform particle size, size distribution, and dispersity. This ensures that the synthetic methodologies like one-pot (chemical) and mixing (physical) will not affect the desired nanoparticles' size, uniform distribution, aggregation, and uneven properties of the nanocomposite. The major advantage of nanocomposites (inorganic–organic) is that they offer both high mass transport and electron transfer as a consequence of their enhanced electrocatalytic properties.

1.6 NANOMATERIALS IN ELECTROCHEMISTRY

Electrochemistry is an important research branch in chemistry; it deals with the interrelation of electrical and chemical effects. Electrochemistry is the study of reaction and discussion of the composition of the electrode and electrolyte/solution interface. The aforesaid system is referred to as an "electrochemical cell"; there are two different types of electrochemical processes that can occur at an electrode surface. The first one is Faradaic processes, which are denoted as the charge transfer reactions occurring at the electrode/electrolyte interface that are governed by Faraday's first law. The second one is non-Faradaic processes; thermodynamically or kinetically unfavorable, the adsorption and desorption reactions can occur at electrode/electrolyte interface without any charge passes. Faraday's first law states that, during electrolysis, the produced electrical current is proportional to the quantity of a chemical species. The electrode process occurs within the double layer and generates the charge unbalance in the electrode/electrolyte interface. The electrochemistry processes depend on the potential of the electrode, which is related to the thermodynamics and kinetics of electrode reactions. Furthermore, the influence of the interfacial potential difference gives an important way to use external control on an electrode reaction. The electrochemistry technique is the most valuable, sensitive, informative, and useful analytical technique in the energy and sensor research field. FC and electroanalytical chemistry techniques play an important role in humans' daily life applications. Most of the techniques, such as batteries, electrocatalysis, and electrochemical sensors, are very attractive for point-of-care applications. In particular, the recent advances in nanoscience and technology in the electrochemical research area will undoubtedly enlarge the possibility of the profound electrochemical process. Various functional nanomaterials and their usage in sustainable electrochemistry applications are shown in Figure 1.8. The electrodeposition in the facile quick method to prepare the readily available electrodes, as mentioned earlier, is electroplating or electrolytic deposition. In this technique, the applied electrical current is a reducing agent for metal ions; as a result, the very thin film formed on the conductive surface. Electrodeposition is the most promising technology, which involves surface phenomena, solid-state processes, and processes occurring in the liquid state, thereby drawing on many scientific disciplines. However, electrochemistry is not a simple process, but it has many applications in the area of energy production and sensor fields.

1.7 SUSTAINABLE ENERGY APPLICATIONS

In sustainable energy development, energy storage plays a vital role, and it is often dependent on inconsistent and impulsive renewable energy sources, such as wind and solar, in terms of constant power supply they are not reliable. In the quest to

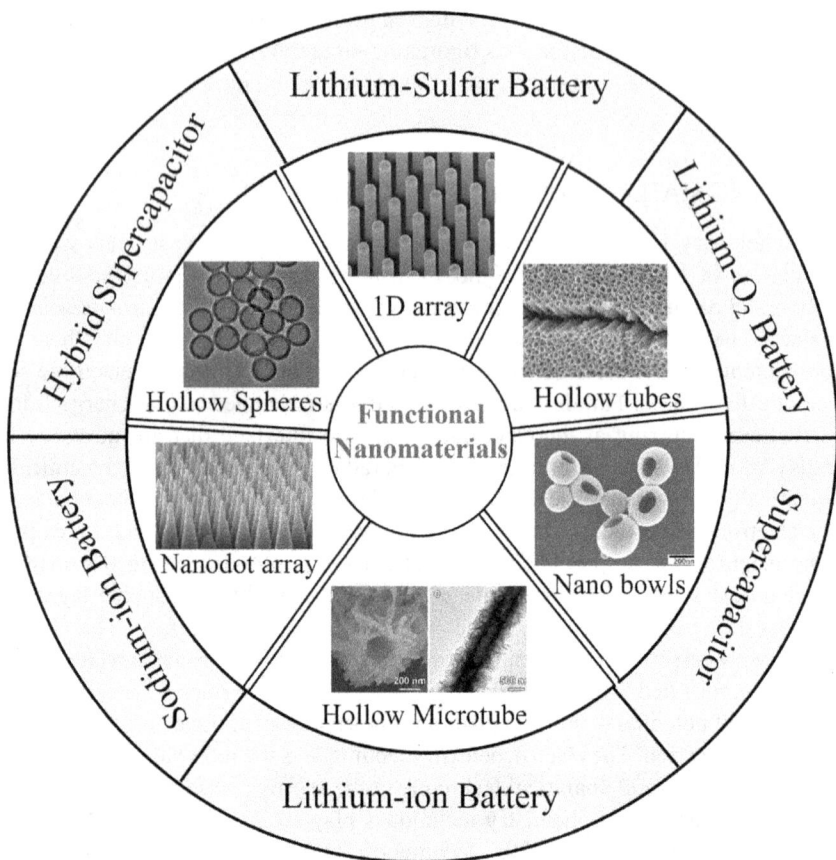

FIGURE 1.8 Various functional nanomaterials and their usage in sustainable electrochemistry applications.

make those sources reliable, technological advancements must be there to store up the energy surplus when the sources are consistent and distribute when energy manufacture is minimal. Such technological advancement should be accompanied by consistency, high accomplishment, competence, large-scale production, and sustainability. Based on chemical, electrochemical, and mechanical, as well as thermal modes, various energy storage systems have been proposed in recent decades in which batteries and supercapacitors are highly attractive for their portability. In this regard, hydrogen is the most suitable alternative fuel for fossil fuels due to its high abundance; carbon-free chemical energy carrier emits water and burned with oxygen, and the chemical energy density is 142 MJ. Kg. However, to be considered, hydrogen must be produced in a sustainable way, such as through water-splitting and photo- and electrocatalytic reactions, by avoiding the production of steam-reforming reactions. For technological advancements, a core-level understanding of electrocatalysis and electrochemical reactions at the electrode surface is required.

In this context, sustainability goals, such as social equity, sanitization, economy, public health, security, and a better environment, can be achieved if we could produce inexpensive and sustainable electricity and fuels to fulfill global energy demands. In the past, human society witnessed and addressed several challenges, such as control over nuclear power, space expeditions, and so on. All these achievements involved a group of intellectuals, but sustainable energy development is different; it requires the active participation of civilians and researchers together to bring about technological advancements. Since energy is a crisis that goes beyond a country's borders and eventually influences the whole human society. Hence, sustainable energy development is of prime importance and is referred to as "development that meets the needs of the present without compromising the ability of future generations to meet their own needs" (Jacobs, 2000). Figure 1.9 depicts the pictorial representation of the advantages of sustainable energy in everyday life. Sustainable energy can help human society in the following ways: (a) boost access to drinking water and sanitization facilities; (b) lower the emission of water, land, and air pollutants; (c) decrease the number of deaths and illnesses related to pollution; (d) diminish poverty, violence, and inequity; (e) expand the environmental impact of transportation systems; and (f) enhances access to clean cooking fuels and technologies.

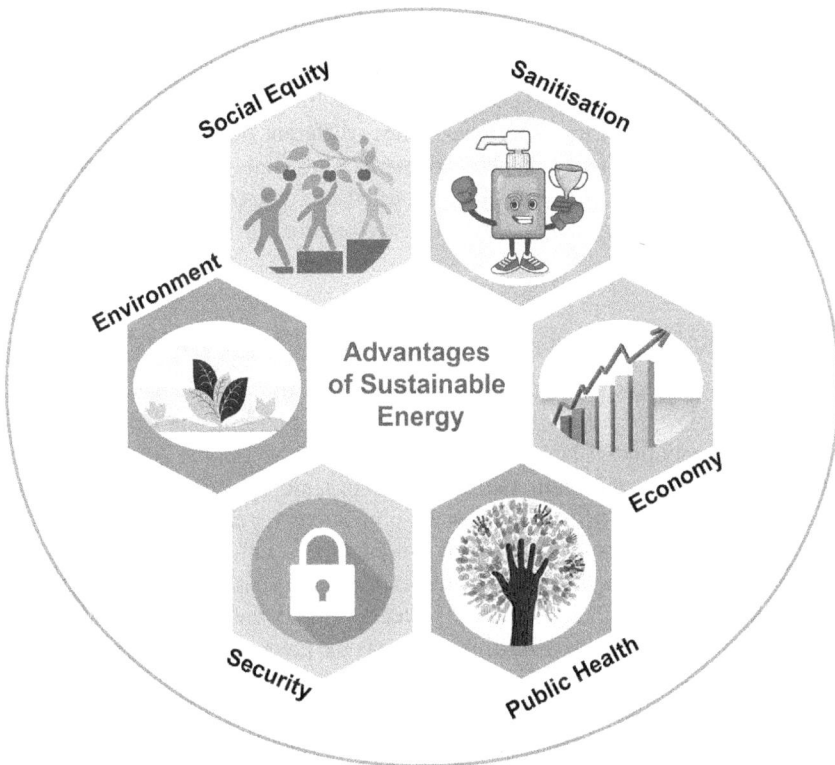

FIGURE 1.9   Pictorial representation of the advantages of sustainable energy in everyday life.

To achieve sustainable energy applications through nanomaterials, researchers should overcome issues such as the complexity of the energy sector, energy production, transformation, distribution, storage, and usage. The eminent nanotechnological advancements would offer feasible solutions to overcome all these issues although with contrasting effects. The following are a few examples of how sustainable energy applications can be developed with the aid of nanomaterials/nanotechnology:

a. Weight reduction: with the increased high volume surface ratio, nanomaterials' weight can be reduced through optimization function; consequently, nanotechnology will help to make airplanes and vehicles out of optimized plastics or metals with CNTs; thereby, fuel consumption can be abridged.
b. Organic light-emitting diodes: traditional light bulbs do offer only 5% of energy to light conversion, whereas by implementing nanolayers of plastic and organic pigments, the conversion rate can be tuned up to 50%.
c. Carbon-based nanomaterials: in the automobile industry, by having control over the strength and rolling resistance of the tires, fuel savings of up to 10% can be reached, and carbon black nanomaterials have recently been used in the modern automobile industries.
d. Self-cleaning nanoscale layers: installing such surfaces would benefit in saving energy and water because frequent cleaning is not required.
e. Nanomaterials for automobiles and industries: nano lubricants, nanoflow agents, and nano-insulating materials are finding wide applicability in the automobile, industry, and construction fields, respectively.
f. Device fabrication for energy harvesting: different types of nanomaterials improve the efficiency of photovoltaic devices, lightweight wind turbine blades, lithium-ion batteries, fuel cells, and catalytic converters in vehicles, thereby improving energy harvesting, conversion, storage, distribution, and usage.

1.7.1 Hybrid Nanomaterials for Sustainable Energy Applications

A foremost task in sustainable energy conversion is the advancement of effective photo(electro)/electrocatalysts for the hydrogen evolution reaction (HER) and oxygen evolution reaction (OER; M. Yan, Jiang, Yang, He, & Huang, 2020). Functional nanomaterials integrated from the inorganic–organic hybrids have proven their encouraging possibility to meet the obscure obligations for photo- and electrocatalysis. The coherent synthesis of functional nanomaterials with well-defined structures has been intensively examined due to the extraordinary properties and captivating applications of the resultant materials. In recent times, inorganic–organic hybrids have been widely emerging as starting materials for preparing distinct hybrid nanomaterials, in which, inorganic and organic species with nano/molecule/atom-scale distribution serve as self-templates and sacrificial agents, respectively, leading materials with tunable morphologies, band gaps, defects, and spatial arrangements. (Jena, 2020; Shahbazi Farahani et al., 2022; Y. Yu, Shi, & Zhang, 2018; Zhang et al., 2022) Synthesizing diverse hybrid nanomaterials with unusual morphologies becomes feasible with the aid of widely available hybrid precursors and abundant transformation tools. Hybrid nanomaterials, especially in the photo- and electrocatalytic processes, do exhibit excellent performance owing to improved photon adsorption, facilitated electron transport, increased active sites, and enhanced intrinsic activity. In general hybrid nanomaterials contain inorganic elements that work for chemical reactions and organic elements that work for controlling the active sites (Goodman, Zhou, & Cargnello, 2020). Furthermore, the large-scale positioning of many types of fuel cells and electrolyzers for sustainable energy applications is presently embarrassing because of the deficiency of sustainable catalysts. In this regard, prominent hetero-atom-doped carbon catalysts have been developed for reactions in fuel cells and water electrolyzers via various synthetic strategies. There is a substantial challenge to the rational design of such hybrid nanomaterials to increase reaction selectivity. Besides, by understanding the key role of these hybrid nanomaterials, structural and mechanistic characterization plays a vital role.

1.7.2 Hybrid Technologies for Harvesting the Sustainability

Sustainable social and economic advancement trusts on novel nanotechnologies that offer maximum efficiency at a marginal environmental cost. (Liu et al., 2010) However, it is very tricky to apply sustainable approaches across the entire life cycle of nanomaterials/nanotechnologies, from design and synthesis to deployment and clearance (Neurock, 2010; Tarafder et al., 2020). Catalysis is at the core of many sustainable energy conversion devices and advanced technologies (Ahmed et al., 2022; Hu et al., 2019). For instance, water electrolysis can be reflected as a green and promising technology to produce green hydrogen, and it plays a vital role in advanced energy technologies, such as solar fuel production, supercapacitors, metal–air batteries, and so on. Hence, the commercialization of water electrolysis technology needs stable, efficient, and cost-effective electrode materials, as well as the construction of green hydrogen; a fed fuel cell device needs hybridization between OER and ORR (X. Li, Liu, & Xiang, 2022) to obtain the hydrogen required to run the fuel cell from a sustainable source, that is, water. Such hybrid technology/device opens the pathway for the development of electrode materials based on nonprecious metal catalysts as well as metal-free catalysts, they mainly overcome sustainability issues. (L. Li et al., 2017) Furthermore, OER and ORR are particularly the most difficult ones to carry out and the most responsible for the performance of water electrolyzers and fuel cells, respectively. Thus, probing for competent, as well as sustainable, catalysts for these two reactions is of dominant importance to make these energy systems efficient and generally available.

The real success of developing sustainable nanomaterials should find the technological advancements in terms of nanotechnology-based products and processes in various industrial sectors as well as everyday needs (Aftab et al., 2020; Jastrzębska & Vasilchenko, 2021). Besides, the adoption of sustainable nano-engineering strategies can reduce the environmental impacts of products and processes by permitting more resourceful use of raw materials and reducing energy consumption, emissions, and leftovers (W. Yu, Wang, Park, & Fang, 2020). In this regard, sustainable nanoarrays, nanoparticle contrast agents for in vivo imaging, biocompatible implant materials, and drug delivery are widely used in the healthcare sector. The improvement

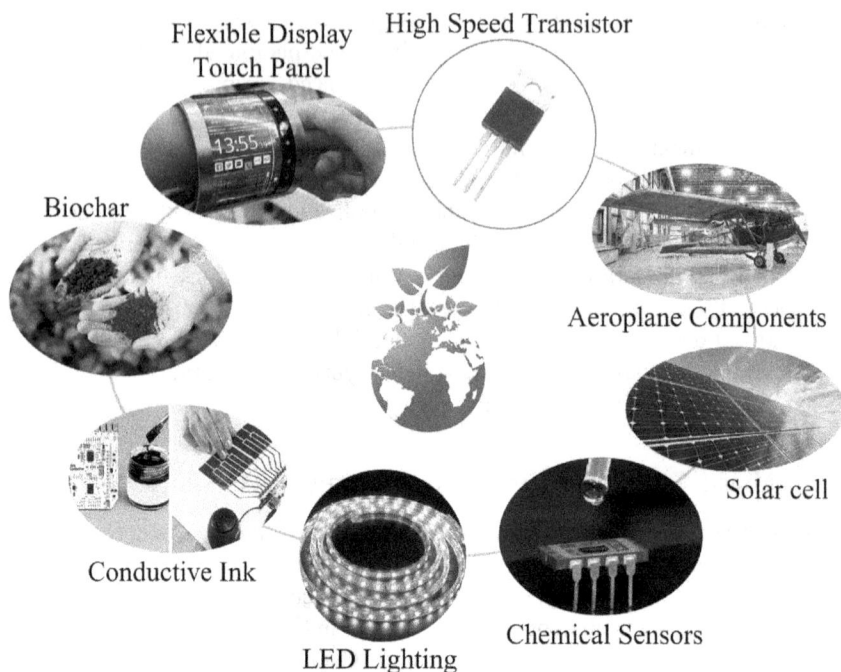

FIGURE 1.10 Various energy applications harvest sustainability.

in the sustainability of healthcare systems not only reduces the cost but will also lead to the latent environmental benefits in terms of bringing the pharmaceutical residues down. In addition, sustainable nanoscale carriers can be employed as controlled delivery systems for fertilizers, pesticides, and herbicides in the agriculture sector. Also, hybrid sensor devices based on sustainable nanomaterials are in use to monitor the crops, soil, and climatic condition thereby resources properly utilized to raise the crop yields. In the food and packaging sectors, sustainable nanomaterials-based devices are employed for monitoring the food quality, moisture, oxygen, and presence of microorganisms. In addition, antimicrobial coatings and biodegradable materials are made from sustainable nanomaterials, thereby food quality and safety in packaging are ensured (Masi, Bollella, Riedel, Lisdat, & Katz, 2020). In lithium-ion batteries, the use of nanomaterials provides shorter diffusion paths for lithium ions, leading to higher charge/discharge rates and much greater energy densities. Besides, lithium ions are also better lodged in the electrode lattice, ensuring the durability of the battery and bringing it fundamentally safer. Various energy applications harvesting sustainability are shown in Figure 1.10.

1.8 RECENT ADVANCEMENTS IN MATERIALS SCIENCE FOR SUSTAINABLE ENERGY APPLICATIONS

Materials science has been contributing progressively to essential parts of our everyday lives with functional nanomaterials as well as their control over their specific synthesis. In particular, synthetic strategies and their optimization play a vital role in determining the assembly, crystallinity, particle size, morphology, porosity, texture, and surface area, along with other properties of materials as well. Regarding the sustainability of the nanomaterials, synthetic strategy, consistency in preparation, and economic aspects are also to be considered. In this regard, advancements in material science in terms of synthesis approaches play a fundamental role in achieving efficient sustainable nanomaterials for energy applications. Recently, several typical material design/architectures were explored, and their electro- and photocatalytic properties strongly depend on their design and execution. However, the prevalent use of fossil fuels is afflicted with the issues, such as the human era experiencing the ever-growing as well as complicated environmental issues associated with unpredictable climate change which we are perceiving now. Crude oil availability for the long term has become a very big question as well. Consequently, developing a library of sustainable nanomaterials to apply with sustainable energy sources is a need of contemporary research in materials science. For instance, the breakthrough in the development of sustainable materials owing to defined sizes, shapes, porosities, crystalline phases, and structures is already reflected in various sustainable energy systems. However, a complete departure from fossil fuels is not possible for at least 15 to 20 years since they are the primary source of energy at present. In parallel, novel catalysts, as well as approaches, are in progress to manage the fossil fuel supply and to bring down its environmental issues. For instance, to promote green Hydrogen as a clean fuel that is produced sustainably for the mobility sector, hydrogen-blended compressed natural gas (HCNG) is emerging as an excellent interim technology for achieving emissions reduction and import substitution. Refueling of HCNG blends in vehicles can be performed with minimum modifications in the infrastructure that is currently under use for dispensing compressed natural gas (CNG). In addition to green hydrogen production, the approaches comprise desulfurization, isomerization, alkylation, gas-to-liquid conversions, adsorption, carbon dioxide conversion, and water-gas shift, and all this strategic research needs sustainable, functional nanomaterials. Besides, the framework of the fuel cell design, membrane, and nonprecious metal catalyst development need to be considered.

The quest for sustainable self-sufficient power sources for micro/nanosystems that can be used as sensors and monitoring electronic devices is always there in sustainable energy development. In this regard, nanogenerators for converting mechanical energy into electricity research are moving ahead. Considering the electrode materials required for the Li-ion batteries, significant progress has been made to achieve functional nanomaterials with controlled porosity and structures on a large scale. Considering the green hydrogen production and its storage and efficient OER and ORR catalysts have been developed, both processes should be merged in a single platform to develop the kind of green hydrogen–fed fuel cell device for the proper utilization of the produced hydrogen from one of the compartments of the device. Furthermore, a solvent-free approach to making materials, referred to as mechanochemistry, has emerged as a feasible strategy for making functional nanomaterials; thereby, environmental consequences of the solvents can be avoided. (Muñoz-Batista, Rodriguez-Padron, Puente-Santiago, & Luque, 2018). MnO is one of the key materials in electrochemical energy storage applications, but its usage is highly limited by its deprived electronic conductivity and a lack of environmentally benign synthesis. A recent study has proposed a steadfast strategy of confining MnO nanowires in pipe-like carbon nanoreactors that exhibited high efficiency in sustainable electrochemical energy storage systems. (L. Li et al., 2017). Besides, in the development of sustainable nanomaterials, the following strategies are being focused on a lot such as alloying noble metals with other sustainable elements, finding the suitable support materials for the noble metal catalysts, implementing noble metal–free catalysts as well as metal-free catalysts, heteroatoms, and multi-element-doped carbon catalysts. Sustainable life cycles of carbon-based devices for electronics and energy applications are shown in Figure 1.11.

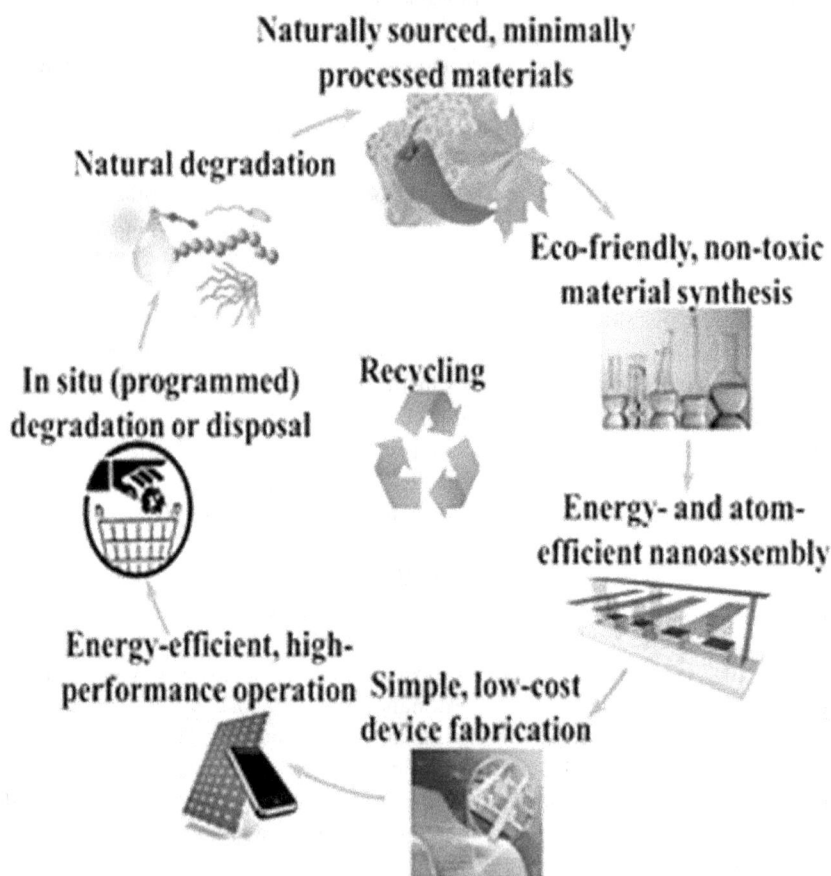

FIGURE 1.11   Sustainable life cycles of carbon-based devices for electronics and energy applications. The figure was taken from the following article: https://pubs.acs.org/doi/10.1021/acs.chemrev.5b00566 with the granted permission from ACS Publications under ACS AuthorChoice.

Source: Bazaka, Jacob, & Ostrikov (2016)

1.9 OPPORTUNITIES AND CHALLENGES

There are opportunities for sustainable nanomaterials to contribute by strengthening industrial modernization and affordability, lowering the utilization of raw materials and energy, and boosting a fresher environment. The emerging nanotechnological advancements have their consequences on all industrial sectors, which include healthcare, agriculture food processing, transport, energy, water, and electronics. Concerning the long term, the currently available fossil fuels to meet energy and transport requirements and their wastes are unsustainable. In this regard, nanotechnology can pave the effective utilization of available resources through hybrid sustainable technologies, finding alternatives for precious materials with abundant ones. Thereby, a sustainable environment can be promised by lowering the emissions and consumption of nonrenewable natural resources. Any scientific revolution carries with it accompanying risks. Nanomaterials' exposure to the environment might occur during synthesis, carriage, storage, supply, custom, and recycling or their disposal. Moreover, existing precautionary measures may be inadequate concerning the hazards and exposure allied with nanomaterials. Hence, prompt action for framing necessary guidelines, measures, and nano-specific risk assessment tools is being developed.

To meet future energy needs in a sustainable way, the following key areas have been identified: solar energy utilization, electrical energy storage, hydrogen production and storage, and catalysis for energy applications. But there is a huge gap that remains between the current science and the technological support to reach it. In addition to identifying the technological advancements, it is essential to create awareness among researchers and common people about the reality of the energy crisis. Such awareness can assist in moving faster toward the changes needed for sustainable energy in the future. In this regard, the enactment of lithium-ion battery technology into electric and hybrid electric vehicles and portable electronic devices, such as smartphones, laptops, and tablets, makes a mandate for efficient, economic, and sustainable nanomaterials for energy storage applications. In the quest for sustainable nanomaterials, the economical aspect should be a decisive screening norm. However, longer reaction times and procedures and reproducibility issues should be overruled. Hence, rational design, control over nucleation and growth, and scalable and reproducible strategies are being developed. Furthermore, for the physiochemical characterizations of the prepared nanomaterials, a local instrumentation facility is more important through which a better understanding and quick optimization of nanomaterials can be carried out. In addition to developing novel nanomaterials, the focus should be on inorganic–organic hybrid materials to boost synergism. Through multidisciplinary efforts from physics, chemistry, materials science, and chemical engineering, the development of sustainable nanomaterials for energy applications could open substantial scientific openings for the groundwork of sustainable nanomaterials with captivating performance soon.

1.10 SUMMARY

Nanomaterials and hybrid nanomaterials-based technologies have contributed to the growth of generous and sustainable activities by making facile and functional nanocatalysts thereby increasing the use of natural resources and energy. We are nearing an era in which functional nanomaterials can be integrated with unique, operational properties to contribute to sustainability through the development of cleaner, less wasteful production processes, and transport systems. In any form of sustainable energy harvesting, low-cost functional nanomaterials can be included to cut down the production cost, energy consumption, emissions, and effluents as well as develop effective and economic energy storage systems. While looking for sustainable nano-materials, nano-specific assessment tools must be developed to track the entire life cycle of the nanomaterials which could bring the perfect balance between the econ-omy and the environmental consequences. For instance, while implementing nano-carriers for pesticides and herbicides in agriculture, we must learn how to control damaging insects without eradicating valuable populations, where control over the development of sustainable nanomaterials comes into the picture. Sustainable nano-materials would bring the solution for energy, environmental and security issues aroused from the consumption of fossil fuels. Besides, the researchers must realize that engaging in sustainable development is not a constraint on innovation but rather is a challenge of creativity.

ACKNOWLEDGMENTS

This work was partially supported by the Incheon National University Research Grant in 2016. RR acknowledges the Raja Ramanna Fellowship Scheme, Department of Atomic Energy, India. Financial support from IoE seed grant-II for new faculty (Scheme No. 6031 (B)) of Banaras Hindu University is gratefully acknowledged.

REFERENCES

Aftab, S., Shah, A., Nisar, J., Ashiq, M. N., Akhter, M. S., & Shah, A. H. (2020). Marketability prospects of microbial fuel cells for sustainable energy generation. *Energy & Fuels, 34*(8), 9108–9136. doi:10.1021/acs.energyfuels.0c01766

Ahmed, M. E., Nayek, A., Križan, A., Coutard, N., Morozan, A., Ghosh Dey, S., . . . Dey, A. (2022). A bidirectional bioinspired [FeFe]-hydrogenase model. *Journal of the American Chemical Society, 144*(8), 3614–3625. doi:10.1021/jacs.1c12605

Asaumi, Y., Rey, M., Vogel, N., Nakamura, Y., & Fujii, S. (2020). Particle monolayer-stabilized light-sensitive liquid marbles from polypyrrole-coated microparticles. *Langmuir, 36*(10), 2695–2706. doi:10.1021/acs.langmuir.0c00061

Bazaka, K., Jacob, M. V., & Ostrikov, K. (2016). Sustainable life cycles of natural-precursor-derived nanocarbons. *Chemical Reviews, 116*(1), 163–214. doi:10.1021/acs.chemrev.5b00566

Edvinsson, T. (2018). Optical quantum confinement and photocatalytic properties in two-, one- and zero-dimensional nanostructures. *Royal Society Open Science, 5*(9), 180387. doi:10.1098/rsos.180387

Gong, X., Liu, G., Li, Y., Yu, D. Y. W., & Teoh, W. Y. (2016). Functionalized-graphene com-posites: Fabrication and applications in sustainable energy and environment. *Chemistry of Materials, 28*(22), 8082–8118. doi:10.1021/acs.chemmater.6b01447

Goodman, E. D., Zhou, C., & Cargnello, M. (2020). Design of organic/inorganic hybrid cata-lysts for energy and environmental applications. *ACS Central Science, 6*(11), 1916–1937. doi:10.1021/acscentsci.0c01046

Hu, C., Qu, J., Xiao, Y., Zhao, S., Chen, H., & Dai, L. (2019). Carbon nanomaterials for energy and biorelated catalysis: Recent advances and looking forward. *ACS Central Science*, *5*(3), 389–408. doi:10.1021/acscentsci.8b00714

Huang, J., & Jiang, Y. (2019). Tailoring resource-efficient catalysts for sustainable energy and chemical processes. *ACS Sustainable Chemistry & Engineering*, *7*(7), 6423–6423. doi:10.1021/acssuschemeng.9b01472

Jacobs, M. (2000). Toward a sustainable future. *Chemical & Engineering News Archive*, *78*(10), 5. doi:10.1021/cen-v078n010.p005

Jastrzębska, A. M., & Vasilchenko, A. S. (2021). Smart and sustainable nanotechnological solutions in a battle against COVID-19 and beyond: A critical review. *ACS Sustainable Chemistry & Engineering*, *9*(2), 601–622. doi:10.1021/acssuschemeng.0c06565

Jena, P. (2020). Clusters and nanomaterials for sustainable energy. *ACS Energy Letters*, *5*(2), 428–429. doi:10.1021/acsenergylett.9b02687

Li, L., Zhu, J., Niu, Y., Chen, Z., Liu, Y., Liu, S., . . . Jiang, J. (2017). Efficient production of coaxial core—shell MnO@Carbon nanopipes for sustainable electrochemical energy storage applications. *ACS Sustainable Chemistry & Engineering*, *5*(7), 6288–6296. doi:10.1021/acssuschemeng.7b01256

Li, X., Liu, Y., & Xiang, Z. (2022). Dithiine bridged phthalocyanine-based covalent organic frameworks for highly efficient oxygen reduction reaction. *The Journal of Physical Chemistry C*, *126*(8), 4008–4014. doi:10.1021/acs.jpcc.1c10846

Liu, C.-J., Burghaus, U., Besenbacher, F., & Wang, Z. L. (2010). Preparation and characterization of nanomaterials for sustainable energy production. *ACS Nano*, *4*(10), 5517–5526. doi:10.1021/nn102420c

Lu, H., Tournet, J., Dastafkan, K., Liu, Y., Ng, Y. H., Karuturi, S. K., . . . Yin, Z. (2021). Noble-metal-free multicomponent nanointegration for sustainable energy conversion. *Chemical Reviews*. doi:10.1021/acs.chemrev.0c01328

Masi, M., Bollella, P., Riedel, M., Lisdat, F., & Katz, E. (2020). Photobiofuel cell with sustainable energy generation based on micro/nanostructured electrode materials. *ACS Applied Energy Materials*, *3*(10), 9543–9549. doi:10.1021/acsaem.0c02169

Muñoz-Batista, M. J., Rodriguez-Padron, D., Puente-Santiago, A. R., & Luque, R. (2018). Mechanochemistry: Toward sustainable design of advanced nanomaterials for electrochemical energy storage and catalytic applications. *ACS Sustainable Chemistry & Engineering*, *6*(8), 9530–9544. doi:10.1021/acssuschemeng.8b01716

Neurock, M. (2010). Engineering molecular transformations for sustainable energy conversion. *Industrial & Engineering Chemistry Research*, *49*(21), 10183–10199. doi:10.1021/ie101300c

Poh, T. Y., Ali, N. A. T. B. M., Mac Aogáin, M., Kathawala, M. H., Setyawati, M. I., Ng, K. W., & Chotirmall, S. H. (2018). Inhaled nanomaterials and the respiratory microbiome: Clinical, immunological and toxicological perspectives. *Particle and Fibre Toxicology*, *15*(1), 46. doi:10.1186/s12989-018-0282-0

Rosso, C., Filippini, G., Criado, A., Melchionna, M., Fornasiero, P., & Prato, M. (2021). Metal-free photocatalysis: Two-dimensional nanomaterial connection toward advanced organic synthesis. *ACS Nano*, *15*(3), 3621–3630. doi:10.1021/acsnano.1c00627

Shahbazi Farahani, F., Rahmanifar, M. S., Noori, A., El-Kady, M. F., Hassani, N., Neek-Amal, M., . . . Mousavi, M. F. (2022). Trilayer metal—organic frameworks as multifunctional electrocatalysts for energy conversion and storage applications. *Journal of the American Chemical Society*, *144*(8), 3411–3428. doi:10.1021/jacs.1c10963

Tarafder, C., Daizy, M., Alam, M. M., Ali, M. R., Islam, M. J., Islam, R., . . . Khan, M. Z. H. (2020). Formulation of a hybrid nanofertilizer for slow and sustainable release of micronutrients. *ACS Omega*, *5*(37), 23960–23966. doi:10.1021/acsomega.0c03233

Turner, C. (2015). Review of chemistry of sustainable energy. *Journal of Chemical Education*, *92*(4), 601–602. doi:10.1021/ed5008298

Yan, M., Jiang, Q., Yang, L., He, H., & Huang, H. (2020). Three-dimensional ternary hybrid architectures constructed from graphene, MoS2, and graphitic carbon nitride nanosheets as efficient electrocatalysts for hydrogen evolution. *ACS Applied Energy Materials, 3*(7), 6880–6888. doi:10.1021/acsaem.0c00975

Yan, Z.-F., Hao, Z.-P., & Lu, M. G. Q. (2010). Perspective on sustainable energy technologies in Asia and Pacific states. *Energy & Fuels, 24*(7), 3713–3714. doi:10.1021/ef100411b

Yip, N. Y., Brogioli, D., Hamelers, H. V. M., & Nijmeijer, K. (2016). Salinity gradients for sustainable energy: Primer, progress, and prospects. *Environmental Science & Technology, 50*(22), 12072–12094. doi:10.1021/acs.est.6b03448

Yu, W., Wang, T., Park, A.-H. A., & Fang, M. (2020). Toward sustainable energy and materials: CO_2 capture using microencapsulated sorbents. *Industrial & Engineering Chemistry Research, 59*(21), 9746–9759. doi:10.1021/acs.iecr.0c01065

Yu, Y., Shi, Y., & Zhang, B. (2018). Synergetic transformation of solid inorganic—organic hybrids into advanced nanomaterials for catalytic water splitting. *Accounts of Chemical Research, 51*(7), 1711–1721. doi:10.1021/acs.accounts.8b00193

Zhang, X., Cui, C., Chen, S., Meng, L., Zhao, H., Xu, F., & Yang, J. (2022). Adhesive iono-hydrogels based on ionic liquid/water binary solvents with freezing tolerance for flexible ionotronic devices. *Chemistry of Materials, 34*(3), 1065–1077. doi:10.1021/acs.chemmater.1c03386

2 Fuel Cells
Engineered Nanomaterials for Improved Fuel Cell Performance and Power Generation

Chandni Singh, Uday Pratap Azad, Ashish Kumar Singh and Sunil Kumar Singh

2.1 INTRODUCTION

With the development of new technologies, new achievements and new ways of living, the world has also proportionately suffered from several critical problems such as environmental pollution, energy crisis, weather irregularity and so on. One of the major problems that the world has experienced, and which will be further aggravated in the near future, is the depletion of energy sources. Energy sources that are environmentally friendly reduce pollution in a greener way. In order to overcome this problem, the concept of fuel cells emerged.

Fuel Cell: A fuel cell is an electrochemical device that converts chemical energy into electrical energy via a redox reaction. It oxidizes fuel at the anode and reduces oxygen from air at the cathode to produce electricity (Vielstich et al., 2003).

Fuel cells mainly consist of three parts, namely, an anode, a cathode and an electrolyte. At the anode, the oxidation of fuel takes place whose e^- are transferred to the cathode, where it reduces half of the O_2, and the ions of fuel are passed to electrolyte and release water, electricity and heat (Figure 2.1).

Redox reaction occurring in the fuel cell:

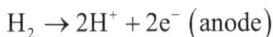

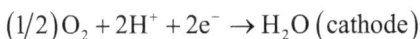

$$H_2 \rightarrow 2H^+ + 2e^- \text{ (anode)}$$

$$(1/2)O_2 + 2H^+ + 2e^- \rightarrow H_2O \text{ (cathode)}$$

$$H_2 + (1/2)O_2 \rightarrow H_2O + W_{ele} + Q_{heat} \text{ (overall reaction)}$$

The concept of fuel cells as energy sources is a greener approach as it is quiet and static, showing higher thermodynamic efficiency and providing excellent load response and a wide range of applications (Sharaf et al., 2014). Mainly fuel cells are

DOI: 10.1201/9781003208709-2

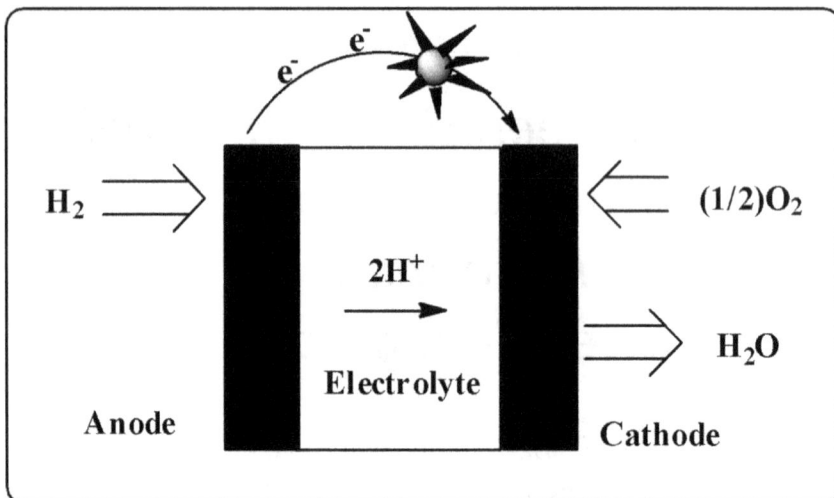

FIGURE 2.1   Schematic representation of fuel cell.

classified into the following types: (1) proton exchange membrane fuel cells (PEM-FCs), including direct methanol fuel cells; (2) alkaline fuel cells; (3) phosphoric acid fuel cells; (4) molten carbonate fuel cells; (v) solid oxide fuel cells; and (6) microbial fuel cells (Sammes, 2006; Blomen & Mugerwa, 1993; Mench, 2008; Herper, 2008; Srinivasan, 2006; Herman et al., 2005). Disadvantages of the fuel cells are that they suffer mainly from the expensive platinum catalyst, long-term durability and stability issues, the high investment cost-per-kW and the relatively large system size and weight (Sharaf et al., 2014)

To overcome the problems faced to some extent, while using fuel cells, nanotechnology is introduced. Therefore, recent developments were undertaken for the synthesis of nanoparticles that led to the formation of various monodisperse nanoparticles with controlled composition interparticle interactions, their shape and size (Xia et al., 2008). To increase the production of current, the oxygen reduction reaction (ORR) should be enhanced at the cathode. The ORR occurs in the cathode part of the fuel cell, and it can be illustrated by the following reaction (Wang et al., 2016).

In acidic electrolytes: two pathways are adopted for ORR:

1. 4-e– pathway

$$O_2 + 4H^+ + 4e^- \rightarrow 2H_2O$$

2. 2-e– pathway

$$O_2 + 2H^+ + 2e^- \rightarrow H_2O_2$$
$$H_2O_2 + 2H^+ + 2e^- \rightarrow 2H_2O$$

In alkaline electrolytes: two possible pathways for ORR:

1. 4-e– pathway

$$O_2 + H_2O + 4e^- \rightarrow 4OH^-$$

2. 2-e– pathway

$$O_2 + H_2O + 2e^- \rightarrow HO^-_2 + OH^-$$

$$HO^-_2 + H_2O + 2e^- \rightarrow 3OH^-$$

In these electrolytes, 4-e– pathways follow direct 4-e– reduction of O_2 to form H_2O in an acidic solution or OH– in basic solutions. Two-e– pathways follow 2-e– reductions of O_2 to form H_2O_2 in an acidic solution or HO_{2-} in basic solution, first, and then another 2-e– gives H_2O or OH–.

To enhance the ORR at the cathode, electrocatalysts are employed that are eco-friendly, give wide active sites for reaction, are resistant to corrosion and have high conductivity.

2.2 VARIOUS TYPES OF NANOMATERIALS USED AS ELECTROCATALYST FOR BOOSTING ORR IN THE CATHODE

2.2.1 Noble Metal-Based Nano-Electrocatalyst for ORR

The following are some noble metal–based nanomaterials for boosting the ORR in the cathode of a fuel cell.

During the ORR process, the surface of Pt was shielded by the electrolyte or hydroxyl ion layer, due to which the active sites of Pt surface get decreased and its catalytic efficiency is also reduced (Markovic et al., 1995). So, the synthesis of Pt nanoparticles is important for fuel cells as their shape and size are easily controlled.

Catalytic efficiency is greatly influenced by the shape and size of Pt nanoparticles (Wang et al., 2008). Pt nanoparticles (NPs) with sizes tunable from 3 nm to 7 nm and controlled polyhedral, truncated cubic or cubic shapes have been studied as catalysis for ORR in PEMFCs.

Monodisperse Pt NPs with a controlled size and shape are synthesized by reaction of Pt $(acac)_2$ and a trace amount of $Fe(CO)_5$ (Wang et al., 2007). The size and shape of Pt NPs were controlled by injecting $Fe(CO)_5$ into the $Pt(acac)_2$ at different temperatures.

The intrinsic catalytic activity of the Pt NPs for the ORR was measured by using a rotating disk electrode (RDE) in 0.5-m H_2SO_4. The current density (J) was calculated and applied into Levich-Koutecky plots at different constant potentials.

$$\frac{1}{J} = \frac{1}{Jk} + \frac{1}{Jdiff} = \frac{1}{Jk} + \frac{1}{B\omega 1/2},$$

in which $B = \dfrac{0.62nFC\,0\,D\,0\,2/3}{\eta\,1/6}$,

j_k = kinetic current density, j_{diff} = the diffusion limiting current density, n = overall number of electrons transferred, $C_0 = O_2$ concentration in the electrolyte (1.26×10^{-3} molL^{-1}), D_0 = diffusion coefficient of O_2 in the H_2SO_4 electrolyte (1.93×10^{-5} cm^{2}s) and η = the viscosity of the electrolyte (1.009×10^{-2} cm^{2}s; Markovic et al., 1995). In the J^{-1}Vs $\omega^{-1/2}$ plot, the slope is $1/B$.

The number of electrons calculated from the slope for 7-nm Pt nanotubes is 3.6 and for 3-nm and 5-nm Pt NPs is 0.7. So, the current density for 7-nm Pt nanotubes is four times that of 3-nm polyhedral, which indicates a nearly complete reduction of O_2 to H_2O, with a 4-e− pathway on the surface of 7-nm Pt.

Due to excellent current density and low overpotential, Pt-based catalysts are widely used for ORR (Bisen & Nanda, 2020a). But it also shows some drawbacks, such as low abundance and high expenses, which limits them for large-scale applications (Bisen & Nanda, 2020b; Xu et al., 2020).

Ru-based nanostructures have been investigated for ORR (Bisen & Nanda, 2021). Synthesized Ru@NC were uniformly dispersed Ru nanoparticles over N-doped carbon nanostructure by one-step pyrolysis of dicyandiamide and ruthenium phthalocyanine at 800 °C as shown in Scheme 2.1.

The electrochemical ORR catalytic activity of synthesized Ru@NC_800 catalyst was investigated and compared with RuPc, Ru@NC_700, Ru@NC_900 and Pt/C catalysis. It suggested that the better performance of Ru@NC is due to the formation of highly efficient 4-e$^-$ transfer via the inner-sphere electron transfer mechanism (ISET) pathway in alkaline medium (Bisen & Nanda, 2021). Nanostructures greatly suppress the HO$_{2-}$ intermediate, which shows its superior stability over 1000 cycles and surpasses the state-of-the-art Pt/C.

Lin et al. (2017) synthesized Rh nanoparticles supported on ultrathin carbon nanosheets by using salt microcrystal (Rh oleate) heated under an inert atmosphere (Scheme 2.2).

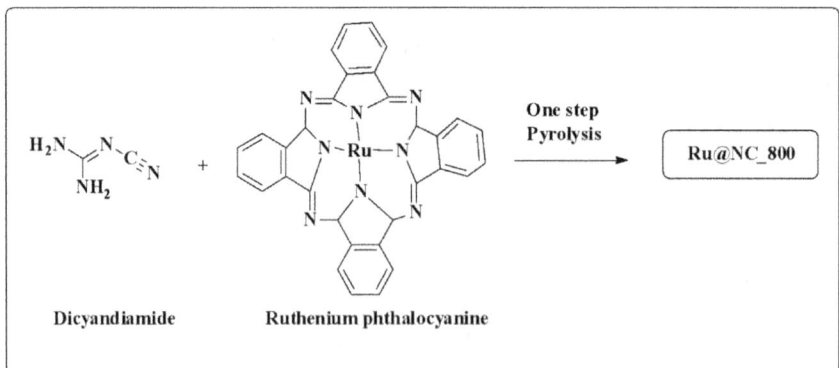

SCHEME 2.1   Schematic synthesis of Ru@NC_800 catalyst by one-step pyrolysis of dicyandiamide and ruthenium phthalocyanine (Bisen & Nanda, 2021).

The electrochemical performance of Rh/C hybrid nanosheets was evaluated. For ORR measurement, a glass carbon RDE is modified. The result shows that the Rh/C nanosheets display an excellent ORR activity and show good stability for long-term cycling.

2.2.2 Metal Nanocluster-Based Catalysts for ORR

To enhance the catalytic activity of nano electrocatalysts for ORR, in the concept of nanoclusters, it is implied that nanoclusters have definitely structure and composition; their size is smaller than NPs, and their smaller size gives them a large surface area for electrocatalysis.

Here, some metal nanoclusters are mentioned for improving the electrocatalytic activity toward ORR.

Hwang et al. (2016) synthesized a hybrid catalyst Pt/FeCo-OMPC (L) for the enhancement of ORR activity (Scheme 2.3).

By using the RDE technique, the ORR polarization curve of the as-prepared catalyst was obtained, which was measured in 0.1 M $HClO_4$. To understand the enhancement of ORR activity by a Pt nanocluster, density functional theory (DFT) was performed. From the obtained result by attempting various methods 5wt% of Pt/FeCo- OMPC (L) electrocatalysts show high-ORR mass activity compared to Pt/C.

Zhao et al. (2016) synthesized an effective catalyst for ORR, an ultrasmall Pd nanocluster. They synthesized a palladium nanocluster that was protected by a 4-tert-butylbenzene thiolate ligand having palladium atoms less than 20 by using a one-phase method. The chemical formulated is $Pd_{13-17}(SR)_{18-22}$ (R= Ph-tBu) (Scheme 2.4).

Then the synthesized nanocluster was evaluated for ORR activity on the RDE in an alkaline solution. From the calculation, the number of electrons transferred in the ORR process for ligand-on and ligand-off Pd nanoclusters is 3.7 and 3.9, respectively. Both ligand-on and ligand-off Pd nanoclusters follow a direct 4-e$^-$ oxygen reduction pathway. This report shows an excellent electrocatalyst in fuel cell technology.

Liu et al. (2013a) synthesized an electrocatalyst (a silver nanocluster supported on carbon nanodots for ORR. Carbon nanodot (CN)–supported silver nanocluster (AgNC/CNs) was synthesized by two steps. In the first step, CNs were synthesized, and in the second step, silver NCs supported on CNs were formed (Scheme 2.5).

They performed an excellent electrocatalyst for ORR and followed four electron pathways. The surfactant-free silver nanocluster supported on CNs also enhances photoluminescence.

2.2.3 Transition Metal–Based Nanocatalyst for Boosting ORR

As noble metals (Pt, Pd, Ag, Au) have low abundance and high cost, which limit their use as catalysts for ORR. In the past few decades, alternatives of precious metal have been developed as non–precious metal catalysts (NPMCs), including metal-free nitrogen-doped carbon (N-doped C) (Chen et al., 2011a; Zheng et al., 2011; Deng et al., 2011), non–precious metal oxides and metal carbides (Su et al., 2012; Esposito & Chen, 2011).

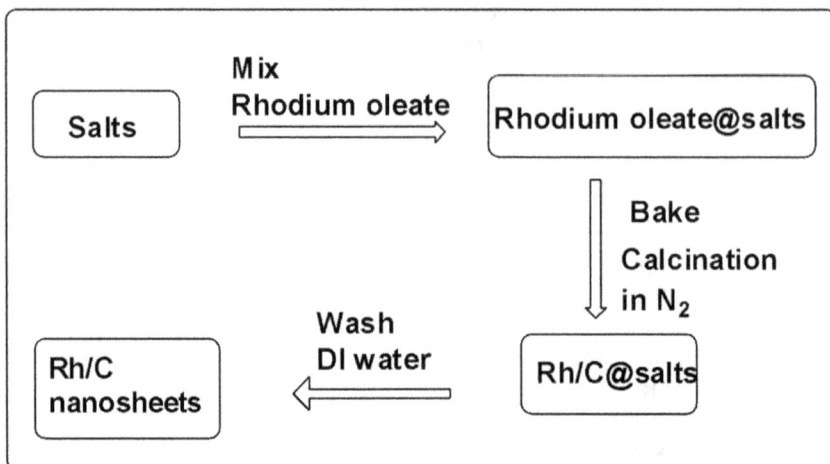

SCHEME 2.2 Schematic preparation of Rh/C nanosheets via salt-template process (Lin et al., 2017).

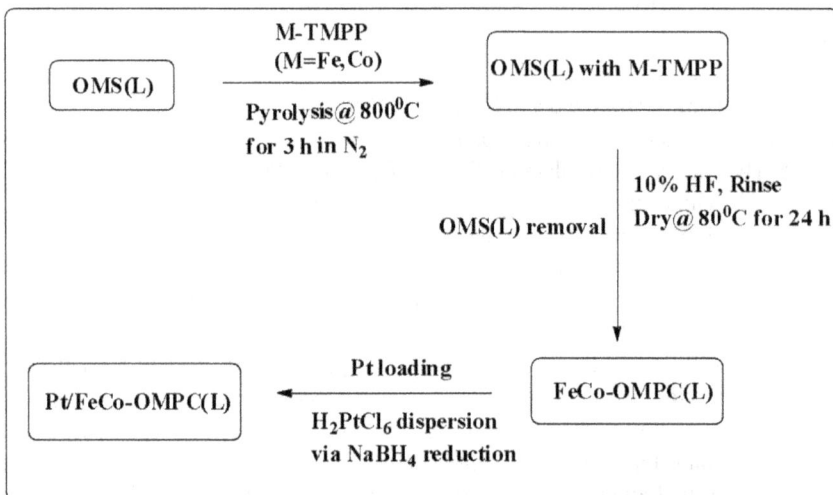

SCHEME 2.3 Schematic synthesis of Pt/FeCo-OMPC (L) (Hwang et al., 2016).

Lin et al. (2014) synthesized an electrocatalyst for ORR in both alkaline and acidic conditions. They reported a new noble metal–free Fe-N/C catalyst (Scheme 2.6).

The synthesized Fe-N/C_800 catalyst shows high ORR activity and major active sites. It follows four-e$^-$ transfer pathways in both alkaline and acidic conditions.

Kim et al. (2018) synthesized a nanocatalyst by using a metal-organic framework (MOF). Its catalytic activity was investigated for ORR and for azide-alkyne Huisgen cycloaddition reaction. The ORR activity of Cu@Cu$_2$O core-shell nanocatalyst was

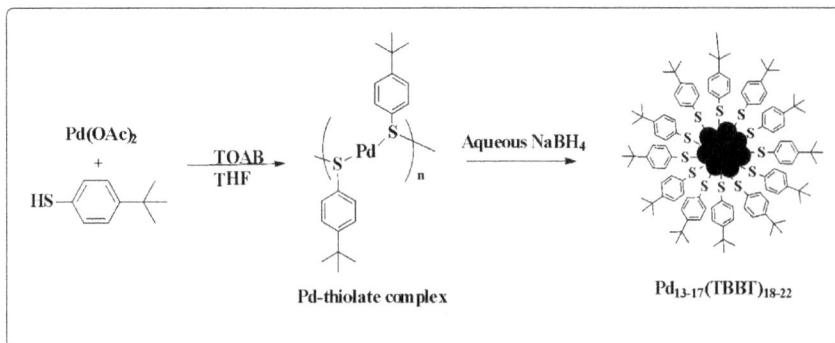

SCHEME 2.4 Synthesis process of $Pd_{13-17}(SR)_{18-22}$ catalyst (Zhao et al., 2016).

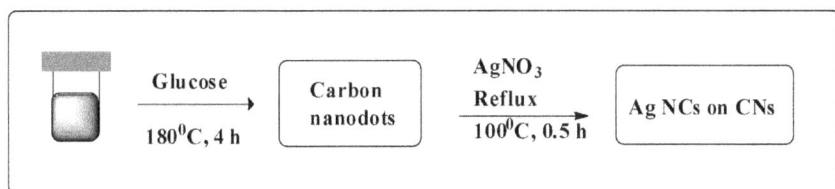

SCHEME 2.5 Schematic representation of AgNC/CNs (Liu et al., 2013a).

evaluated by plotting Koutecky-Levich (K-L) plots versus reversible hydrogen electrode (RHE). The electron transfer number of $Cu@Cu_2O$ core-shell is on average 3.9 V at 0.45–0.75 V, which suggests that the $Cu@Cu_2O$ nanocatalyst follows a direct 4-e– pathway. This catalyst also shows good catalytic activity for azide-alkyne Huisgen cycloaddition under mild reaction conditions.

To improve ORR activity (Dong et al., 2017) reported a novel high-performance electrocatalyst metal nitride/N-rGO. This catalyst was successfully synthesized by mounting transition metal nitride nanoparticles on nitrogen-doped reduced graphene oxide (RGO) (Scheme 2.7).

The synthesized $TiCoN_x$/N-rGO were investigated for ORR activity and compared with TiN/N-rGO and commercial Pt/C. It performed excellent ORR activity in an alkaline medium with good stability and methanol tolerance. The outstanding performance of the catalyst toward ORR is due to the synergy between the binary transition metal nitride and N-rGO

2.2.4 MIXED-METAL NANOCATALYSTS FOR IMPROVING ORR

To improve ORR activity in the cathode of fuel cells, more and more active catalysts have been explored, one of which is the mixed metal–based nanocatalyst. Attempts are made for Pt-based bi- and trimetallic electrocatalyst, alloying it with transition metals (Stamenkovic et al., 2007; Yano et al., 2007) and a monolayer coating of Pt with metal NPs (Gong et al., 2010; Sasaki et al., 2010)

Wang et al. (2013a) synthesized a stable electrocatalyst for ORR. This work presents an intermetallic core-shell $Pt_3Co@Pt/C$ nanoparticle. The ORR activity of the catalyst is obtained by using an RDE O_2 saturated in 0.1 M $HClO_4$ solution. Pt/C, Pt_3Co/C-400 and Pt_3Co/C-700 electrocatalysts are thin films on RDE. Pt_3Co/C-700 shows higher mass activity than that of Pt_3Co/C-400 and pure Pt.

Noh et al. (2015) proposed a high electrochemical durable ternary PtCuNi NP electrocatalyst for ORR.

To evaluate the ORR activity, d-band center energy module (Hammer et al., 2000) is applied. So, it roughly estimates the relative ORR activity of synthesized electrocatalyst $Pt_{skin}CuNi$ with respect to $Pt_{skin}Cu$, $Pt_{skin}Ni$ and pure Pt NPs. The $Pt_{skin}CuNi$ shows high ORR activity among them all.

To enhance the ORR activity in acidic media, Jiang et al. (2020) synthesized a trimetallic AuPdPb nanowire electrocatalyst. They synthesized core-shell trimetallic catalyst with anisotropic 1D nanostructure (Au@PdPb nanowires) using a two-step water bath method at 60°C.

The ORR activity of Au@PdPb nanowires was investigated by using RDE in an O_2-saturated 1.0 M $HClO_4$ solution. The mass activity for Au@PdPb nanowires at

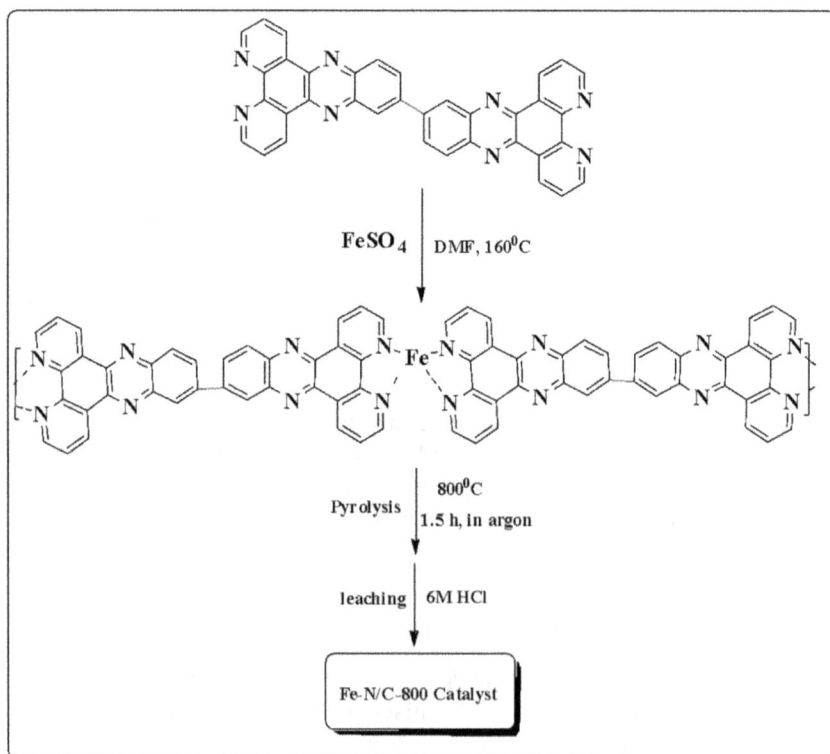

SCHEME 2.6   Schematic synthesis of the catalyst Fe-N/C_800 (Lin et al., 2014).

0.85 V is 1.68 higher than that of commercial Pd black and Au@Pd nanowires. The Au@PdPb nanowires exhibit excellent ORR activity and stability.

2.2.5 METAL OXIDE–BASED NANOCATALYSTS FOR ORR

To achieve the goal of cost-effectiveness, noble metal–free and highly active catalyst for ORR, recently, a study has investigated transition metal compounds, such as transition metal oxide (Menezes et al., 2015), chalcogenides (Wu et al., 2017, Balamurugan et al., 2017) and carbides (Huang et al., 2016). Here we mention some transition metal oxides that are easier to obtain and control but have some drawbacks (mostly they are semiconductors).

Tong et al. (2017) synthesized an efficient bifunctional catalyst, an electrocatalyst that was composed of cobalt oxide (CoO_xNPs/BNG) (Scheme 2.8).

By using a rotating ring-disk electrode (RRDE) in 0.1 M KOH solution, the ORR activity of an electrocatalyst was measured. The number of electrons transferred of the electrocatalyst (CoO_xNPs/BNG) is 4.0 from 0.2–0.6 V, which suggests that it follows 4-e– pathways under alkaline conditions. This work provides an idea for developing a non–noble metal hybrid for the synthesis of an electrocatalyst (Scheme 2.9).

Yan et al. (2012) reported an efficient electrocatalyst Cu_2O nanoparticle dispersed on RGO with a size of about 4 nm (Scheme 2.10).

The ORR process of the Cu_2O/RGO electrode was measured by using RRDE voltammograms. The transferred electron number per oxygen molecule is 3.2–3.6 at the potential range from −0.4 to −1.0 V. This suggests that the electrocatalyst follows a combined pathway of two and four e^-.

To enhance ORR in alkaline media, Raghavendra et al. (2018) reported a RGO-supported bimetallic nano electrocatalyst. Pd@Au core-shell nanoparticles supported

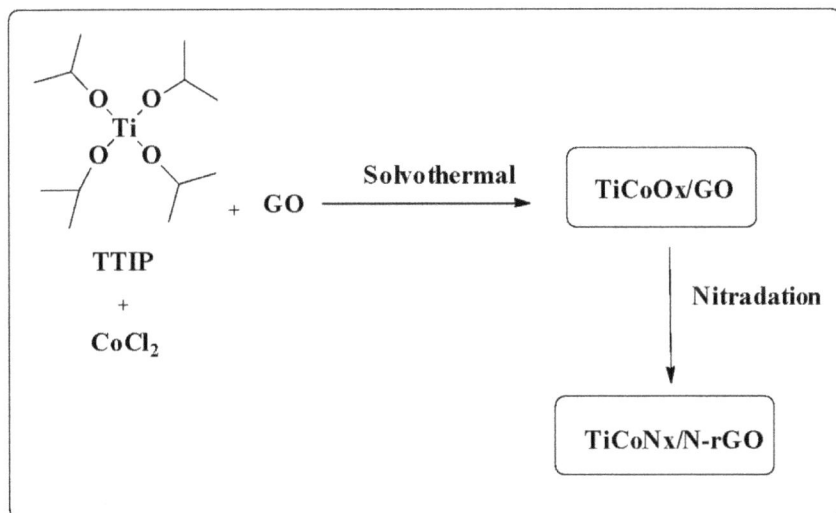

SCHEME 2.7 Synthesis of $TiCoN_x/N$-rGO (Dong et al., 2017).

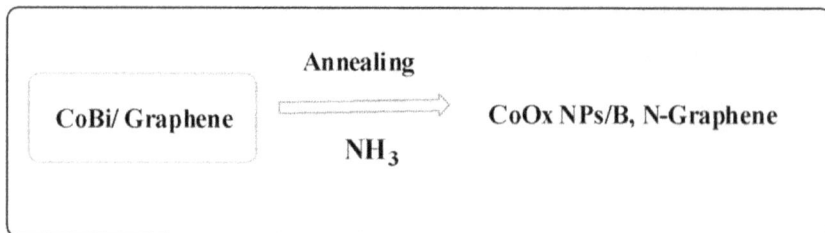

SCHEME 2.8   Synthesis of (CoO$_x$NPs/BNG) (Tong et al., 2017).

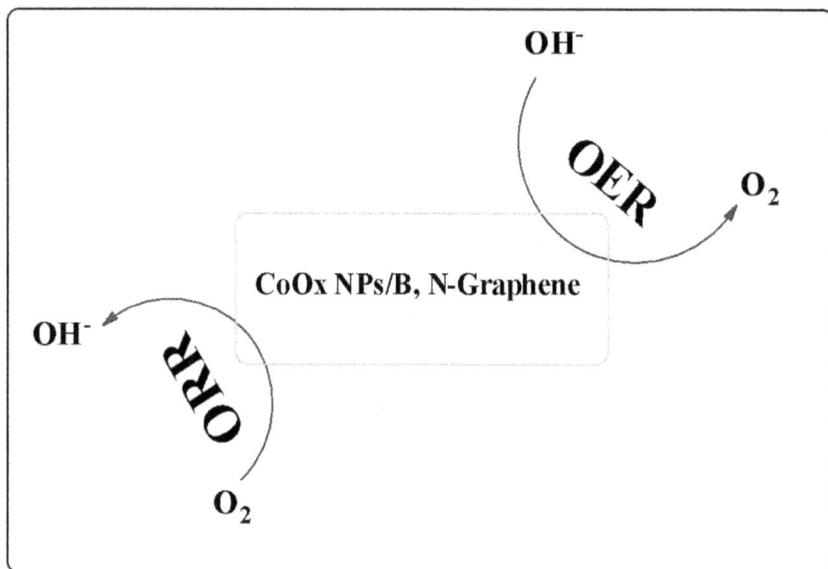

SCHEME 2.9   Graphical representation of (CoO$_x$NPs/BNG) working as a catalyst for ORR and OER (Tong et al., 2017).

on RGO were synthesized through a two-step synthetic route. In the first step, the formation of Pd@Ag/RGO core-shell NPs was done. In the second step, the pure Ag shells are replaced with Au particles (Scheme 2.11).

The electrocatalytic activities of Pd/RGO, Au/RGO and commercial Pt/C catalysts were studied for ORR. The onset potential of the catalyst for ORR activity Pd@Au/RGO (0.183 V) > Au/RGO (0.167 V) > Pt/C (0.164 V) > Pd/RGO (0.156 V).

The high ORR activity of Pd@Au/RGO catalyst is due to the electronic mediation of Pd by Au, which favors facile adsorption of oxygen and its corresponding reduction process.

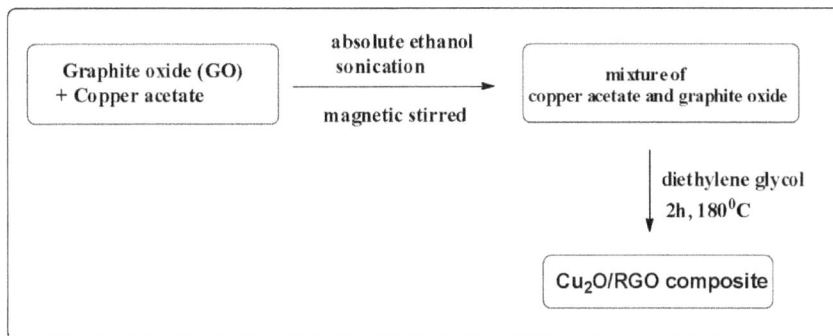

SCHEME 2.10   Representation of synthesis of the Cu_2O/RGO composite (Yan et al., 2012).

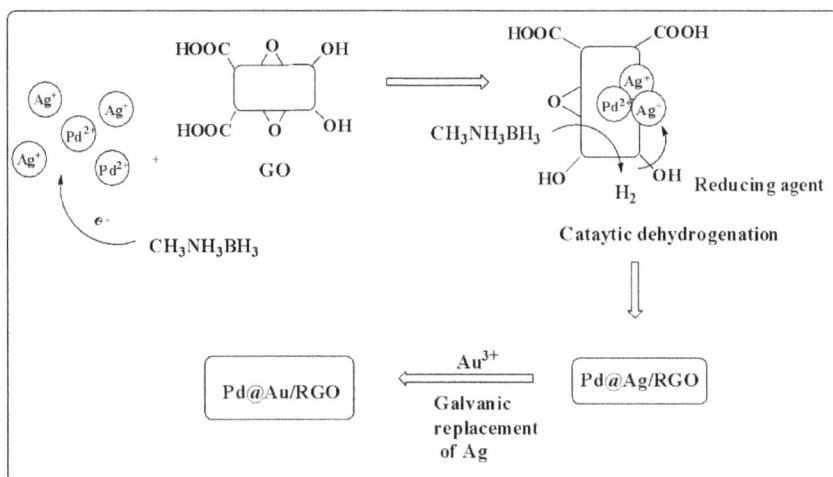

SCHEME 2.11   Schematic representation of synthesis of Pd@Au/RGO (Raghavendra et al., 2018).

2.2.6   NANODENDRITE-BASED ELECTROCATALYSTS FOR ORR

Attempts are made to synthesize nanostructure with different morphologies, such as nanoflowers (Liu et al., 2021), nanodendrites (Cheng et al., 2019), nanoprism (Abkhalimov et al., 2018) and nanorod (Amirtharaj & Mariappan, 2021). Dendritic structure exhibited large surface area, more active sites than other shapes, higher porosity and better interconnectivity in the particles (Maniam et al., 2016). Here we mention some nanodendrite-based catalysts for improving ORR.

Xie et al. (2017) designed an amino acid–assisted strategy for constructing a uniform dendrite-like PtAu porous nanocluster (PtAuPNCs) (Scheme 2.12).

The catalytic activity of PtAuPNCs for ORR was studied using a linear sweep voltammogram (LSV). The electron transfer number was calculated to be 3.87 and 3.62 for PtAuPNCs and Pt black catalysts, respectively, according to the slopes of

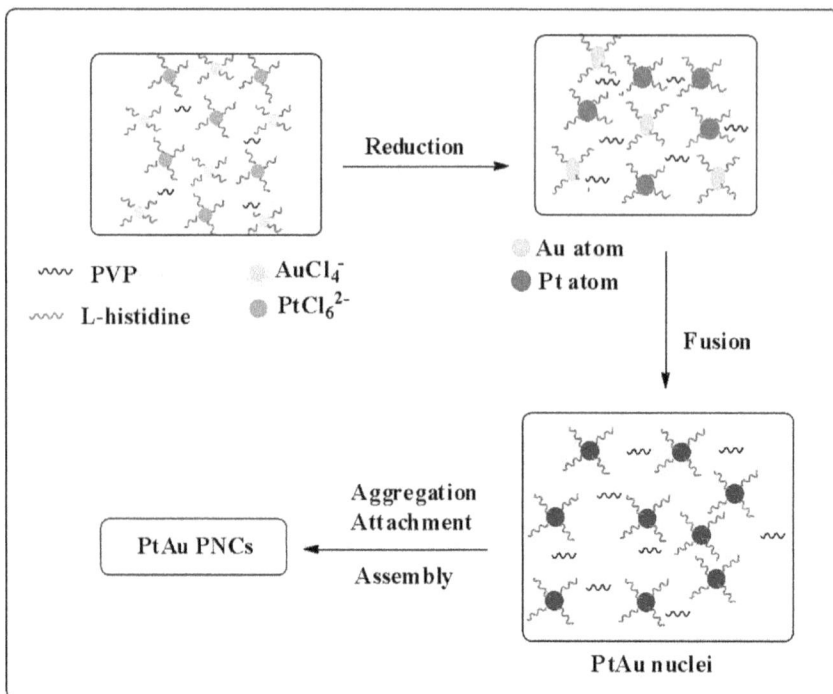

SCHEME 2.12 Illustration of the formation of PtAuPNCs (Xie et al., 2017).

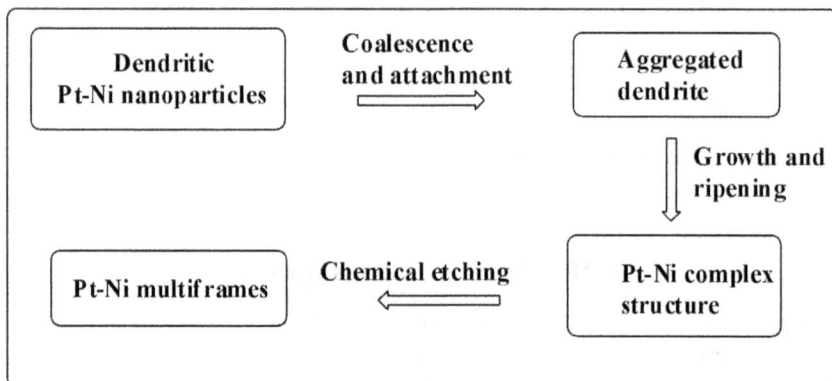

SCHEME 2.13 Schematic representation of formation of the Pt-Ni multiframes (Kwon et al., 2018).

the K-L plots at 0.3V. It shows that for the ORR direct 4-e⁻ transfer pathway was followed by a PtAuPNC catalyst.

Kwon et al. (2018 successfully synthesized dendrite-embedded bimetallic Pt-Ni multiframes for ORR (Scheme 2.13).

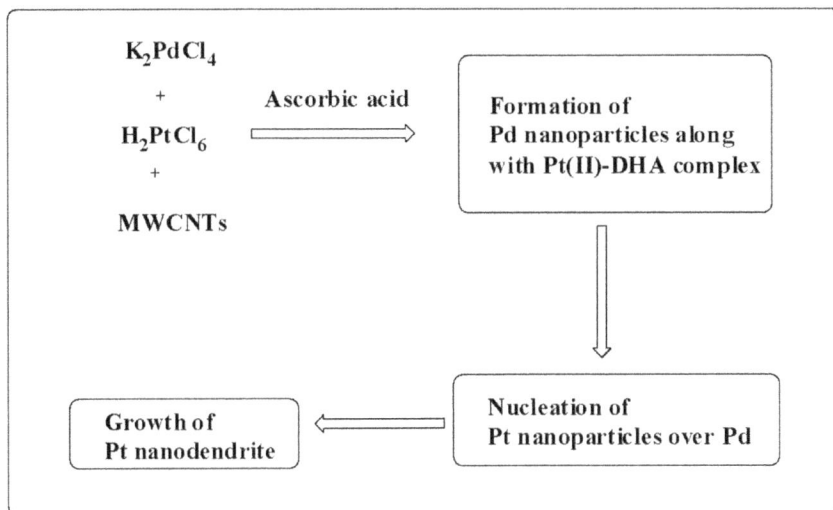

SCHEME 2.14   Schematic synthesis of Pt-on-Pd nanodendrite (Ghosh et al., 2014).

The Pt-Ni multiframe shows a 30-fold higher electrochemically active surface area (ECSA) and mass ORR activity than that of the state-of-the-art Pt/C catalyst. This bimetallic Pt-Ni multiframe showed high catalytic activity and durability even after 1000 cycles.

Ghosh et al. (2014) reported Pt-on-Pd dendritic nanostructures of different compositions and their catalytic activity toward ORR. The growth of bimetallic nano-structure of different compositions ($Pt_{64}Pd_{36}$, $Pt_{52}Pd_{48}$ and $Pt_{21}Pd_{79}$) over multiwalled carbon nanotubes (MWCNTs) were synthesized (Scheme 2.14).

The electroactivity of the Pt-on-Pd electrocatalyst for ORR was studied using an RRDE. Mass activity increases in the order $Pt_{52}Pd_{48}$ < $Pt_{21}Pd_{79}$ < $Pt_{64}Pd_{36}$ are due to the change in the electronic state of the electrocatalyst. It shows high activity for ORR.

2.2.7   2D Nanomaterial-Based Catalysts for ORR

For improving electrocatalytic performances, it is important to maximize the surface of the catalyst as the reaction occurs on it. So various attempts are taken to fabricate nanostructures (1D, 2D and 3D).

Here we mention some fabricated 2D nanomaterials for enhancing ORR.

Lyu et al. (2020) synthesized a 2D PtPd bimetallic nano-electrocatalyst for ORR (Scheme 2.15).

The ORR activity of the 2D nanocatalyst was studied using the K-L equation; the calculated n value obtained from the K-L equation is approximately 4.0, which suggests that it follows 4-e$^-$ pathways. The aspect, which enhances the catalytic activity of a nano-catalyst for ORR, is the porous structure and the synergistic effect of Pt alloying with Pd.

To boost the ORR activity, Sahoo et al. (2020) reported a 2D nanocube Pd-based catalyst. It is a facile one-pot synthesis of PdNCs from halide ions and polyvinylpyr-rolidone (PVP) in water under the precise control of acid etching (Scheme 2.16).

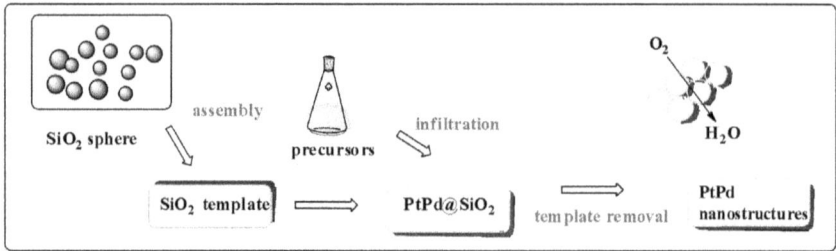

SCHEME 2.15 Systematic synthetic scheme of 2D PtPd bimetallic nanostructure (Lyu et al., 2020).

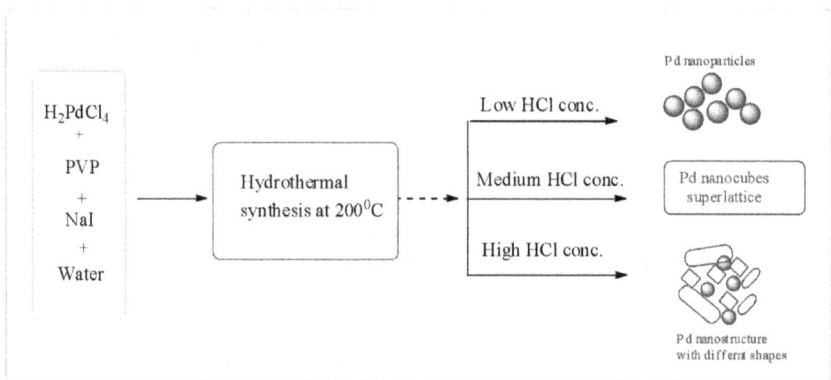

SCHEME 2.16 Illustration of synthesis of 2D Pd nanotubes (Sahoo et al., 2020).

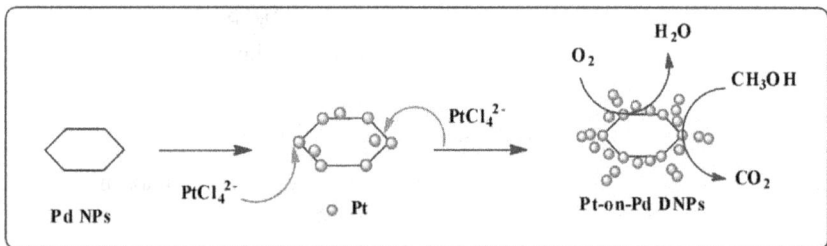

SCHEME 2.17 Schematic representation of 2D Pt-on-Pd DNPs (Peng et al., 2020).

LSVs recorded in O_2-saturated 0.1 M KOH were used to investigate the activity of catalysts toward ORR. It had excellent electrocatalytic activity, with stability for ORR and methanol oxidation reaction (MOR).

To enhance fuel cell performance, Peng et al. (2020) synthesized a bifunctional 2D Pt-on-Pd dendritic nanosheet (Scheme 2.17). This bimetallic electrocatalyst has a hexagonal Pd nanosheet core with spatially separated Pt branches by seed-mediated growth.

2.2.8 3D-NANOMATERIAL-BASED CATALYSTS FOR ORR

Here we mention some fabricated 3D nanomaterials for enhancing ORR.

Wu et al. (2012) reported a 3D nanocatalyst for ORR. They synthesized a 3D N-doped graphene aerosol supported on Fe_3O_4 NPs (Fe_3O_4/NGAs) (Scheme 2.18).

By studying the ORR activity of (Fe_3O_4/NGAs) catalyst, they found that the catalyst showed excellent catalytic activity for ORR, with a high current density and a lower H_2O_2 yield in the alkaline medium.

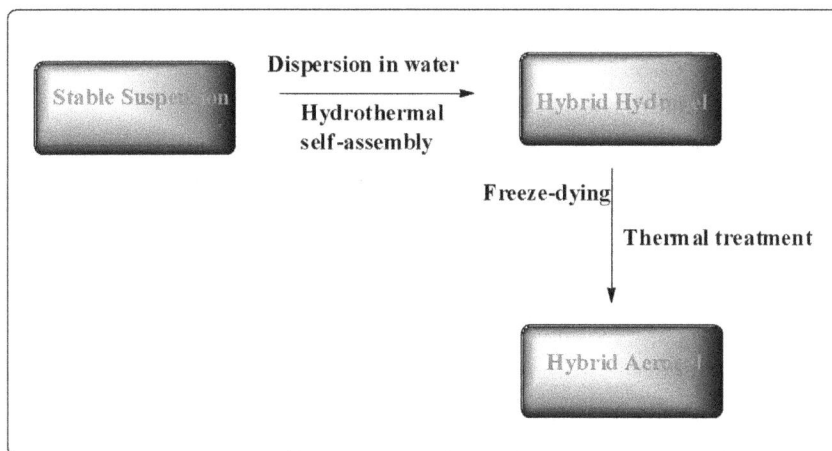

SCHEME 2.18 Fabrication of Fe_3O_4/NGAs catalyst stable suspension of GO, iron acetate and polypyrrole (ppy) was formed and then treated with hydrothermal self-assembly. Fe and ppy supporting graphene hybrid hydrogel were prepared. Then by freeze-drying and thermal treatment, a monolithic (Fe_3O_4/NGAs) hybrid aerogel was obtained (Wu et al., 2012).

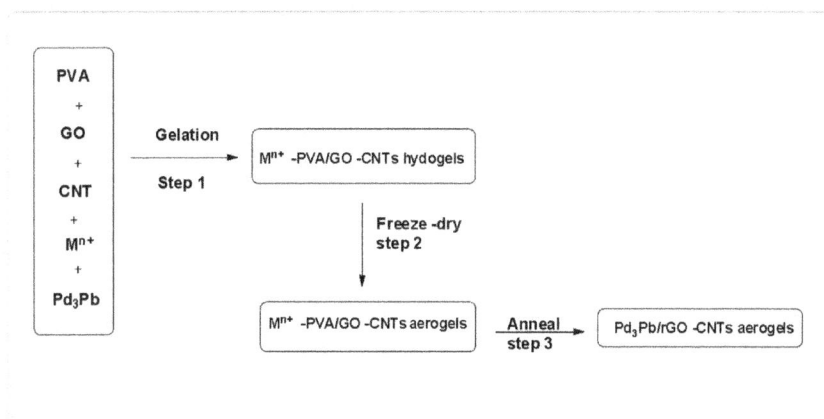

SCHEME 2.19 Schematic representation of 3D hydrogel PVA/GO CNTs (Fu et al., 2018).

Fu et al. (2018) reported an aerogel for the high activity of electrocatalyst for ORR. The novel electrocatalyst contains 3D hydrogel PVA/GO CNTs, which allows efficient capture of Pd_3Pb NPs (Scheme 2.19).

To evaluate the electrocatalytic process of catalyst for ORR. RDE in O_2-saturated 0.1 M KOH solution was investigated and compared with commercial catalysts. From the slopes of the K-L plots, the average electron transfer number of Pd_3Pb/rGOCNTs is 3.84 in the range of 0.2–0.7 V.

It has high activity and stability for ORR. The 3D interconnected network prevents it from aggregation, leaching and provides fast accessibility of reactants to the Pd_3Pb active sites.

2.2.9 GRAPHENE-SUPPORTED NOBLE METAL ELECTROCATALYSTS AND THEIR APPLICATIONS

Graphene, an allotrope of carbon, has unique properties which show its wide application in various fields. Here we mention some graphene-based electrocatalysts that performed excellently in fuel cell reactions.

TABLE 2.1

Some Electrocatalyst with Their Application (Liu et al., 2014b)

Material	Application	Ref
Pt/GNs	MOR	(Zhang et al., 2010)
	ORR	(Jafri et al., 2010)
	MOR/ORR	(Li et al., 2009a; Ha et al., 2011)
	MOR	(Li et al., 2010b)
	MOR/ORR	(Qiu et al., 2011)
	MOR	(Qiu et al., 2013)
	MOR	(Maiyalagan et al., 2012)
	MOR	(Liu et al., 2011)
	MOR	(Kundu et al., 2011)
	MOR/HOR	(Yoo et al., 2009, 2011)
PtPd/GNs	MOR	(Lu et al., 2013)
	MOR	(Guo et al., 2010)
	MOR/EOR	(Ji et al., 2012a; Yang et al., 2012)
	MOR	(Lu et al., 2014)
PtAu/GNs	FAOR	(Rao et al., 2011a)
	FAOR	(Zhang et al., 2011a)
	MOR/ORR	(Hu et al., 2011)
	MOR	(Yang et al., 2013)
PtRu/GNs	MOR/EOR	(Dong et al., 2010)
	MOR	(Wang et al., 2013b)
	MOR	(Lv et al., 2011)
	MOR	(Li et al., 2010a)
	MOR	(Woo et al., 2013)

Material	Application	Ref
PtCo/GNs	MOR	(Huang et al., 2012)
	ORR	(Yan et al., 2013)
	ORR	(Ma et al.,2012)
	MOR/ORR	(Shen et al., 2012; Yue et al., 2010)
	ORR	(Nam et al., 2012)
PtNi/GNs	MOR	(Luo et al., 2013)
	MOR	(Luo et al., 2012)
	ORR	(Zhang et al., 2011)
	MOR	(Hu et al., 2012b)
	MOR	(Hu et al., 2012c)
PtFe/GNs	ORR	(Guo & Sun, 2012)
	MOR	(Ji et al., 2012b)
PtCu/GNs	MOR	(Liu et al., 2012)
PtSn/GNs	MOR	(Anandan et al., 2012)
	MOR	(Han et al., 2012)
PtCr/GNs	ORR	(Rao et al., 2011b)
PtAg/GNs	MOR	(Feng et al., 2011)
PtPdAu/GNs	MOR	(Zhang et al., 2011b)
PtPdCu/GNs	EOR	(Hu et al., 2012a)
Pd/GNs	FAOR/EOR	(Chen et al., 2011b)
	FAOR	(Yang et al., 2011b)
	MOR	(Zhao et al., 2011)
	ORR/FAOR	(Giovanni et al., 2012; Bong et al., 2010)
	FAOR	Jiang et al., 2012)
Pd-MnO$_2$/RGO	MOR	(Liu et al., 2013c)
AuPd/GNs	ORR	(Yang et al., 2011a)
PdRu/GNs	MOR	(Awasthi & Singh, 2013)
PdSn/GNs	MOR	(Awasthi & Singh, 2012a)
PdAu/GNs	FAOR	(Chai et al., 2011)
PdAg/GNs	ORR	(Liu et al., 2013b)
PdRuSn/GNs	MOR	(Awasthi & Singh, 2012b)
Pd$_3$Y/GNs	ORR/EOR	(Seo et al., 2013)
Au/GNs	ORR	(Zhang et al., 2012)
Au/GNs	ORR	(Li et al., 2009b)
Au/GNs	ORR	(Hu et al., 2010)

2.2.10   GRAPHENE-SUPPORTED NONPRECIOUS METAL ELECTROCATALYSTS AND THEIR APPLICATIONS

TABLE 2.2
Various Electrocatalysts and Their Application for Oxygen Reduction Reaction (Liu et al., 2014b)

Catalyst	Application	Ref
Fe_3O_4/N-graphene	ORR in alkaline	(Wu et al., 2012)
Co_3O_4/(N-)rGO	ORR/EOR in alkaline	(Liang et al., 2011)
Cu_2O/rGO	ORR in alkaline	(Yan et al., 2012)
CuO/N-rGO	ORR in alkaline	(Zhou et al., 2013)
MnO_2/rGO	ORR in alkaline	Qian et al., 2011)
Ir_xV/rGO	ORR in alkaline	(Zhang et al., 2013)

2.3   CONCLUSION

To develop sustainable energy sources for the future, different steps have been taken in a greener way. Fuel cells are an excellent approach to becoming a pollution-free world. Fuel cells show some drawbacks that are reduced by applying nanotechnology to their working. Various nanomaterials are synthesized in this respect to improve the reaction in the fuel cell. ORR, which occurs in the cathode part of the fuel cell, is one of the important reactions in fuel cells, and to increase it, nano electrocatalysts are applied. By using nanocatalysts, fuel production increases in a greener way, with a low cost, excellent performance and better durability. They provide wide active sites for reactions and increase conductivity and resistance to corrosion. As discussed earlier, nano electrocatalyst (noble metal–based nano-electrocatalysts, metal nano-clusters, transition metal–based and mixed metal nanocatalysts, metal oxide–based nanocatalysts, nanodendrite-based electrocatalysts, 2D and 3D nanomaterials-based catalyst) have excellent performances in an eco-friendly manner.

ACKNOWLEDGEMENTS

One of the authors, Chandni Singh, gratefully acknowledges the University Grant Commission, India, for awarding a fellowship.

REFERENCES

Abkhalimov, E. V., Timofeev, A. A., Ershov, B. G. (2018) Electrochemical mechanism of silver nanoprisms transformation in aqueous solutions containing the halide ions. *J Nanopart Res. 20, 26.*

Amirtharaj, S. N., Mariappan, M. (2021) Rapid and controllable synthesis of Mn_2O_3 nanorods via a sonochemical method for supercapacitor electrode application. *Appl. Phys. A. 127, 607.*

Anandan, S., Manivel, A., Ashokkumar, M. (2012) One-step sonochemical synthesis of reduced graphene oxide/Pt/Sn hybrid materials and their electrochemical properties. *Fuel Cells 12, 956–962.*

Awasthi, R., Singh, R. N. (2012a) Optimization of the Pd-Sn-GNS nanocomposite for enhanced electrooxidation of methanol. *Int. J. Hydrogen Energy 37, 2103–2110.*

Awasthi, R., Singh, R. N. (2012b) Synthesis and structural characterization of a ternary palladium—ruthenium—tin nanoalloy supported on graphene nanosheets for methanol electrooxidation in alkaline medium. *Catal. Sci. Technol. 2, 2428–2432.*

Awasthi, R., Singh, R. N. (2013) Graphene-supported Pd-Ru nanoparticles with superior methanol electrooxidation activity. *Carbon 51, 282–289.*

Balamurugan, J., Peera, S. G., Guo, M., Nguyen, T. T., Kim, N. H., Lee, J. H. (2017) A hierarchical 2D Ni-Mo-S nanosheet@nitrogen doped graphene hybrid as a Pt-free cathode for high-performance dye sensitized solar cells and fuel cells. *J Mater Chem A 5, 17896–17908.*

Bisen, O. Y., Nanda, K. K. (2020a) Alkaline earth metal based single atom catalyst for the highly durable oxygen reduction reaction. *Appl. Mater. Today 21, 100846.*

Bisen, O. Y., Nanda, K. K. (2020b) Unique one-step strategy for nonmetallic and metallic heteroatom doped carbonaceous materials. *ACS Omega 5(51), 32852–32860.*

Bisen, O. Y., Nanda, K. K. (2021) Uniform distribution of ruthenium nanoparticles on nitrogen doped carbon nanostructure for oxygen reduction reaction. *ACS Appl. Energy Mater. 4, 12191–12200.*

Blomen, L. J. M. J., Mugerwa, M. N. (1993) *Fuel Cell Systems*, Plenum Press, New York.

Bong, S., Uhm, S., Kim, Y. R., Lee, J., Kim, H. (2010) Graphene supported Pd electrocatalysts for formic acid oxidation. *Electrocatalysis 1, 139–143.*

Chai, J., Li, F.H., Hu, Y. W., Zhang, Q. X., Han, D. X., Niu, L. (2011) Hollow flower-like AuPd alloy nanoparticles: One step synthesis, self-assembly on ionic liquid-functionalized graphene, and electrooxidation of formic acid. *J. Mater. Chem. 21, 17922–17929.*

Chen, T., Cai, Z., Yang, Z., Li, L., Sun, X., Huang, T., Yu, A., Kia, H. G., Peng, H. (2011a) Nitrogen-doped carbon nanotube composite fiber with a core—sheath structure for novel electrodes. *Adv. Mater. 23, 4620–4625.*

Chen, X. M., Wu, G. H., Chen, J. M., Chen, X., Xie, Z. X., Wang, X. R. (2011b) Synthesis of "clean" and well-dispersive Pd nanoparticles with excellent electrocatalytic property on graphene oxide. *J. Am. Chem. Soc. 133, 3693–3695.*

Cheng, Z.-Q., Li, Z.-W., Xu, J.-H., Yao, R., Li, Z.-L., Liang, S., Cheng, G.-L., Zhou, Y.-H., Luo, X., Zhong, J. (2019) Morphology-controlled fabrication of large-scale dendritic silver nanostructures for catalysis and SERS applications. *Nanoscale Res. Lett. 14, 89.*

Deng, D., Pan, X., Yu, L., Cui, Y., Jiang, Y., Qi, J., Li, W-X., Fu, Q., Ma, X., Xue, Q., Sun, G., Bao, X. (2011) Toward N-doped graphene via solvothermal synthesis. *Chem. Mater. 23, 1188–1193.*

Dong, L. F., Gari, R. R. S., Li, Z., Craig, M. M., Hou, S. F. (2010) Graphene-supported platinum and platinum—ruthenium nanoparticles with high electrocatalytic activity for methanol and ethanol oxidation. *Carbon 48, 781–787.*

Dong, Y., Deng, Y., Zeng, J., Son, H., Liao, S. (2017) A high-performance composite ORR catalyst based on the synergy between binary transition metal nitride and nitrogen-doped reduced graphene oxide. *J. Mater. Chem. A 5, 5829–5837.*

Esposito, D. V., Chen, J. G. (2011) Monolayer platinum supported on tungsten carbides as low-cost electrocatalysts: Opportunities and limitations. *Energy Environ. Sci. 4, 3900–3912.*

Feng, L. L., Gao, G., Huang, P., Wang, A. S., Zhang, C. L., Zhang, J. L., Guo, S. W., Cui, D. X. (2011) Preparation of Pt Ag alloy nanoisland/graphene hybrid composites and its high stability and catalytic activity in methanol electro-oxidation. *Nanoscale Res. Lett. 6, 551.*

Fu, G., Liu, Yu., Wu, Z., Lee, J.-M. (2018) 3D robust carbon aerogels immobilized with Pd$_3$Pb nanoparticles for oxygen reduction catalysis. *ACS Appl. Nano Mater. 1, 4, 1904–1911.*

Ghosh, S., Mondal, S., Retna Raj, C. (2014) Carbon nanotube-supported dendritic Pt-on-Pd nanostructures: Growth mechanism and electrocatalytic activity towards oxygen reduction reaction. *J. Mater. Chem. A 2, 2233–2239.*

Giovanni, M., Poh, H. L., Ambrosi, A., Zhao, G. J., Sofer, Z., Sanek, F., Khezri, B., Webster, M. Pumera, R. D. (2012) Noble metal (Pd, Ru, Rh, Pt, Au, Ag) doped graphene hybrids for electrocatalysis. *Nanoscale 4, 5002–5008.*

Gong, K. P., Su, D., Adzic, R. R. (2010) Platinum-monolayer shell on $AuNi_{0.5}Fe$ nanoparticle core electrocatalyst with high activity and stability for the oxygen reduction reaction. *J. Am. Chem Soc. 132, 14364–14366.*

Guo, S. J., Dong, S. J., Wang, E. K. (2010) Three-dimensional Pt-on-Pd bimetallic nanodendrites supported on graphene nanosheet: Facile synthesis and used as an advanced nano-electrocatalyst for methanol oxidation. *ACS Nano, 4, 547–555.*

Guo, S. J., Sun, S. H. (2012) FePt nanoparticles assembled on graphene as enhanced catalyst for oxygen reduction reaction. *J. Am. Chem. Soc. 134, 2492–2495.*

Ha, H. W., Kim, I. Y., Hwang, S. J., Ruoff, R. S. (2011) One-pot synthesis of platinum nanoparticles embedded on reduced graphene oxide for oxygen reduction in methanol fuel cells. *Electrochem. Solid-State Lett. 14, B70.*

Hammer, B., Nørskov, J. K. (2000) Theoretical surface science and catalysis—calculations and concepts. *Adv. Catal. 45, 71–129.*

Han, F., Wang, X. M., Lian, J., Wang, Y. Z. (2012) The effect of Sn content on the electrocatalytic properties of Pt-Sn nanoparticles dispersed on graphene nanosheets for the methanol oxidation reaction. *Carbon 50, 5498.*

Herman, A., Chaudhuri, T., Spagnol, P. (2005) Bipolar plates for PEM fuel cells: A review. *Int. Journal of Hydrogen Energy 30, 1297–1302.*

Herper, G. D. J. (2008) *Fuel Cell Projects for the Evil Genius*, McGraw-Hill, New York, NY.

Hu, C. G., Cheng, H. H., Zhao, Y., Hu, Y., Liu, Y., Dai, L. M., Qu, L. T. (2012a) Newly-designed complex ternary Pt/PdCu nanoboxes anchored on three-dimensional graphene framework for highly efficient ethanol oxidation. *Adv. Mater. 24, 5493–5498.*

Hu, Y. J., Jin, J. A., Wu, P., Zhang, H., Cai, C. X. (2010) Graphene—gold nanostructure composites fabricated by electrodeposition and their electrocatalytic activity toward the oxygen reduction and glucose oxidation *Electrochim. Acta, 56, 491–500.*

Hu, Y. J., Wu, P., Yin, Y. J., Zhang, H., Cai, C. X. (2012b) Effects of structure, composition, and carbon support properties on the electrocatalytic activity of Pt-Ni-graphene nanocatalysts for the methanol oxidation. *Appl. Catal. B 111, 208–217.*

Hu, Y. J., Wu, P., Zhang, H., Cai, C. X. (2012c) Synthesis of graphene-supported hollow Pt-Ni nanocatalysts for highly active electrocatalysis toward the methanol oxidation reaction. *Electrochim. Acta 85, 314.*

Hu, Y. J., Zhang, H., Wu, P., Zhang, H., Zhou, B., Cai, C. X. (2011) Bimetallic Pt-Au nanocatalysts electrochemically deposited on graphene and their electrocatalytic characteristics towards oxygen reduction and methanol oxidation. *Phys. Chem. Chem. Phys. 13, 4083–4094.*

Huang, H., Du, C., Wu S., Song, W. (2016) Thermolytical entrapment of ultrasmall MoC Nanoparticles into 3D frameworks of nitrogen-rich graphene for efficient oxygen reduction. *J Phys Chem C 120, 15707–15713.*

Huang, H. J., Sun, D. P., Wang, X. (2012) PtCo alloy nanoparticles supported on graphene nanosheets with high performance for methanol oxidation. *Chin. Sci. Bull. 57, 3071–3079.*

Hwang, S.-M., Choi, Y., Kim, M. G., Sohn, Y. J., Cheon, J. Y., Joo, S. H., Yim, S. D., Kuttiviel, K. A., Sasaki, K., Adzic, R. R., Park, G. G.(2016) Enhancement of oxygen reduction reaction activities by Pt nanoclusters decorated on ordered mesoporous porphyrinic carbons. *J. Mater. Chem. A. 4, 5869–5876.*

Jafri, R. I., Rajalakshmi, N., Ramaprabhu, S. (2010) Nitrogen doped graphene nanoplatelets as catalyst support for oxygen reduction reaction in proton exchange membrane fuel cell. *J. Mater. Chem. 20, 7114–7117.*

Ji, H. Q., Li, M. G., Wang, Y. L., Gao, F. (2012a) Electrodeposition of graphene-supported PdPt nanoparticles with enhanced electrocatalytic activity. *Electrochem. Commun. 2012, 24, 17–20.*

Ji, Z. Y., Zhu, G. X.,. Shen, X. P., Zhou, H., Wu, C. M., Wang, M.(2012b) Reduced graphene oxide supported FePt alloy nanoparticles with high electrocatalytic performance for methanol oxidation. *New J. Chem. 36, 1774–1780.*

Jiang, X., Xiong, Y., Zhao, R., Zhou, J., Lee, J.-M., Tang, Y. (2020) Trimetallic Au@PdPb nanowires for oxygen reduction reaction. *NanoRes. 13(10), 2691–2696.*

Jiang, Y. Y., Lu, Y. Z., Li, F. H., Wu, T. S., Niu, L., Chen, W. (2012) Facile electrochemical codeposition of "clean" graphene—Pd nanocomposite as an anode catalyst for formic acid electrooxidation. *Electrochem. Commun. 19, 21–24.*

Kim, A., Muthuchamy, N., Yoon, C., Joo, S. H., Park, K. H. (2018) MOF-derived Cu@Cu$_2$O nanocatalyst for oxygen reduction reaction and cycloaddition reaction. *Nanomaterials 8, 138.*

Kundu, P., Nethravathi, C., Deshpande, P. A., Rajamathi, M., Madras, G., Ravishankar, N. (2011) Ultrafast microwave-assisted route to surfactant-free ultrafine Pt nanoparticles on graphene: Synergistic co-reduction mechanism and high catalytic activity. *Chem. Mater. 23, 2772–2780.*

Kwon, H., Kabiraz, M. K., Park, J., Oh, A., Baik, H., Choi, S.-I., Lee, K. (2018) Dendrite-embedded platinum—nickel multiframes as highly active and durable electrocatalyst toward the oxygen reduction reaction. *Nano Lett. 18, 2930–2936.*

Li, F. H., Yang, H. F., Shan, C. S., Zhang, Q. X., Han, D. X., Ivaska, A., Niu, L. (2009a) The synthesis of perylene-coated graphene sheets decorated with Au nanoparticles and its electrocatalysis toward oxygen reduction. *J. Mater. Chem. 19, 4022–4025.*

Li, H. Y., Zhang, X. H., Pang, H. L., Huang, C. T., Chen, J. H. (2010a) PMo$_{12}$-functionalized graphene nanosheet-supported PtRu nanocatalysts for methanol electro-oxidation. *J. Solid State Electrochem. 14, 2267–2274.*

Li, Y. J., Gao, W., Ci, L. J., Wang, C. M., Ajayan, P. M. (2010b) Catalytic performance of Pt nanoparticles on reduced graphene oxide for methanol electro-oxidation. *Carbon 48, 1124–1130.*

Li, Y. M., Tang, L. H., Li, J. H. (2009b) Preparation and electrochemical performance for methanol oxidation of Pt/graphene nanocomposites. *Electrochem. Commun. 11, 846–849.*

Liang, Y. Y., Li, Y. G., Wang, H. L., Zhou, J. G., Wang, J., Regier, T., Dai, H. (2011) Co$_3$O$_4$ nanocrystals on graphene as a synergistic catalyst for oxygen reduction reaction. *J. Nat. Mater. 10, 780–786.*

Lin, C., Wu, G., Li, H., Geng, Y., Xie, G., Yanga, J., Liu, B., Jin, J. (2017) Rh nanoparticles supported on ultrathin carbon nanosheets for high-performance oxygen reduction reaction and catalytic hydrogenation. *Nanoscale 9, 1834–1839.*

Lin, L., Zhu, Q., Xu, A-W. (2014a) Noble-metal-free Fe-N/C catalyst for highly efficient oxygen reduction reaction under both alkaline and acidic conditions. *J. Am. Chem. Soc. 136, 11027–11033.*

Liu, M., Chen, W. (2013a) Green synthesis of silver nanoclusters supported on carbon nanodots: Enhanced photoluminescence and high catalytic activity for oxygen reduction reaction. *Nanoscale 5, 12558–12564.*

Liu, M., Zhang, R., Chen, W. (2014b) Graphene-supported nanoelectrocatalysts for fuel cells: Synthesis, properties and applications. *Chem. Rev. 114(10), 5117–5160.*

Liu, M. M., Lu, Y. Z., Chen, W. (2013b) PdAg nanorings supported on graphene nanosheets: Highly methanol-tolerant cathode electrocatalyst for alkaline fuel cells. *Adv. Funct. Mater. 23, 1289–1296.*

Liu, R., Zhou, H. H., Liu, J., Yao, Y., Huang, Z. Y., Fu, C. P., Kuang, Y. F. (2013c) Preparation of Pd/MnO$_2$-reduced graphene oxide nanocomposite for methanol electro-oxidation in alkaline media. *Electrochem. Commun. 26, 63–66.*

Liu, X. W., Mao, J. J., Liu, P. D., Wei, X. W. (2011) Fabrication of metal-graphene hybrid materials by electroless deposition. *Carbon 49, 477–483.*

Liu, Y. J., Huang, Y. Q., Xie, Y., Yang, Z. H., Huang, H. L., Zhou, Q. Y. (2012) Preparation of highly dispersed CuPt nanoparticles on ionic-liquid-assisted graphene sheets for direct methanol fuel cell. *Chem. Eng. J. 197, 80–87.*

Liu, Y. J., Niu, S., Hu, R. (2021) $FeCo_2S_4/NH_2$-GQDs nanoflowers supported on Ni foam as novel binder-free electrode for high-performance supercapacitors. *Appl. Phys. A. 127, 417.*

Lu, Y. Z., Jiang, Y. Y., Chen, W. (2014) Graphene nanosheet-tailored PtPd concave nanocubes with enhanced electrocatalytic activity and durability for methanol oxidation. *Nanoscale 6, 3309–3315.*

Lu, Y. Z., Jiang, Y. Y., Wu, H. B., Chen, W. (2013) Nano-PtPd cubes on graphene exhibit enhanced activity and durability in methanol electrooxidation after CO stripping—cleaning. *J. Phys. Chem. C 117, 2926–2938.*

Luo, B. M., Xu, S., Yan, X. B., Xue, Q. J. (2012) Synthesis and electrochemical properties of graphene supported PtNi nanodendrites. *Electrochem. Commun. 23, 72–75.*

Luo, B. M., Xu, S., Yan, X. B., Xue, Q. J. (2013) PtNi alloy nanoparticles supported on polyelectrolyte functionalized graphene as effective electrocatalysts for methanol oxidation. *J. Electrochem. Soc. 160, F262.*

Lv, R. T., Cui, T. X., Jun, M. S., Zhang, Q., Cao, A. Y., Su, D. S., Zhang, Z. J., Yoon, S. H., Miyawaki, J., Mochida, I., Kang, F. Y. (2011) Open-ended, N-doped carbon nanotube—graphene hybrid nanostructures as high-performance catalyst support. *Adv. Funct. Mater. 21, 999–1006.*

Lyu, X., Zhang, W.-N., Li, G., Shi, B-W., Zhang, Y.-N., Chen, H., Li, S.-C., Wang, X. (2020) Two-dimensional porous PtPd nanostructure electrocatalysts for oxygen reduction reaction. *Appl. Nano Mater. 3, 8586–8591.*

Ma, Y. W., Liu, Z. R., Wang, B. L., Zhu, L., Yang, J. P., Li, X. A. (2012) Preparation of graphene-supported Pt-Co nanoparticles and their use in oxygen reduction reactions. *New Carbon Mater. 27, 250–257.*

Maiyalagan, T., Dong, X. C., Chen, P., Wang, X. (2012) Electrodeposited Pt on three-dimensional interconnected graphene as a free-standing electrode for fuel cell application. *J. Mater. Chem. Commun., 22, 5286–5290.*

Maniam, K. K., Muthukumar, V., Chetty, R. (2016) Electrodeposition of dendritic palladium nanostructures on carbon support for direct formic acid fuel cells *Int. J Hyd Ener 41, 18602–18609.*

Markovic, N. M., Gasteiger, H. A., Ross, P. N. (1995) Oxygen reduction on platinum low-index single-crystal surfaces in sulfuric acid solution: Rotating ring-Pt(hkl) disk studies *J. Phys. Chem. 99, 3411–3415.*

Mench, M. M. (2008) *Fuel Cell Engines,* John Wiley and Sons Ltd., London.

Menezes, P. W., Indra, A., González-Flores, D., Sahraie, N. R., Zaharieva, I., Schwarze, M., Strasser, P., Dau, H., Driess, M. (2015) High-performance oxygen redox catalysis with multifunctional cobalt oxide nanochains: Morphology-dependent activity. *ACS Catalysis, 5, 2017–2027.*

Nam, K. W., Song, J., Oh, K. H., Choo, M. J., Park, H., Park, J. K., Choi, J. W. (2012) Monodispersed PtCo nanoparticles on hexadecyltrimethylammonium bromide treated graphene as an effective oxygen reduction reaction catalyst for proton exchange membrane fuel cells. *Carbon, 50, 3739–3747.*

Noh, S. H., Han, B., Ohsaka, T. (2015) First-principles computational study of highly stable and active ternary PtCuNi nanocatalyst for oxygen reduction reaction. *Nano Research, 8(10), 3394–3403.*

Peng, X., Lu, D., Qin, Y., Li, M., Guo, Y., Guo, S. (2020) Pt-on-Pd dendritic nanosheets with enhanced bifunctional fuel cell catalytic performance. *ACS Appl. Mater. Interfaces 12, 27, 30336–30342.*

Qian, Y., Lu, S. B., Gao, F. L. (2011) Synthesis of manganese dioxide/reduced graphene oxide composites with excellent electrocatalytic activity toward reduction of oxygen. *Mater. Lett. 65, 56–58.*

Qiu, H. J., Dong, X. C., Sana, B., Peng, T., Paramelle, D., Chen, P., Lim, S. (2013) Ferritin-templated synthesis and self-assembly of Pt nanoparticles on a monolithic porous graphene network for electrocatalysis in fuel cells. *ACS Appl. Mater. Interfaces 5, 782–787.*

Qiu, J. D., Wang, G. C., Liang, R. P., Xia, X. H., Yu, H. W. (2011) Controllable deposition of platinum nanoparticles on graphene as an electrocatalyst for direct methanol fuel cells. *J. Phys. Chem. C 115, 15639–15645.*

Raghavendra, P., Reddy, G. V., Sivasubramanian, R., Chandana, P. S., Sarma, L. S. (2018) Reduced graphene oxide-supported Pd@Au bimetallic nano electrocatalyst for enhanced oxygen reduction reaction in alkaline media. *Int J. Hyd. Ener, 22, 43(8), 4125–4135.*

Rao, C. V., Cabrera, C. R., Ishikawa, Y. (2011a) Graphene-supported Pt-Au alloy nanoparticles: A highly efficient anode for direct formic acid fuel cells. *J. Phys. Chem. C 115, 21963–21970.*

Rao, C. V., Reddy, A. L. M., Ishikawa, Y., Ajayan, P. M. (2011b) Synthesis and electrocatalytic oxygen reduction activity of graphene-supported Pt_3Co and Pt_3Cr alloy nanoparticles *Carbon 49, 931–936.*

Sahoo, L., Gautam, U. K. (2020) Boosting bifunctional oxygen reduction and methanol oxidation electrocatalytic activity with 2D superlattice-forming Pd nanocubes generated by precise acid etching. *ACS Appl. Nano Mater. 3(8), 8117–8125.*

Sammes, N. (2006) *Fuel Cell Technology: Reaching Towards Commercialization*, Springer, London.

Sasaki, K., et al. (2010) Core-protected platinum monolayer shell high-stability electrocatalysts for fuel-cell cathodes. *Angew. Chem. Int. Ed. 49, 8602–8607.*

Seo, M. H., Choi, S. M., Seo, J. K., Noh, S. H., Kim, W. B., Han, B. (2013) The graphene-supported palladium and palladium—yttrium nanoparticles for the oxygen reduction and ethanol oxidation reactions: Experimental measurement and computational validation. *Appl. Catal. B 129, 163–171.*

Sharaf, O. Z., Orhan, M. F. (2014) An overview of fuel cell technology: Fundamentals and applications *Renew Sust Ener Rev 32 810–853.*

Shen, J. F., Yan, B., Shi, M., Ma, H. W., Li, N., Ye, M. X. (2012) Fast and facile preparation of reduced graphene oxide supported Pt-Co electrocatalyst for methanol oxidation. *Mater. Res. Bull. 47, 1486–1493.*

Srinivasan, S. (2006) *Fuel Cells: From Fundamentals to Applications*, Springer Verlag, Springer, New York, NY.

Stamenkovic, V. R. et al. (2007) Trends in electrocatalysis on extended and nanoscale Pt-bimetallic alloy surfaces. *Nature Mater. 6, 214–247.*

Su, H.-Y., Gorlin, Y., Mann, I. C., Calle-Vallejo, F., Nørskov, J. K., Jaramillo, T. F., Rossmeisl, J. R. (2012) Identifying active surface phases for metal oxide electrocatalysts: A study of manganese oxide bi-functional catalysts for oxygen reduction and water oxidation catalysis. *Phys. Chem. Chem. Phys 14, 14010–14022.*

Tong, Y., Chen, P., Zhou, T., Xu, K., Chu, W., Wu, C., Xie, Y. (2017) A bifunctional hybrid electrocatalyst for oxygen reduction and evolution: Cobalt oxide nanoparticles strongly coupled to B,N-decorated graphene. *Angew. Chem. Int. Ed. 56, 7121–7125.*

Vielstich, W., Lamm, A., Gasteiger, H. A. (Eds.) (2003) *Handbook of Fuel Cells*, Vol. 3, Wiley, Chichester.

Wang, C., Daimon, H., Lee, Y., Kim, J., Sun, S. (2007) Synthesis of monodisperse Pt nanocubes and their enhanced catalysis for oxygen reduction. *J. Am. Chem. Soc. 129, 6974–6975.*

Wang, C., Daimon, H., Onodera, T., Koda, T., Sun, S. (2008) A general approach to the size- and shape-controlled synthesis of platinum nanoparticles and their catalytic reduction of oxygen. *Angew. Chem. 120, 3644–3647.*

Wang, D., Xin, H. L., Hovden, R., Wang, H., Yu, Y., Muller, D. A., DiSalvo, F. J., Abruña, Héctor, D. (2013a) Structurally ordered intermetallic platinum—cobalt core—shell nanoparticles with enhanced activity and stability as oxygen reduction electrocatalysts. *Nature Materials 12, 81–87.*

Wang, W., Lei, B., Guo, S. (2016) Engineering multimetallic nanocrystals for highly efficient oxygen reduction catalysts. *Adv. Energy Mater 6, 1600236.*

Wang, Y. S., Yang, S. Y., Li, S. M., Tien, H. W., Hsiao, S. T., Liao, W. H., Liu, C. H., Chang, K. H., Ma, C. C. M., Hu, C. C. (2013b) Three-dimensionally porous graphene—carbon nanotube composite-supported PtRu catalysts with an ultrahigh electrocatalytic activity for methanol oxidation. *Electrochim. Acta 87, 261–269.*

Woo, S., Lee, J., Park, S. K., Kim, H., Chung, T. D., Piao, Y. (2013) Enhanced electrocatalysis of PtRu onto graphene separated by Vulcan carbon spacer. *Power Sources 222, 261–266.*

Wu, C., Zhang, Y., Dong, D., Xie, H., Li, J. (2017) Co_9S_8 nanoparticles anchored on nitrogen and sulfur dual-doped carbon nanosheets as highly efficient bifunctional electrocatalyst for oxygen evolution and reduction reactions. *Nanoscale 9, 12432–12440.*

Wu, Z. S., Yang, S. B., Sun, Y., Parvez, K., Feng, X. L., Mullen, K. (2012) 3D nitrogen-doped graphene aerogel-supported Fe_3O_4 nanoparticles as efficient electrocatalysts for the oxygen reduction reaction. *J. Am. Chem. Soc. 134, 9082–9085.*

Xia, Y., Xiong, Y., Lim, B., Skrabalak, S. E. (2008) Shape-controlled synthesis of metal nanocrystals: Simple chemistry meets complex physics? *Angew. Chem. Int. Ed. 48, 60–103.*

Xie, X.-W, Lv, J.-J., Liu, L., Wang, A.-J., Feng, J.-J., Xu, Q.-Q. (2017) Amino acid-assisted fabrication of uniform dendrite-like PtAu porous nanoclusters as highly efficient electrocatalyst for methanol oxidation and oxygen reduction reactions. *Int J Hyd Ener 42, 4, 2104–2115.*

Xu, Z., Zhao, H., Liang, J., Wang, Y., Li, T., Luo, Y., Shi, X., Lu, S., Feng, Z., Wu, Q., Sun, X. (2020) Noble-metal-free electrospun nanomaterials as electrocatalysts for oxygen reduction reaction. *Mater. Today Phys. 15, 100280.*

Yang, G. H., Li, Y. J., Rana, R. K., Zhu, J. J. (2013) Pt—Au/nitrogen-doped graphene nanocomposites for enhanced electrochemical activities. *J. Mater. Chem. A 1, 1754–1762.*

Yan, X.-Y., Tong, X.-L., Zhang, Y.-F., Han, X.-D., Wang, Y.-Y., Jin, G.-Q., Qin, Y., Guo, X.-Y. (2012) Cuprous oxide nanoparticles dispersed on reduced graphene oxide as an efficient electrocatalyst for oxygen reduction reaction. *Chem. Commun. 48, 1892–1894.*

Yang, J., Deng, S. Y., Lei, J. P., Ju, H. X., Gunasekaran, S. (2011a) Electrochemical synthesis of reduced graphene sheet-AuPd alloy nanoparticle composites for enzymatic biosensing. *Biosens. Bioelectron. 29, 159–166.*

Yang, J., Tian, C. G., Wang, L., Fu, H. G. (2011b) An effective strategy for small-sized and highly-dispersed palladium nanoparticles supported on graphene with excellent performance for formic acid oxidation. *J. Mater. Chem. 21, 3384–3390.*

Yang, X., Yang, Q. D., Xu, J., Lee, C. S. (2012) Bimetallic PtPd nanoparticles on Nafion— graphene film as catalyst for ethanol electro-oxidation. *J. Mater. Chem. 22, 8057–8062.*

Yano, H., Kataoka, M., Yamashita, H., Uchida, H., Watanabe, M. (2007) Oxygen reduction activity of carbon-supported Pt-M (M = V, Ni, Cr, Co, and Fe) alloys prepared by nanocapsule method. *Langmuir, 23(11), 6438–6445.*

Yan, Z. H., Wang, M., Huang, B. X., Liu, R. M., Zhao, J. S. (2013) Graphene supported Pt-Co alloy nanoparticles as cathode catalyst for microbial fuel cells. *Int. J. Electrochem. Sci. 8, 149–158.*

Yoo, E., Okada, T., Akita, T., Kohyama, M., Honma, I., Nakamura, J. (2011) Sub-nano-Pt cluster supported on graphene nanosheets for CO tolerant catalysts in polymer electrolyte fuel cells. *J. Power Sources 196, 110–115.*

Yoo, E., Okata, T., Akita, T., Kohyama, M., Nakamura, J., Honma, I. (2009) Enhanced electrocatalytic activity of Pt subnanoclusters on graphene nanosheet surface. *Nano Lett. 9, 2255–2259.*

Yue, Q. L., Zhang, K., Chen, X. M., Wang, L., Zhao, J. S., Liu, J. F., Jia, J. B. (2010) Generation of OH radicals in oxygenreduction reaction at Pt-Co nanoparticles supported on graphene in alkaline solutions. *Chem Commun. 46, 3369–3371.*

Zhang, K., Yue, Q. L., Chen, G. F., Zhai, Y. L., Wang, L., Wang, H. S., Zhao, J. S., Liu, J. F., Jia, J. B., Li, H. B. (2011) Effects of acid treatment of Pt–Ni alloy Nanoparticles@ Graphene on the kinetics of the oxygen reduction reaction in acidic and alkaline solutions. *J. Phys. Chem. C. 115, 379–38958.*

Zhang, L. S., Liang, X. Q., Song, W. G., Wu, Z. Y. (2010) Identification of the nitrogen species on N-doped graphene layers and Pt/NG composite catalyst for direct methanol fuel cell. *Phys. Chem. Chem. Phys. 12, 12055–12059.*

Zhang, Q. X., Ren, Q. Q., Miao, Y. Q., Yuan, J. H., Wang, K. K., Li, F. H., Han, D. X., Niu, L. (2012) One-step synthesis of graphene/polyallylamine—Au nanocomposites and their electrocatalysis toward oxygen reduction *Talanta 89, 391–395.*

Zhang, R. Z., Chen, W. (2013) Non-precious Ir-V bimetallic nanoclusters assembled on reduced graphene nanosheets as catalysts for the oxygen reduction reaction. *J. Mater. Chem. A 1, 11457–11464.*

Zhang, S., Shao, Y. Y., Liao, H. G., Liu, J., Aksay, I. A., Yin, G. P., Lin, Y. H. (2011a) Graphene decorated with PtAu alloy nanoparticles: Facile synthesis and promising application for formic acid oxidation. *Chem Mater. 23, 1079–1081.*

Zhang, Y. Z., Gu, Y. E., Lin, S. X., Wei, J. P., Wang, Z. H., Wang, C. M., Du, Y. L., Ye, W. C. (2011b) One-step synthesis of PtPdAu ternary alloy nanoparticles on graphene with superior methanol electro oxidation activity. *Electrochim. Acta 56, 8746–8751.*

Zhao, Y. C., Zhan, L., Tian, J. N., Nie, S. L., Ning, Z. (2011) Enhanced electrocatalytic oxidation of methanol on Pd/polypyrrole—graphene in alkaline medium. *Electrochim. Acta 56, 1967–1972.*

Zheng, Y., Jiao, Y., Chen, J., Liu, J., Liang, J., Du, A., Zhang, W., Zhu, Z., Smith, S. C., Jaroniec, M., Lu, G. Q., Qiao. S. Z. (2011) Nanoporous graphitic-C_3N_4@Carbon metal-free electrocatalysts for highly efficient oxygen reduction. *J. Am. Chem. Soc. 133, 20116–20119.*

Zhoa, S., Zhang, H., House, S. D., Jin, R., Yang, J. C., Jin, R. (2016) Ultrasmall palladium nanoclusters as effective catalyst for oxygen reduction reaction. *Chem Electro Chem 3, 1225–1229.*

Zhou, R. F., Zheng, Y., Hulicova-Jurcakova, D., Qiao, S. Z. (2013) Enhanced electrochemical catalytic activity by copper oxide grown on nitrogen-doped reduced graphene oxide. *J. Mater. Chem. A 1, 13179–13185.*

3 Polyoxometalate-Induced Nano-Engineered Composite Materials for Energy Storage
Supercapacitor Applications

Neeraj K. Sah, Shankab J. Phukan, Dasnur Nanjappa Madhusudan, Dr. Kamatchi Sankaranarayanan, Dr. Manas Roy and Dr. rer. nat. Somenath Garai

3.1 INTRODUCTION

In the present social and economic circumstances, with rapid industrialization, the obligation for sustainable and green energy is the foremost requirement for humankind. The two prime factors that we need to centralize in our research domain are the harvesting and storage of energy. The green and sustainable energy harvested by the means of solar, thermal and hydrothermal methods, among others, are stored with the assistance of principal energy storage devices, which presumes the conversion of electrical or nonmechanical energy into the form of chemical energy/potential energy.

Until now, the Li-ion batteries and supercapacitor (SC)-based devices are playing a major role in energy storage devices; the SC stores lesser quantities of energy but generates a very high power density. These are also known as ultra-capacitors, as well as electrochemical capacitors, which use electrode materials with large surface areas and thin electrolytic dielectrics to obtain capacitances that are many orders of magnitude higher than normal capacitors. SCs (electrical double-layer capacitors) have a twofold increase in energy storage density over ordinary electrolytic capacitors. SCs are capable of achieving greater energy densities that can be charged and discharged up to 10^6 times without significant power loss with a high specific power density (Sagadevan et al., 2021). SCs store electrical charge, and that charge–discharge cycle can be repeated several times because this process is extremely reversible. At the contact between a carbon electrode with a large surface area and a liquid electrolyte, SCs store electrical charge in the form of an electrical double layer. Electrical double layers are generated and released when electrolyte ions are

DOI: 10.1201/9781003208709-3

oriented at the electrolyte/electrolyte contact in a specific manner. SCs possess critical components like electrodes, electrolytes and the separator; the electrodes are typically thin coatings that are applied and electrically linked to a conductive metal current collector. The electrode materials must exhibit some characteristics such as good conductivity, high-temperature stability, large surface area, high cyclability, long-term stability and electrochemical oxidation/reduction resistance, among others (Da Silva et al., 2020). On the basis of the principal electrode materials and the principle of operation, electrical double-layer capacitors, pseudocapacitors and hybrid capacitors are the three types of SCs known to date. In electrical double-layer capacitors, the charge storage mechanism is based on the separation of electronic and ionic charges at the interfaces, but in the case of pseudocapacitors and hybrid capacitors, the reversible solid-state redox processes that occur concurrently with electrostatic contribution regulate the charge–storage process. The introduction of nanomaterials is revolutionizing supercapacitors allowing for large storage capacity and a lot of new applications. In recent years, based on the application level, asymmetric supercapacitors are designed that are constructed by connecting two capacitors in series, one of which is a capacitor and the other is a pseudocapacitor or a battery that is achievable by varying electrode capacity ratio (Miller & Simon, 2008). In the future, asymmetric SCs will play a major role in the energy storage field. Although the major problems are that most of them are extremely massive, with hard/huge sizes, they suffer from efficiency (wastage of energy 10–30%) and environmental-compatible issues. So it is quite impossible to fix the specific demands of flexible electronic devices with this parameter (Zhou et al., 2014). Consequently, the creation of light, flexible and compact units, along with shape conformability, visual variety, outstanding mechanical characteristics, good fabrication, elevated capacity and excellent conductivity, are some essential prerequisites for the energy storage devices of the future. Interestingly, flexible energy storage technologies, namely, flexible lithium-ion battery and SCs, have received a lot of scientific as well as commercial attention in recent years (X. Wang et al., 2014).

Recent research disciplines have focused more acutely on the consolidated energy storage systems than those of the general conventional devices because of their various established functions like micro-electromechanical systems, self-powering systems and the like. Xie *et al.* (2014) have demonstrated an electrochromic smart window with a self-contained power source with tunable transmittance powered by dye-sensitized solar cells that also functions like a photo-charged electrochromic supercapacitor with greater areal capacitance along with reversible color shifts.

SC-based composite electrodes with double-layer capacitance and pseudocapacitance have excellent energy and power densities compared to ordinary capacitors and batteries. The charge is stored *via* charge polarization at the electrical double layer in the electrochemical double-layer capacitors (EDLCs) and the pseudocapacitors, which store charge generated from specific redox activities, are the two major forms of supercapacitors (Y. Yang et al., 2019). The use of pseudocapacitance materials instead of EDLC materials can boost specific capacitance; under the same electrode area circumstances, the specific capacitance of pseudocapacitance capacitors can be 10–100 times higher than that of EDLCs. In addition, the electrode's operating voltage is also an important component in increasing the energy density of devices;

for example, EDLCs may charge and discharge in seconds for tens of thousands of cycles and possess a near-immediate response, in addition to having a higher power density. EDLCs have a poor energy density as compared to batteries, whereas the energy density (E) of an SC can be increased by raising the specific capacitance (C) or widening the operating voltage (V), according to the equation $E = \frac{1}{2}CV^2$ (Y. Wang et al., 2016). Recently, Rani *et al.* (2019) synthesized N-incorporated, thiol-functionalized reduced-graphene oxide scrolls (NTGS) in concern to enhance the energy density of the electrical double-layer capacitors. Zhang *et al.* (2020) have developed a high-performance SC with two species that are self-redox active: Fe_2O_3 and layered Ni-Co hydroxide/oxy-hydroxide. Both materials have high surface areas, adequate reaction sites, low ion diffusion lengths and increased rate capability due to the inclusion of reduced graphene oxide. Nevertheless, their low energy density and specific capacitance, however, preclude them from being used in many applications that need a lengthy duration. Graphene SCs with a high energy density have been developed by Yang *et al.*, who have utilized a scalable nano-porous graphene manufacturing technique including a simple strengthening procedure in a hydrogen environment; with extremely porous graphene electrodes, they have obtained ultra-high specific gravimetric and volumetric capacitances (H. Yang et al., 2013). By orderly adjusting the quantity of graphene-related additives during mass balancing the positive and negative electrodes, Redondo *et al.* (2020) suggested a new approach for increasing the energy density of supercapacitor cells. For various electrodes with varied additive loadings, the capacitance has been determined, along with an electrochemically stable potential window. There are primarily three ways of enhancing the energy density of supercapacitors:

a. by fine-tuning the surface properties of the electrode materials, the system's capacitance may be improved.
b. by using electrodes and electrolytes with a broad variety of electrochemically stable potential windows to boost the device's voltage.
c. alternatively, we can boost capacitance and voltage simultaneously by combining a Faradaic electrode along with a non-Faradaic electrode, as in a hybrid method.

(Demarconnay et al., 2011)

The improvement of some energy storage devices still lies in its primitive stages due to the insufficiency of expedients, environmental questionings and others. Viable endeavors are much required for the development of renewable resources of energy production for reducing the emissions of toxic substances and hazardous environmental impacts. Undoubtingly, the upcoming research on energy storage devices stores many favorable circumstances for modernization and upgradation in the energy storage mechanisms. In some opportunities, polyoxometalates (POMs) have proved to be an optimistic electrode material for supercapacitor appliances.

The renowned polynuclear metal-oxo anionic compounds, which are called POMs, are widely celebrated due to their discrete and chemically modifiable structures. The formation of structurally diversified species with high anionic characteristics and differences in nuclearity ranging from 6 to 368 metal centers per molecule is very common in most POM aqueous solutions. Beneficially, the synthetic pathways of the POMs are based on economical and effortlessly accessible precursors, due to which the large-scale synthesis of the POMs is very promising. The electrode materials are based on POMs managed to procure considerable recognition because of their previously mentioned simplistic synthetic routes, which in turn reduces the inclined issues like rapid capacity decay, volume change and phase transition.

The Keplerate-type POM Mo_{132} has been utilized by Pakulski *et al.* (2019) to design a new Mo_{132}-DTAB-EEG hybrid as the SC electrode material. Hou *et al.* reported two metallo-calixarene compounds with POM included inside their nano-cavity $[Ag_5(C_2H_2N_3)_6]$ $[H_5\subset SiMo_{12}O_{40}]$ (**1**) and $[Ag_5(C_2H_2N_3)_6]$ $[H_5\subset SiW_{12}O_{40}]$ (**2**), $(C_2H_2N_3 = 1H$-1,2,4-triazole); due to the greater reduction potential and flexibility

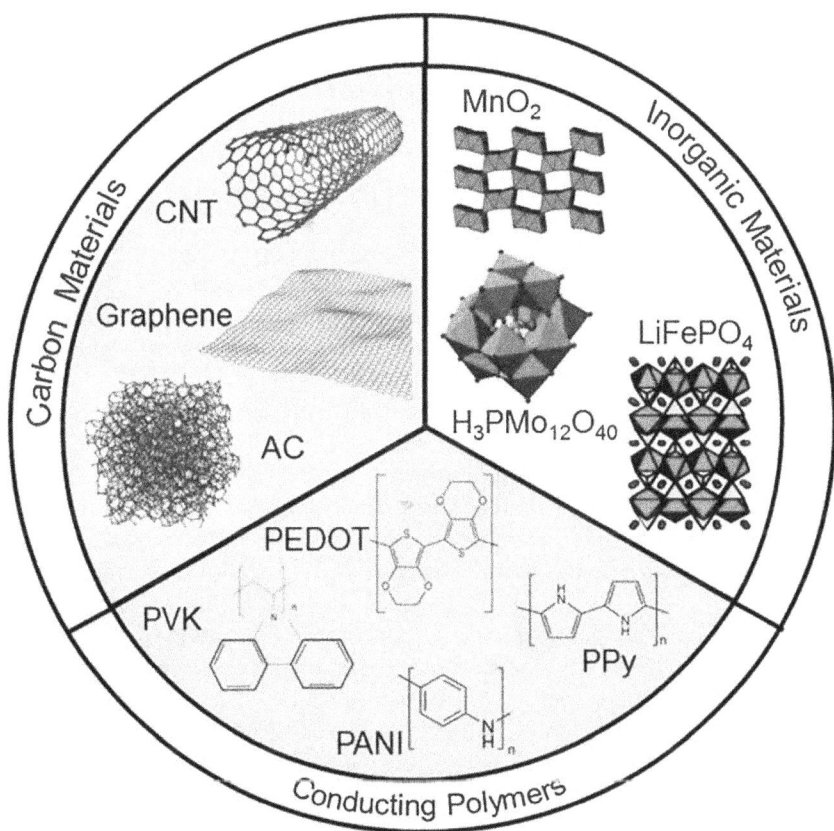

FIGURE 3.1 The carbon materials, conducting organic polymers and a range of inorganic electro-active species are shown schematically as three types of components utilized in the creation of hybrid materials for energy storage.

Source: Reproduced from (Dubal et al., 2015), Copyright 2015, with permission from the Royal Society of Chemistry.

of the Mo-based Keggin ions, compound-**1** has a higher specific capacitance as well as cyclic voltammetry stability than compound-**2** (Hou et al., 2019). Because of the wide negative voltage window and distinct electrochemical behavior, POMs are used as pseudocapacitor materials as negative electrode. Wang *et al.* developed a novel asymmetric supercapacitor device with a negative electrode made of a POM-based coordination polymer; two novel coordination polymers based on POM: [Ni(itmb)$_4$ (HPMo$_{12}$O$_{40}$)]·2H$_2$O (**1**) and [Zn(itmb)$_3$(H$_2$O)(HPMo$_{12}$O$_{40}$)]·4H$_2$O (**2**) (itmb=1-(imidazo-1-ly)-4-(1,2,4-triazol-1-ylmethyl)benzene) (G. Wang et al., 2020). The newly formed POM-based coordination polymers feature Keggin-type {PMo$_{12}$} polyanions with high specific capacitance and redox reactivity, which suggests that POM-based coordination polymers might be viable negative electrode materials toward high-potential SCs. Three types of components are shown in Figure 3.1 in the design of hybrid materials for energy storage.

This chapter mainly highlights the prime points of extensive presentations of POM-based nanostructured composite systems for energy storage applications. A descriptive configuration of presumed challenges inclining in the field of energy storage processes and SC applications that are endeavored by POM-based composite systems along with various synthetic routes of the POM chemistry is set forth.

3.2   SYNTHESIS METHODOLOGY FOR POM AND ITS COMPOSITE MATERIALS

The synthetic methodologies of POMs were first discovered by Berzelius in 1826 when he obtained a yellow precipitate of ammonium phosphomolybdate "(NH$_4$)$_3$[PMo$_{12}$O$_{40}$]$_{aq}$" by adding excess amount of ammonium molybdate to the phosphoric acid solution (Gouzerh & Che, 2006). After that, several synthetic methodologies were employed to synthesize different classes of POMs. In 1864, Galissard de Marignac synthesized heteropoly 12:1 tungstosilicic acids and their salts; however, the composition of the heteropoly anion was not clear until it was described by Warner in 1907 based on his idea of coordination chemistry (Baker & Glick, 1998), followed by the structural hypothesis as proposed by Miolati and Pizzighelli (Gouzerh & Che, 2006). According to their theory, the hypothetical acid "H$_7$[PO$_6$]" was regarded as the parent acid from which 12-molybdophosphoric acid can be derived *via* the substitution of O-atom with Mo$_2$O$_7$ units. In 1929, Pauling postulated another hypothesis to govern the structural features of ionic crystals, which was mostly based on ionic radii, should also hold to heteropoly anions, but this proposed structure had limitation as the edge-sharing options between MO$_6$ octahedra were not taken into account (Baker & Glick, 1998). Ultimately, [PMo$_{12}$O$_{40}$]$^{3-}$ anion and isostructural anions were termed as Keggin anions after the structure of H$_3$[PW$_{12}$O$_{40}$].5H$_2$O that was characterized by X-ray crystallography in 1934 by James F. Keggin (1934). Therefore, the itinerary for the early development of the POM chemistry postulated the generation of sites on the composition and framework of heteropoly anions of the 12-tungstophosphoric acid, which was historically noted down as H$_7$[P(W$_2$O$_7$)$_6$] (Miolati-Rosenheim), then as H$_3$[PO$_4$W$_{12}$O$_{18}$(OH)$_{36}$] (Pauling) and finally as H$_3$[P(W$_3$O$_{10}$)$_4$], often written as H$_3$[PW$_{12}$O$_{40}$] (Keggin). Later, Souchay *et al.* (1945) synthesized numerous POM compounds, including anions consisting of both isopoly and heteropoly acids, and focused mainly on the disclosure of the formulas and the condensation reaction in solution. He was primarily interested in tungstates, but surprisingly, he was also interested in the reduction of Mo^{VI} in aqueous solutions to comprehend the framework of "molybdenum blue". After that, numerous reduced derivatives were prepared and characterized to validate the unparalleled capability of heteropoly anions to behave as an electron pool (Gouzerh & Che, 2006). Within an edge-sharing trio of WO_6 octahedra, reduced heteropolyanions containing several electrons (multiples of six) were discovered to undergo intramolecular dissociation (essential factors like pH, temperature, solvent type, counter-ion type, heteroatom stereochemistry) caused by the positioning of W^{IV} centers and the creation of three W-W bonds. The optical and EPR spectroscopy has revealed electron movement phenomena in polyanions with mixed valences. It was reported that the Keggin- and Dawson-type compounds exhibit both configurational/structural and conformational isomerisms and in fact the chemistry of lacunary polyoxotungstates self-assemble to larger polyanions, *e.g.*, $[As_4W_{40}O_{140}]^{28-}$ and $[P_8W_{48}O_{184}]^{40-}$. During the development of polyoxothiometalate chemistry, the researchers started using the precursor "$[Mo_2S_2(H_2O)_6]^{2+}$", as a "magic building block", in order to create cyclic network structures that showed prominent host–guest chemistry (Gouzerh & Che, 2006; Müller et al., 2001; Müller & Serain, 2000). In this section, we have insight into the different methodologies to synthesize different classes of POMs.

3.2.1 Synthesis of Lindqvist Structure $[M_6O_{19}]^{n-}$ (M = Nb, Ta, Mo, W, n = 2, 8)

The first isopoly-Lindqvist anion $[Nb_6O_{19}]^{8-}$, which is also known as hexaniobate, was first introduced by Lindqvist in 1953. It can be synthesized by fusing Nb_2O_5 in an alkaline solution (pH of 11) at temperature less than 100°C. It was believed that in alkaline solutions the $[Nb_6O_{19}]^{8-}$ ion exist as the major niobate form which may be easily segregated (M. Nyman et al., 2006). The niobate-based POMs could bind inorganic and organic cationic moieties as they have stable biocompatible pH levels and strong negative charges. The alkali salts of $[Nb_6O_{19}]^{8-}$ ion were known as good water-soluble precursors to synthesize heteropolyniobates and considered as the sole way to make several heteropolyniobates phases. The heteropolyniobates may be precipitated from pH 11–12.5 and separated *via* hydrothermal treatment at temperatures above 150°C. M. Nyman *et al.* (2006) reported a sequence of alkali (K, Rb, Cs) salts of $[Nb_6O_{19}]^{8-}$ and postulated the formation mechanism of heteropolyanions from the Lindqvist ion. The crystals of alkali salts of hexaniobate were obtained by straight crystallization from alkali hydroxide solution in high concentration followed by slow diffusion of nonsolvents into aqueous precursor solutions. The common step for the synthesis of $[Nb_6O_{19}]^{8-}$ was the fusing of amorphous Nb_2O_5. xH_2O with excess alkali followed by dissolution in an aqueous medium and cooled to room temperature. Similarly, the Lindqvist ion of Ta, $[Ta_6O_{19}]^{8-}$, was the only polyanion found to form when tantalum oxide Ta_2O_5 was dissolved in an aqueous alkali hydroxide solution. But its form was unknown until 1977 (Keggin, 1934) and a practical, solution-based method was reported ten years later (Souchay, 1945). Later on, tetrabutylammonium (TBA) salts of protonated $[H_2Ta_6O_{19}]^{6-}$ cluster were discovered (M. D. Nyman et al., 2007). Afterward, Anderson *et al.* introduced a new technique for the production of $[Ta_6O_{19}]^{8-}$ salts; $Na_8[Ta_6O_{19}]$ ·$15H_2O$, $K_7Na[Ta_6O_{19}]\cdot14H_2O$ (M. D. Nyman et al., 2007). It was reported that $Rb_3Ta(O_2)_4$ and $K_3Ta(O_2)_4$ could be synthesized with the help of $TaCl_5$, hydrogen peroxide and alkali RbOH and KOH. Initially, $TaCl_5$ reacted with hydrogen peroxide to form tantalum peroxide, which was used as the precursor for synthesizing the salts of tantalum Lindqvist ion with the help of alkali hydroxide (KOH, RbOH), K_3VO_4 and Na_3VO_4 by condensation method. Here, $[VO_4]^{3-}$ were used for the catalytic decomposition of peroxide ligands, and their use deduced the boiling time. Moreover, salts (alkylammonium) of Lindqvist ion $[Mo_6O_{19}]^{2-}$ and $[W_6O_{19}]^{2-}$ were prepared by Jahr, Fuchs, and Oberhauser in 1968 for the first time, using controlled hydrolysis of molybdic and tungstic acid esters (Klemperer, 1990). Later on, more convenient preparation methods and synthesis of other salts were available like the synthetic procedure for tetrabutylammonium hexamolybdate was demonstrated by W. G. Klemperer and his coworkers, which can be explained by the following equation (Klemperer, 1990):

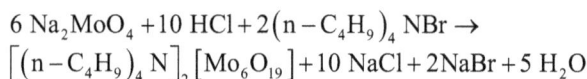

$$6\, Na_2MoO_4 + 10\, HCl + 2\left(n - C_4H_9\right)_4 NBr \rightarrow$$
$$\left[\left(n - C_4H_9\right)_4 N\right]_2 \left[Mo_6O_{19}\right] + 10\, NaCl + 2NaBr + 5\, H_2O$$

3.2.2 SYNTHESIS OF KEGGIN ION $[XM_{12}O_{40}]^{N-}$

The first Keggin structure "$(NH_4)_3[PMo_{12}O_{40}]_{aq}$" was reported by J.J. Berzelius (Gouzerh & Che, 2006). The synthetic route for the preparation of Keggin ion–containing molybdenum and tungsten can be given by the following equations (Izumi et al., 1992; Pope & Müller, 1991):

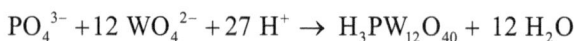

$$23\, H^+ + \left[HPO_4\right]^{2-} + \left[12\, MoO_4\right]^{2-} \rightarrow \left[PMo_{12}O_{40}\right]^{3-} + 12\, H_2O$$

$$PO_4^{3-} + 12\, WO_4^{2-} + 27\, H^+ \rightarrow H_3PW_{12}O_{40} + 12\, H_2O$$

Notably, the core heteroanion stereochemistry (trigonal: B(III); pyramidal: X(III), X=As, Sb, Bi; tetrahedral: Si (IV), Ge(lV), P(V), As(V); octahedral: $M^{2+/3+}$ cations of transition metals) substantially influenced the generation and structure of the hetero-polyanions. In 1864, the dodecatungstosilicic acid $H_4[SiW_{12}O_{40}]$ was initially synthe-sized by Marignac in an acid medium, which was later considered a Keggin β -isomer after the X-ray structure obtained in 1975 (Contant & Herveb, 2002). Tézé *et al.* reported the synthetic procedure and the crystal structure of the γ -isomer of the $[SiW_{12}O_{40}]^{4-}$, which involved the reaction of γ -decatungstosilicate and perchloric acid in the pres-ence of HCl and ethanol, followed by addition of sodium tungstate for the removal of insoluble material. When the filtrate is treated with tetrabutylammonium bromide followed by recrystallization from organic solvent produced dodecatungstosilicate polyanion (Teze & Herve, 1990). Furthermore, they synthesized and demonstrated the synthetic procedure of the salts of α -$[SiW_{12}O_{40}]^{4-}$ and β -$[SiW_{12}O_{40}]^{4-}$ isomers of the Keggin ion. Recently, a new mineral, Kegginite, $Pb_3Ca_3[AsV_{12}O_{40}(VO)].20H_2O$, was found from the Packrat mine, near Gateway, Mesa County, Colorado, U.S.A., and the crystal structure of this mineral contains $[As^{5+}V_{12}^{5+}O_{40}(VO)]^{12-}$, an ε -isomer of Keggin with a mono-capped moiety (Kampf et al., 2017).

3.2.3 SYNTHESIS OF LACUNARY KEGGIN ION

The plenary-Keggin or α -Keggin anions are usually stable under acidified conditions, but the controlled base hydrolysis leads to the elimination of one, two, or three [M=O] fragments from the plenary structure resulting in the formation of lacunary Keggin ions (Sarwar et al., 2021). These lacunary structures are referred to as monolacunary $[XW_{11}O_{39}]^{n-}$, dilacunary $[XW_{10}O_{36}]^{n-}$ or trilacunary $[XW_9O_{34}]^{n-}$ depending on the number of metal-polyhedra eliminated. The synthesis of the sodium salts of monolacunary $[PW_{11}O_{39}]^{7-}$ Keggin was reported first time in 1983, and its first crystal structure was elucidated in 2002 (Honma et al., 2002). The formation of different isomeric forms of lacunary-Keggin derivatives was investigated by recording the ^{31}P-NMR of freshly prepared aqueous solutions of W(VI) and P(V) with different acidities. The monolacunary $[PW_{11}O_{39}]^{7-}$ appeared in the pH range of 3.5–5, and the further addition of OH- ions, that is, an increase in alkalinity of solution (pH; 5.5–7.5), converted the $[PW_{11}O_{39}]^{7-}$ to $[PW_9O_{34}]^{9-}$ and finally decomposed to tungstates and phosphates (Sarwar et al., 2021). According to the number of empty sites, lacunary polyanions can function as ligands with a wide range of metal cations, leading to mono-, di-, or tri-capped complexes. These complexes were the earliest mixed oxide model materials and were essential for catalytic applications, notably the oxidation of organic substances. Tézé et al. described the preparation method of salts of various lacunary Keggin ions (α -, β - and γ -isomer) containing tungsten as addenda moieties (Kampf et al., 2017). They reported the synthetic procedure for sodium α -nonatungustosilicate $Na_{10}[$ α -Si $W_9O_{34}]$, sodium β -nonatungustosilicate $Na_9[$ β -$SiW_9O_{34}H].23H_2O$, potassium γ -decatungustosilicte $K_8[$ γ -$SiW_{10}O_{36}].12H_2O$, potassium α -undecatungustosilicate $K_8[$ α -$SiW_{11}O_{39}].13H_2O$, sodium β_1-undecatungustosilicate $Na_8[$ β_1-$SiW_{11}O_{39}]$ solvent, potassium β_2-undecatungustosilicate $K_8[$ β_2-$SiW_{11}O_{39}].14H_2O$ and potassium β_3-undecatungustosilicate $K_8[$ β_3-$SiW_{11}O_{39}].14H_2O$. The synthetic route can be represented by the following equations:

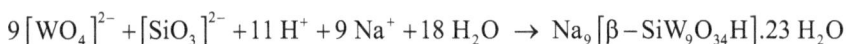

$$9\left[WO_4\right]^{2-} + \left[SiO_3\right]^{2-} + 10\ H^+ + 10\ Na^+ \rightarrow Na_{10}\left[\alpha - SiW_9O_{34}\right] + 5\ H_2O$$

$$9\left[WO_4\right]^{2-} + \left[SiO_3\right]^{2-} + 11\ H^+ + 9\ Na^+ + 18\ H_2O \rightarrow Na_9\left[\beta - SiW_9O_{34}H\right].23\ H_2O$$

$$11\left[WO_4\right]^{2-} + \left[SiO_3\right]^{2-} + 16\ H^+ \rightarrow \left[\alpha - SiW_{11}O_{39}\right]^{8-} + 8\ H_2O$$

$$11 \left[WO_4 \right]^{2-} + \left[SiO_3 \right]^{2-} + 16 \, H^+ + 8 \, K + 6 \, H_2O \ \rightarrow \ K_8 \left[\beta_2 - SiW_{11}O_{39} \right].14 \, H_2O$$

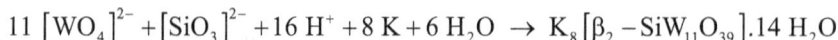

The reaction of transition metal ions (Cr, Mn, Fe, Co, Cu) with $[PW_{11}O_{39}]^{7-}$ produced an enormous range of structures with the formula $[PMW_{11}O_{39}]^{n-}$. Additionally, the polyanion also produced monometallic sandwich-type complex units $[M(\alpha\text{-}PW_{11}O_{39})_2]^{10-}$ with M = Zr (IV) or Hf (IV). The reaction of trilacunary $[PW_9O_{34}]^{9-}$ with series of lanthanide (III) (La, Pr, Nd, Sm, Eu, Gd, Tb, Dy, Er, Tm, Yb, and Y) salts generated (1:2) sandwich-type enantiomeric $[Ln(\alpha\text{-}PW_{11}O_{39})_2]^{11-}$ compounds (Sarwar et al., 2021).

3.2.4   Synthesis of Anderson-Evans {[XM$_6$O$_{24}$]$^{M-}$, M = W, Mo, V; X = Mn (III) or Fe (III)}

In 1937, Anderson reported the framework for the 6-molybdoperiodate ion, $[I^{7+}Mo_6O_{24}]^{5-}$ and additional anhydrous 6-heteropoly compounds that combine to create normal salts along with the anticipated number of monovalent cations, based on Pauling principles. Evans validated Anderson's theory by utilizing single-crystal X-ray measurements to determine the orientations of the central heteroatom along with the Mo-sites in ammonium and potassium normal salts of $[Te^{6+}Mo_6O_{24}]^{6-}$ (Baker & Glick, 1998; Evans, 1948).

In order to make inorganic Anderson-Evans POMs, an aqueous phase containing metal-oxide anions (MoO_4^{2-}, WO_4^{2-}) and a salt of a templating heteroatom was acidified. The number and nature of metal-oxide anions, the pH and kind of acid employed, the nature and concentration of heteroatoms, the presence of any extra ligands, the employment of a reducing agent, the reaction mixture's temperature and the solvent used had a crucial role for the synthesis of Anderson-Evans POM. In order to obtain inorganic clusters, two typical synthesis procedures were followed: synthesis in open-aqueous and hydrothermal environments. Different counter-cations were generally introduced during synthesis; however, in some cases, to create the Anderson-Evans structure with the necessary counter-cation, cation exchange of previously synthesized Anderson-Evans clusters were utilized (Blazevic & Rompel, 2016). In the same line, Yan et al. obtained crystal structures of series of the Anderson-Evans Type lanthanopolyoxometalates: $[Ln(H_2O)_n]_2[TeMo_6O_{24}]\cdot 6H_2O$ [n = 6 or 7, Ln = La, Ce, Pr and Sm] by using sodium molybdate, telluric acid dihydrate and $Ln(NO_3)_3\cdot 6H_2O$ as precursors. The synthetic route involved the aqueous synthesis under reflux conditions, along with the pH maintenance at 5.5 (Yan et al., 2008). Piştea *et al.* (2017) reported green techniques for the synthesis of Anderson-Evans polyanions [{XMo_6O_{24}}, where X = Cr(III), Mn(II), Co(II), Ni(II), and Cu(II),] by using aqueous solutions of precursors followed by varying pH. The authors proposed two green techniques that were simple, quick, repeatable and affordable, with crystalline molecules and high yields.

3.2.5   Synthesis of Wells-Dawson Structure [X$_2$M$_{18}$O$_{62}$]$^{N-}$

A new class of POMs currently known as Wells-Dawson structure $[X_2M_{18}O_{62}]^{n-}$ was initially reported by Rosenheim and his coworkers by preparing dimeric ammonium 9-molybdophosphate-(V) (*i.e.*, 18 molybdodiphosphate) in 1915. After that,

A. F. Wells proposed a precise structure for the tungstate isomorph, the dimeric (2:18) 9-tungstophosphate anion, in 1945 while Tsigdinos established the formula $[P_2W_{18}O_{62}]^{6-}$ for the molybdic complex in 1952, based on the structure suggested by Wells for the tungstate complex. Within a year, Dawson determined the positions of the W atoms in $[P_2W_{18}O_{62}]^{6-}$ through the single-crystal X-ray investigation that was in harmony with the structure postulated by Wells which constituted the basis of that POM cluster as the Wells-Dawson structure (Baker & Glick, 1998).

The preparation procedure for potassium salt of α, β -isomeric forms of $K_6P_2W_{18}O_{62}$ and Well-Dawson structure was demonstrated by R. G. Finke and coworkers. The synthesis of α , β mixture of $K_6P_2W_{18}O_{62}$ involved the heating of $Na_2WO_4.2H_2O$ solution, followed by the addition of phosphoric acid to the solution. Then, the solution was refluxed for 5–13 hours and subjected to filtration. The filtrate was further treated with solid KCl to obtain the crystals. The α -form was synthesized using the α, β mixture, by oxidizing it with bromine followed by bicarbonate and acid workup and subsequently recrystallized from hot solution (Finke et al., 1987). Additionally, Contant *et al.* explained the preparation method of potassium octadecatungustodiphosphates which involved heating the mixture of sodium tungstate and excess phosphoric acid at reflux conditions; the synthetic route can be given by the following equation (Contant et al., 1990):

$$18 \left[WO_4\right]^{2-} + 32\ H_3PO_4 \rightarrow \left[P_2W_{18}O_{62}\right]^{6-} + 30\ \left[H_2PO_4\right] - +18\ H_2O$$

3.2.6 FORMATION OF UNIQUE TOPOLOGICALLY IMPORTANT PENTAGONAL BUILDING BLOCKS AND THEIR SELF-ASSEMBLY FOR NANOSCOPIC MOLECULAR INORGANIC CONTAINERS

The idea for the introduction of curvature in any of the building blocks of the inorganic clusters pioneered the formation of nanosized inorganic clusters having concave surfaces. The introduction of central pentagonal bipyramidal $\{MoO_7\}$-subunit harnessed the generation of the arch-shaped pentagonal $\{Mo (Mo)_5\}$-type unit, a transient member of the combinatorial dynamic library (CDL) formed on the acidification of the molybdate solution. The incomplete reduction of the acidified molybdate solution influenced the generation of different building blocks, which can self-assemble among themselves to form molybdenum-oxide-based giant clusters. The pentagonal units, the major building blocks of those giant clusters were also transferable for the fabrication of orthorhombic Mo(V)Sb and Mo(V)Te oxide catalytic phases (Sadakane et al., 2009). Corresponding pure tungsten-based or mixed Mo and W based–pentagonal units were also common (Müller & Gouzerh, 2012). The representative structures of the container molecules were the following:

3.2.6.1 Icosahedral Inorganic Superfullerenes or the Keplerates

The self-assembly of $\{Mo^{VI}(Mo^{VI})_5O_{21}(H_2O)_6\}^{6-}$-type pentagonal and $\{Mo^V_2O_4(H_2O)_2\}^{2+}$-type dumbbell-shaped dinuclear building blocks in the carboxylate buffer (pH about 4) led to the formation of icosahedral shaped $[\{Mo^{VI}(Mo^{VI})_5O_{21}(H_2O)_6\}_{12}\{Mo^V_2O_4(RCO_2)\}_{30}]^{42-}$ nanocluster (where R = H or alkyl), the topology of which matched with the capsid of the most elementary virus, "*Satellite Tobacco Necrosis Virus*" (STNV; Müller & Gouzerh, 2012). As a comparison, similar to the 12 capsomers of the virus, the central pentagonal bipyramidal $\{MoO_7\}$ subunit of 12-$\{Mo(Mo)_5\}$-type pentagons, were also found at the icosahedron's vertices and were linked by 30-$\{Mo^V_2\}$-type dinuclear linkers to form the prototypical capsule with an approximate diameter of 3 nm. The linkers of the capsule spanned the vertices of a truncated icosahedron, being comparable in shape with C_{60}-Buckminsterfullerene but much larger in size justifying its name of "Inorganic Superfullerenes". However, the combination of pentagonal and the dinuclear linkers of the Mo_{132} capsule can also be considered in the following disposition: $\{Mo_{11}\}_{12} \equiv \{(Mo^{VI})(Mo^{VI}_5)(Mo^V_5)\}$ reminiscent to Kepler's early hypothetical representation of the cosmos, as reported in his work *Mysterium Cosmogrphicum* (Kemp, 1998).

As a general phenomenon, the introduction of preformed $\{Mo^V_2O_4(H_2O)_2\}^{2+}$ or $\{Mo^V_2O_2S_2(H_2O)_2\}^{2+}$ building blocks into the potent reference center of molybdates or tungstates at pH = 4, resulted in the generation of $\{M_{72}Mo_{60}O_{372}\}^{n-}$ or $\{M_{72}Mo_{60}S_{60}O_{312}\}^{n-}$ (M = Mo/W and n depended on the numerals of the cavity internal ligands present; Müller & Gouzerh, 2012). Similarly, the presence of certain trivalent metal ions, for example, Fe^{3+}, Cr^{3+} or oxo-species like $(VO)^{2+}$ in an acidified aqueous solution of molybdate or tungstate at pH ≈ 2, led to the evolution of the smaller Keplerate of the type $\{M_{72}M'_{30}\}^{n-}$; M = Mo/W and M' = Fe^{III}, Cr^{III} or VO^{II}, n depended on the type and the number of the cavity internal ligands present and the presence of extra Mo fragments inside the cavity), where the M' atoms spanned the vertices of an icosidodecahedron (Müller & Gouzerh, 2012). The $\{M_{72}M'_{30}\}$-clusters consisting of 20 $\{M_3M'_3O_6\}$-pores (pore outer aperture 0.9 pm) can be plugged by ammonium, partially aquatic potassium, barium or gadolinium ions. Although the smaller Keplerates are mostly neutral, they tend to behave as "polyprotic nanoacids" due to higher *Lewis's* acidity of the M' centers (when M'= Fe/ Cr) by the deprotonation of water molecules coordinated to M' centers.

3.2.6.2   The Giant Big Wheel or the Bielefeld Wheel: Synthesis

Molybdenum blue solutions were known to exist naturally by the oxidation of plentiful Mo-minerals, especially with its disulfides. In fact, the intervalence charge transfer (IVCT) related deep blue coloration is the most abundant species of molybdenum in a mild reducing environment, and previously, it could also be synthesized in laboratory by simple reduction of the molybdate solution in acidic medium. Several distinct compounds based on wheel-shaped nanocontainer molecules were described by single-crystal X-ray crystallography and other physical methods since the last 20 years, and they may be divided into two classes: Mo_{154}, the large wheel and Mo_{176} or the Bielefeld wheel (Müller & Gouzerh, 2012). Subsequently, Müller and co-workers used NH_2OH as a reductant; just a few single crystals were separated from a Mo-blue solution. The formula was first published with a negative charge error limit (owing to ambiguity in the number of protons), but the Bielefeld team labored for months and finally arrived to the accurate formula to be $[Mo_{154}(NO)_{14}O_{448}H_{14}(H_2O)_{70}]^{28-} \equiv [Mo^{VI}_{126}Mo^V_{28}] \equiv \{Mo_{154}\}$ as available today (Müller & Serain, 2000). At low pH levels, the Mo_{36}-type cluster was recognized as the dominating member of the CDL in aqueous solution, which is composed of $\{(Mo)Mo_5\}$-units (one pentagonal MoO_7 unit distributing edges with five MoO_6 octahedra; Gouzerh & Che, 2006). The species like $\{Mo^V_2O_4\}^{2+}$, Fe^{3+} or $(VO)^{2+}$ were being considered as linkers for the formation of spherical/wheel-shaped species by the addition of these linkers to the $\{Mo_{36}\}$-cluster's aqueous solution and thereby were utilized as a promoter for the efficient nucleation of Mo_{36} plugs into the inner cavity of the Mo_{154} wheel structure, which was considered to unveil the central dogma of nanoscopic coordinative nucleation under confinement, resembling the different modes of molybdenum storage inside the *Azatobactor vinelandii*. Interestingly, they were also able to replicate similar scenarios in the laboratory conditions as well and therefore, justifying the role of the POM systems as the "Nano-Labs", which will also be benchmarked here for energy storage applications.

3.2.6.3 Nano-Hedgehog or the Blue Lemon

Apart from these "Keplerates", the Bielefeld researchers also created a protein-sized massive POM cluster of 368 molybdenum atoms, branded as "nano-hedgehog or blue lemon", due to its peculiar shape with metal-oxo spikes facing outward (Müller et al., 2001). The self-assembly pathway was not only dependent on the building blocks of the molybdenum dynamic library but also on the pH of the medium and the existence of foreign cations as exemplified in the template formation of $\{M(SO_4)\}_{16}$ (where M = K^+, NH_4^+) ring inside the Mo_{146}-type wheel. However, in the absence of those cations, the same reaction mixture results in the construction of an extremely large (comparable to the size of the proteins like hemoglobin) mixed-valent nanocluster containing 368 Mo atoms. This largest inorganic nanocluster also belongs to the Robin Day classification III and is the intermediate regarding its degree of reduction (~30%) compared to the capsules (higher reduced) and the wheels (lower reduced).

3.2.7 METHODOLOGY OF POM-FABRICATED INORGANIC/ORGANIC COMPOSITES

Organic–inorganic hybrid materials established on polymers with conjugation received considerable attention in recent years. These materials were used as electrodes for batteries, SCs, fuel cells and solar cells among other energy conversion, production and storage technologies. Their importance relied on the wide variety of organic and inorganic counterparts from which these hybrids can be constructed. A significant amount of new structural polyoxometalates were synthesized, resulting in the possibilities of sophisticated nanoscale devices (sensors, nanocontainers, and nanoreactors). Controlling the surface textures and characteristics of polyoxometalates, as well as the capacity to build two- and three-dimensionally organized structures using efficient organization techniques, are critical in the polyoxometalate-based applications (Fan et al., 2012).

The conductive organic polymers (COPs) bear reversible redox as well as conducting properties; also, their low cost and good processability assisted them in being promising candidates for the fabrication of POM-based composites. Various COPs and their derivatives were utilized to fabricate hybrid materials with the POMs, among those some common COPs are polythiophene (PT), polyaniline (PANI) and polypyrrole (PPy). Typically, the POM can be incorporated into the COP matrix in two ways: (1) the deposition of polymer film on the substrate by electropolymerization technique or spin coating followed by the dipping of polymer film into the POM-containing solution for its immobilization into the matrix of polymers and (2) the chemical or electrochemical oxidation of the monomer molecules in the vicinity of POM solution resulted in the creation of a polymeric membrane (generally used as the electrolyte in this method). The polymerization of monomers like thiophene, pyrrole and aniline is supported by the high ionic conductivity, strong oxidizing power and acidic character of heteropoly acids. This technique involved depositing of a polymer film on the working electrode and doping it with POM molecules, leading to the establishment of a nanostructured hybrid material in which the bulky POM molecules were confined in a COP matrix (Genovese & Lian, 2015). Gomez-Romero and coworkers reported a hybrid nanocomposite material consisting of PANI and polyoxomolybdate, $H_3[PMo_{12}O_{40}]$ {PMo_{12}}, as electrodes for electrochemical capacitors in a solid state. There are two methods that were employed for the production of PANI–POM hybrid materials: (1) combined chemical–electrochemical method which involved treatment of aniline with {PMo_{12}} followed by the transfer of the resulting product into an acidic solution which was finally electrodeposited on a carbon foil, and (2) the direct electrochemical polymerization method in which direct electro-polymerization occurred from aniline and {PMo_{12}} solution. The authors found thicker, more porous films by a direct method which exhibited higher Faradic current than the films prepared by the combined method. Similarly, in 2006, Gomez-Romero and co-workers further reported another hybrid POM composite PEDOT/{PMo_{12}}, for its applications in electrochemical supercapacitors. They used a modified synthetic procedure and employed hydrogen peroxide as an external oxidant during the synthesis, which was found to have a significant impact on the material's morphology. The layer-by-layer (LBL) self-assembling is a promising method, typically used in order to fabricate multilayer-nanocomposites, in which oppositely charged species like POM anions, cationic surfactants, polymers, polyelectrolytes and biological molecules were deposited based on the electrostatic interaction. This method may be used to immobilize polyoxometalate clusters into the ultrathin films, allowing for a great degree of control over the homogeneity, thickness and inclination of each layer at the nanoscale size and the formed composite films showed excellent properties like electrolytic and photochromic activities. In this process, the first step was the immersion of negatively charged substrate like polystyrene-sulfonate (PSS) form or a charged monolayer form of an alkyl-silane or alkyl-thiol into a solution of polycation due to which adsorption of polycation occurred and formation of a complete layer happened. Thereafter, immersing the sample in POM solution caused its next film to adsorb the nanoclusters, and repeating this alternating deposition led to the formation of the multilayers. Thus, every deposition stage took only a few moments due to the high electrostatic interactions; the only need is that the constituents be suitably charged in order to adhere irreversibly at the interfaces (Liu et al., 2003). Aside from the cationic connectors, there were a number of additional factors that can influence the architecture of LbL thin films, including anion charge, deposition duration, ionic strength, pH and coating concentration of the solution. In 2003, Dong and co-workers at Changchun Institute of Applied Chemistry fabricated metalloporphyrin-polyoxometalate hybrid film of [Tetrakis (*N*-methylpyridyl) porphyrinato] cobalt (CoTMPyP) and 1:12 silicotungstic acid (SiW_{12}) by LBL technique. In acidic conditions, the SiW_{12}-containing multilayer screens exhibited exceptional electrocatalytic performance for the hydrogen evolution reaction (HER) at significantly higher positive potential.

The Langmuir, as well as Langmuir-Blodgett methods, were some promising routes to organize polyoxometalate clusters into films. The hollow form of Keplerate-type polyoxometalate $\{Mo_{132}\}$ was modified by the cationic surfactant, dimethyl-dioctadecylammonium (DODA), to obtain surfactant-encapsulated clusters (SECs); for example, $(DODA)_{40}(NH_4)_2[(H_2O)_n@Mo_{132}O_{372}(CH_3COO)_{30} (H_2O)_{72}]$ (n = 50), a single anionic Keplerate cluster residing in a hydrophobic shell of 40 DODA$^+$ cations (Volkmer et al., 2000). The surfactant encapsulated Keplerates were prepared by the treatment of the aqueous solution of $\{Mo_{132}\}$ with the trichloromethane solution of surfactant DODA-bromide through phase transfer. The cationic part of the surfactant molecule was facing toward the negatively charged surface of the cluster due to the electrostatic attractions. Similarly, mixed-valence gigantic inorganic ring, $[Mo_{154}O_{462}H_{14}(H_2O)_{70}]^{14-}$ ($\{Mo_{154}\}$) and DODA were combined by a cation-metathesis to form $(DODA)_{20}[Mo_{154}O_{462}H_8(H_2O)_{70}]$-hybrid based Langmuir–Blodgett film. The chemical stability of the wheels encapsulated by the DODA cationic surfactants in CHCl$_3$ was found to be higher than that of the "native" sodium salt of the $\{Mo_{154}\}$-ring in water owing to the semi-hydrophobic confinement derived restricted access of water molecules to the nanocluster interior (Akutagawa et al., 2008).

3.2.8 SYNTHESIS OF THE POM@NANO DOMAIN CARBON PARTICULATE–BASED COMPOSITES

So far, we have discussed a lot about POMs that were composed of d-block transition elements and nano-sized metal anionic oxide, and because of their various redox characteristics, they constituted sustainable targets for electrocatalysis. The Keggin-type POMs comprised of an encapsulated central (XO$_4$) tetrahedral arrangement and three octahedral connecting corners, edges and end sites made by 12 fragments of MO$_6$ octahedral addenda atoms, which provided favorable accessibility of electron transfer through the empty d-orbitals along the metal–oxygen bonds. Photochemically, radiolytically and electrochemically Keggin polyoxometalates were known to execute stepwise multi-electron reversible redox processes with minimal structural alterations; hence, the replacements of addenda or heteroatoms can readily change the redox characteristics. It had a broad range of electrochemical applications, including energy conversion and storage system fuel cells, battery applications, pseudocapacitors and sensors, as well as possible electrode material in batteries (Y. Kim & Shanmugam, 2013). In acidic media, POMs undergo reversible one- or even multi-electron accumulation to create a differentially reduced form of POMs, as indicated in Equations 3.1–3.3.

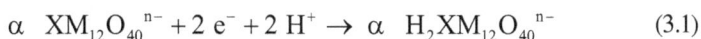

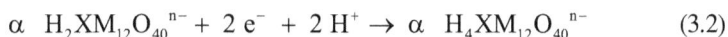

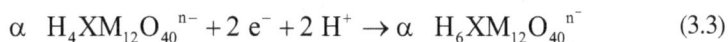

$$\alpha \ XM_{12}O_{40}^{n-} + 2\ e^- + 2\ H^+ \rightarrow \alpha \ H_2XM_{12}O_{40}^{n-} \qquad (3.1)$$

$$\alpha \ H_2XM_{12}O_{40}^{n-} + 2\ e^- + 2\ H^+ \rightarrow \alpha \ H_4XM_{12}O_{40}^{n-} \qquad (3.2)$$

$$\alpha \ H_4XM_{12}O_{40}^{n-} + 2\ e^- + 2\ H^+ \rightarrow \alpha \ H_6XM_{12}O_{40}^{n-} \qquad (3.3)$$

Keggin POMs come in three different crystalline forms: primary, secondary and tertiary. The primary form was widely distributed on support and acted as a nano-Bronsted acid, in an aqueous solution, protons are ionized to produce polyanions. However, due to their unique crystal structure, POMs degrade in very dilute aqueous media, prompting researchers to immobilize the POMs on apposite base materials like silica, metal cations (Fe, Co, Cs, *etc.*), polymer chains with positive charges and on nano-carbon supports. However, there were still some drawbacks, such as a strong electrostatic reaction between positively charged polymer matrix and negatively charged POMs were also known to indulge structural or, more important, functional modifications. Illustratively, the silica was not suited for electrochemical investigations due to its poor conductivity while the POM's natural stacking interactions with carbon materials provided it an excellent support characteristic. POMs were dispersed using a variety of materials made of carbon; comprising activated carbon, carbon nanotubes, highly ordered pyrolytic graphite, glassy carbon and graphene sheets (Schwegler et al., 1992).

3.2.9   PREPARATION OF GRAPHENE–POMS HYBRID SHEET

Graphene was a one-of-a-kind contender for a POM base material because of its small area of surface and high dispersibility in the nano-dimension. Due to its variable characteristics like a large area of the surface, high mechanical and thermal characteristics, as well as good electrical properties, graphene, which was made up of a monolayer of sp^2 carbon atoms, was an appealing material in several study domains, comprising electrocatalytic approaches of various kind. However, the intensity of the contact between graphene oxide and POMs was not well investigated in nature. Tessonnier *et al.* used the POMs to improve the dispersion of a thermally reduced graphene oxide (rGO) sheet by using functional groups in alkyl chains over the graphene layers (Schwegler et al., 1992). Non-reduced oxygen functional moieties, like hydroxyl and carboxyl groups, were partially visible on the graphene oxide surface when reduced using the hydrazine reduction technique. The electrostatic reaction between phosphomolybdic acid and graphene oxide was enhanced by these functional groups. Because this contact was significantly more powerful than the charge transfer reaction between pure carbon and POMs, it can improve the binding affinity between the two-component materials, resulting in an immobilized hybrid catalyst. A modified Hummer technique (Schwegler et al., 1992) was used to oxidize graphene oxide (GO) made of graphite powder. The mixture was then neutralized and dried in an oven after being reduced with hydrazine hydrate in a nitrogen environment. Using an ultrasonicator, the rGO is disseminated and ultra-sonicated in deionized water (DI) water for 30 minutes, which often induces exfoliation. The POM and the rGO solution were used because they can behave as a Bronsted acid and base, respectively, reacting *via* electrostatic contact (Scheme 3.1). The graphene solution was diluted with ethylene glycol in this phase to spread the POM over the graphene layer; the diol solvent also helped to maintain a suitable, moderate hydrophilic–hydrophobic balance and the pH for the interaction with POMs like phosphomolybdic acid. Ultimately, to eliminate any physisorbed POMs on the rGO, the filtered contents were rinsed thoroughly

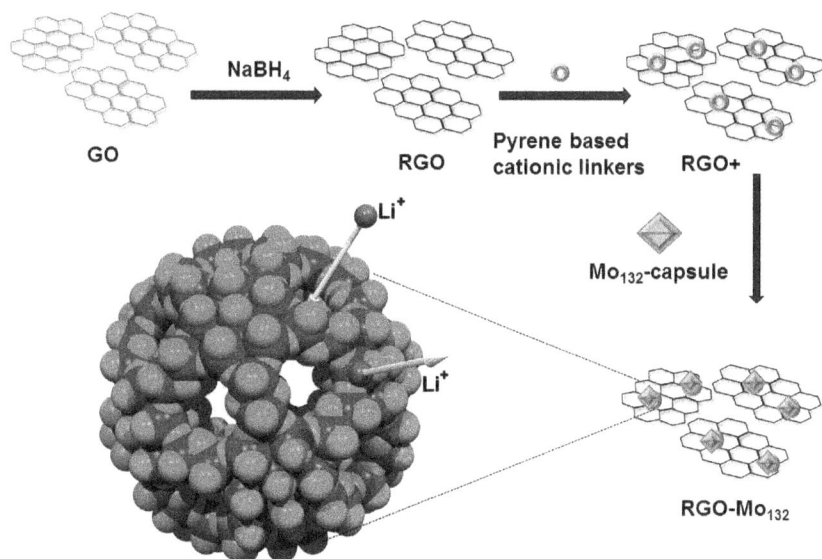

SCHEME 3.1 Schematic representation of the synthesis of nanohybrid of rGO-Mo$_{132}$ nano-container as the cathode material.

with DI water, before being dried overnight in a vacuum oven to yield charge-driven homogenous and aggregation-free nanocomposites.

3.2.10 PREPARATION OF CARBON NANOTUBE–POM AND CARBON NANOFIBER–POM NANOCOMPOSITE

Chemisorption was a simple and effective method for making nanostructured carbon-POM composites because the carbon surface and POM molecules can interact strongly. The followings were some of the most typical stages in the production of carbon-POM composites: (1) the oxidation of a carbon substrate with a strong acid to provide surface functional groups that can be utilized like a POM binding site; (2) the carbon substance was subsequently dispersed in an organic-POM or aqueous solution, which was disturbed at room temperature by stirring or ultrasonication; (3) the solid result was then washed multiple times to eliminate any loose surplus species before being dried to produce the surface customized carbon-POM composite (Genovese & Lian, 2015). Multiwalled carbon nanotubes (MWCNTs), carbon nanofibers (CNFs), mesoporous carbon, activated carbon and graphene may all be treated with the aforementioned synthetic methodologies. Cuentas-Gallegos *et al.* (2006) used this method to synthesize POM-MWCNT, as well as POM-CNF, composites, emphasizing the relevance of carbon oxidation. They oxidized both CNF and MWCNT materials by refluxing them for 7 hours in 2M HNO_3, 2M HNO_3 with 0.5 M H_2SO_4 or 2M H_2SO_4 and then sonicating the oxidized carbon in a 50 or 75 weight percent solution of the Keggin POM $[PMo_{12}O_{40}]^{3-}$ {PMo_{12}}. Because of the acid treatment, on the carbon surface, different oxygen functional groups, such as carbonyl, hydroxyl and carboxylic acid were formed, as indicated by the Fourier transform infrared (FTIR) results. The carbon specimens which were reacted with 2M H_2SO_4, had the largest concentration of carboxylic acid moieties, which resulted a composite matter that was the most distributed, homogeneously ordered {PMo_{12}} on the carbon surfaces (Cuentas-Gallegos et al., 2006).

In research of CNT-$[P_2W_{18}O_{62}]^{6-}$ composites, Kang *et al.* (2004) also demonstrated that the chemisorption process resulted in the delivery of oxygen to its surface; they discovered that CNTs that were not oxidized earlier had practically negligible POM adhesion. However, the HNO_3-refluxed CNTs represented a homogeneous and finely dispersed surface ornamentation of {P_2W_{18}} nanoparticles after sonication in the $[P_2W_{18}O_{62}]^{6-}$ {P_2W_{18}} solution. As a consequence of these findings, it was concluded that oxidized carboxylic groups provided a distinct binding location for POM chemisorption and this hypothesis as well as the general process for POM chemisorption over the carbon surface, was already widely identified by the FTIR method (Cuentas-Gallegos et al., 2006; Kang et al., 2004). The distinctive peak at 1700 cm^{-1} (matching to the C=O stretching form of carboxylic groups) in FTIR did not vanish for oxidized carbon and carbon–POM composites, but it was moved to somewhat higher frequencies, which implied a POM–carboxylic site interaction that the carbonyl component of the group was not consumed. Cuentas-Gallegos proposed the following mechanism for strong carbon-POM chemisorption that occurs at the hydroxyl end of the group, resulting in electron transfer along with the POM reduction (Cuentas-Gallegos et al., 2006). The concentration of the POM solution, as well as the level of carbon oxidation, was an essential specification in the chemisorption operation; higher POM solution concentrations result in more POM molecules being loaded onto the carbon surface, however, greater quantities of POM might cause agglomeration. Because the carbon material was dispersed in a more dilute polyoxometalate solution, the average POM crystallite size was lower, and the surface modification is more homogeneous and well-dispersed along the whole surface.

3.3 APPLICATION OF POM AS ELECTROCHEMICAL SUPERCAPACITORS

Electrochemical capacitors (ECs), aka SCs, deliver superior advancements, practically in providing greater power density in contrast to batteries; owing to the dependency of their charge storage, the reaction occurs on the surface of the electrode materials, in the absence of any diffusion of ions in the bulk materials. Concerning the fast harvesting of energy or elevated power distribution capacity, SCs can engage as an optimistic substitute for rechargeable batteries. The earliest EC prototype has been patented in the year 1957 (Becker, 1957) and the technology has advanced ahead to explore the pulsed current system for the fulfillment of the large current density together with high specific power (10,000 W Kg^{-1}) for a short period. Therefore, it can also be employed to distribute high power pulses while averting the major energy storage units (*e.g.*, battery) from rapid performance deterioration. Based on the mechanism of charge storage and functionalization of active materials, electrochemical capacitors can be sorted into three basic classifications, namely, EDLCs, pseudo-capacitors and EDLC-pseudo hybrid-SCs. EDLC is the most prevalent supercapacitor and is used most frequently in commercial markets, where it is also categorized as a wet electrolyte capacitor incorporated with a liquid-phase electrolyte containing ions to provide optimum charge transportation. The majority of the liquid phase electrolytes utilize aprotic solvents like diethyl carbonate, propylene carbonate, ethylene carbonate or dimethyl carbonate, where the presence of solvated salts like lithium hexafluoroarsenate ($LiAsF_6$) or tetraethylammonium tetrafluoroborate ($TEABF_4$) and others, are desirable to increase the specific conductivity. Electrostatic interaction is being used to stockpile the energy in the Helmholtz double layers at the point of interface of the electrode surface and the electrolyte by EDLC, whereas the derivation of the double-layer capacitance is due to the dependency on the potential of electrostatically reserved surface energy at the interface of capacitor electrodes. The second class of SCs, which are known as the pseudocapacitors or Faradaic supercapacitors, are commercially more inert than EDLC and have much resemblance with batteries than capacitors themselves, according to their operational mechanism. The phenomenon of pseudocapacitance occurs when the materials of the electrodes moderate the transfer of electrons to undergo a redox reaction on the electrode's surface, where the reactions enable the crossing of energy across the double layer, consonant to the charging and discharging of batteries. With the combination of EDLC and pseudocapacitors, hybrid supercapacitors have been fabricated, which are the newest and most advanced third type of capacitors. A higher amount of specific capacitance is displayed by the hybrid SCs in comparison to the conventional EDLCs and the rare pseudocapacitors individually, where the combination of these two kinds of technology accounts for the asymmetric behavior of these capacitors, which is an enhancing factor in its corresponding capacitance values. In this type of SC, the negative electrode contains the composites of the same materials as that of the pseudocapacitive electrode and due to the occurrence of Faradaic reactions on the negative electrode's surface; the energy density of these type of SCs are very high.

One of the major obstacles faced by SCs is the low energy density along with other soaring obstructions like high production cost, surpassing self-discharge with low voltage per cell and the most fruitful effort to overcome the low energy density challenge can be realized by fabricating novel and more effective SC electrode materials (Saubanère et al., 2016). Carbon materials in their diversified formations are the most preferred electrode materials for SCs as they possess a large surface area, a lower level of toxicity, cost-effectiveness and incorporated electrode fabrication technologies. The EDLCs are generally equipped with electrodes of carbon materials like activated carbon, where they showcase many beneficial attributes, namely, promising electrical properties and high activated surface area at an average cost as well. The activated carbon materials exemplifying porous structures can be obtained using activation processes, often exhibiting an extended pore size sequence with micropores (<2 nm), mesopores (2–50 nm) and macropores (>50 nm) (Shi, 1996). Current predilections in research propound the fact of increasing aspirations of the

CNTs as an electrode material for EDLC. With a two-dimensional layered structure possessing the thickness of an atom, graphene has materialized as a prolific carbon material that conveys a high prospect for energy storage devices. The 2D architecture of graphene has been executed to attain a capacitance up to 550 F·g^{-1}, and this value justifies the excellent aspects of graphene: chemical stability and high electrical conductivity along with an immensely high specific surface area of 2630 m^2·g^{-1} (T. Y. Kim et al., 2011). Metal oxides also confer a potent substitution for electrode materials due to their low resistivity and high specific capacitance and are generally applied on pseudocapacitors. The most famous metal oxides are MnO_2, NiO, RuO_2 and IrO_2 owing to their profound characteristics like environment amiability with an exceptional capacitive performance of MnO_2 (S. Chen et al., 2010), endured specific capacitance of NiO, which is electrochemically converted from $Ni(OH)_2$ (Wu & Wang, 2012), and extended stability for the higher number of cycles of RuO_2, fabricated *via* the electrodeposition method (Gujar et al., 2007). Conducting polymers have been proven to be another profound element for pseudocapacitor electrode materials with fairly higher values of conductivity, capacitance and equivalent series resistance where augmented capacitance values have been observed for a synergistically stable system of PANI nanofiber composites integrated with graphene nanosheets (Li et al., 2011). Poly(3,4-ethylenedioxythiophene) (PEDOT), a conjugative polymer with conducting properties, exhibits authoritative capacitance performances and sets forth standard effective potentials to be asserted as an electrode material. In 2014, a novel approach has been grasped to merge poly(3,4-ethylenedioxythiophene): poly(styrenesulfonate) (PEDOT: PSS) with graphitic carbon nitride (g-C_3N_4) by LBL assembly to obtain a constructive material for the electrode of SCs. The resultant hybrid material has displayed a twofold increment in specific capacitance in contrast to the conventional PEDOT electrodes, where a fixed 96.5% capacitance stability has been accomplished even after the completion of 1000 cycles (X. Chen et al., 2015). In the case of hybrid SCs, both polarizable (*e.g.*, carbon) and non-polarizable (*e.g.*, conducting polymer or metal) electrodes are employed for the storage of charges, where the utilization of both Faradic and non-Faradic processes are involved and procured high energy storage capacity while taking advantages of the particular properties of both the capacitor and the battery-type electrodes. The three significant classifications of hybrid SCs, considering the differences in their designs and operational mechanisms, are asymmetric hybrid supercapacitors, composite hybrid supercapacitors (CHSs) and hybrid supercapacitors that can be recharged like batteries. The asymmetric-type hybrid SCs are equipped with two disparate electrodes, where one acts as a capacitive electrode, while the other acts as a Faradic electrode. The applications of carbon derivative materials (Karthikeyan et al., 2010) in the capacitor-type electrode instigate harvesting the charge like an electrochemical capacitor and employing metal or metal oxides (Ma et al., 2017) for the Faradic electrodes. However, the principal motive of the CHSs is to synergistically enhance the output yield of cycling stability, specific capacitance and conductivity *via* synergistic affiliation of the properties of both carbon and metal oxides. The carbon or carbon-derived materials will promote the channeling through charge transfer processes and by the assistance of redox reactions. The storage of charges can be accomplished through the variable valency of metal-oxide materials, contributing to higher specific capacitance and high energy density. Whereas, subjected to asymmetric hybrid SCs, rechargeable battery–type hybrid SCs are also assembled *via* fusion of two separate electrodes, such as the union of a SC electrode along with a battery electrode. This unique hybridization of two electrodes assembly has proven to satisfy the higher energy demand of a SC and higher power need hours through a battery, assisted by the synergistic combination of energy prospects of batteries with cycle life, power, long periodic stability and short recharge time of SCs. Typically, PbO_2, $Ni(OH)_2$ and $LTO(Li_4Ti_5O_{12})$ are the most researched electrode materials to fabricate the anode and activated carbon for cathode associated with the battery-like hybrid SCs (Du Pasquier et al., 2003).

Furthermore, the compounds of POMs are also preferred as the electrode material for the SCs, as a number of cationic substances like conducting polymers and polymeric ionic liquids are incorporated as nano-hybrids with anionic POMs to construct electrodes. These negatively charged inorganic entities with highly porous architecture and tunable solubility and with other exclusive characteristics of facile redox chemistry, charge transfer and photo- and electrochemistry stand out as an obligatory successor of nano-ingredients for the construction of nanocomposites for a broad range of energy applications like electrochemical/photochemical catalysis, electrochemical sensors, energy conversion and storage utility (Ueda, 2018). The introductory hybridization of POMs with conducting polymers to functionalize the system as a solid-state electrode material for SCs has been conducted in Barcelona, in 2003 where the $H_3[PMo_{12}O_{40}]$/PANI nanocomposite, constructed by electrodeposition method, exhibited an energy density value of 24.4 mJ·cm^{-2} and an excellent capacitance of 195 mF·cm^{-2} with retaining the cycle stability up to 500 cycles (Gómez-Romero et al., 2003). After establishing the name of POM as a potent candidate for SC electrode material the researchers from Barcelona have reported two novel hybrid composite systems with convenient energy storage applications: an amalgamated electrode prepared *via* assembling phosphotungstic acid ($H_3PW_{12}O_{40}$), {PW_{12}} with activated carbon (AC), in 2013 and another hybrid material fabricated by merging {PW_{12}} with rGO with the help of hydrothermal treatment, in 2014 (Suárez-Guevara et al., 2014a, 2014b). The {PW_{12}}/AC composite with the integrated properties of the redox-active POM clusters have been added with the double-layer capacitance that guides to an inherent incrementation of specific capacitance (254 F·g^{-1} for PW_{12}/AC) and increment in the operating voltage by a factor of 60%, whereas the specific capacitance value for the {PW_{12}}/RGO composite has been recorded to be 276 F·g^{-1}. The organic materials, specifically nano-carbon and its derivatives along with organic conducting polymers have validated themselves as the most influential companions of POMs to fabricate electrodes for energy storage purposes. This is also because of high conductivity performances and effective chemical, along with physical, stability of the organic materials as well as the POM skeleton. For instance, two Keggin-type heteropolyanions, $[PMo_{12}O_{40}]^{3-}$ {PMo_{12}} and $[PMo_{10}V_2O_{40}]^{5-}$ {$PMo_{10}V_2$}, have been fused with MWCNTs with the aid of LBL deposition for the acquiring incremental in charge storage capacity, for the surface modification along with obtaining pseudocapacitance characteristics (Bajwa et al., 2013). The capacity of charge storage has been heightened by an immense factor of five, along with each coating layer of either {PMo_{12}} or {$PMo_{10}V_2$}. With the substitution of two vanadium oxides, the exhibition of storage capacity of charge for $\{PMo_{10}V_2\}$ has been similar to that of $\{PMo_{12}\}$, although the potential window got broad. Generally, the conventional chemisorption method is adopted for the fabrication of nanohybrids of POMs and nano-carbon composites, where the noncovalent interactions are commonly predominant and a direct synergistic interaction prevails between the POMs and the oxidized surface of carbon composites. Besides the Keggin structures, Dawson-type POM merged with AC to assemble SC electrode materials has been first reported in 2015, where Dawson-like $(NH_4)_6[P_2Mo_{18}O_{62}]\cdot14.2H_2O$ $\{P_2Mo_{18}\}$ massed with AC, have exhibited profound electrochemical properties, thanks to the cooperative characteristics of the pseudocapacitance of $\{P_2Mo_{18}\}$ and the double-layer capacitance of AC (Mu et al., 2015). With a specific capacitance of 275 F·g^{-1} at a high value of current density of 6 A·g^{-1}, the $\{P_2Mo_{18}\}$/AC composite electrode also maintains a unique rate capability of 89%, even in the conditions of incremental current density from 2 to 6 A·g^{-1}. Under other conditions, the interaction of conducting polymers (CPs) with POMs is majorly electrostatic, where, after the fabrication of CP *via* the spin-coating technique or the electro-polymerization technique, the films of CP are soaked in the solution of POMs and the cationic–anionic electrostatic interaction between the CPs and POMs facilitates the immobilization of the latter into the polymer matrix. In recent years, many research works have been devoted to the new generation of metal-organic framework (MOF)–derived electrodes that can be stabilized kinetically *via* the functionalization of the ligand or by the formation of greatly conductive composites with two-dimensional planar composition (Sheberla et al., 2017). Since 2012, MOFs have been distinguished as the potential candidates for electrochemical capacitor applications, owing to their many efficient features like versatile structural architecture, higher specific surface area and embodied pseudocapacitive redox centers. However, owing to the overall mediocre chemical stability and electrical conductivity, many undesired restrictions are observed in the case of MOFs for electrochemical applications. So, as a progressive step towards the synergistic improvement of energy storage devices, integration of POMs with MOFs has been proposed to fabricate pristine crystalline POM-based MOFs (POMOFs), which consist of an architecture of well-distributed active sites (Chai, Gómez-García, et al., 2019; Roy et al., 2018). An integration of arsenic-based Keggin-type POM $(AsW_{12}O_{40})$ with Ag-MOF to submit two composites of three-dimensional POMOFs, namely, $(imi)_2[\{Ag_3(tpb)_2\}_2(H_2O)\{AsW_{12}O_{40}\}_2]\cdot6H_2O$ **(a)** and $[(Ag_7bpy_7Cl_2)\{AsW^V_2W^{VI}_{10}O_{40}\}]\cdot H_2O$ **(b)**, where imi = imidazole, tpb = 1,2,4,5-tetrakis(4-pyridyl)benzene and bpy = 4,4′-bipyridyl, have been reported in 2020 (Cui et al., 2020). With intricated three-dimensional networks, novel topologies and intersecting channels, the resultant POMOFs showcased higher order of capacitances: 929.7 and 986.1 F·g^{-1} for **(a)** and **(b)**, respectively, at a current density of 3 A·g^{-1}, along with greater capacity retention rates, better rate capabilities, improved conductivities and electrochemical recyclability as compared to their individual components. Further, two more hybrid Mo-based POMOF composites have been synthesized over hydrothermal reaction, which has been functionalized as effective capacitor electrode materials: $[Cu^IH_2(C_{12}H_{12}N_6)(PMo_{12}O_{40})]\cdot[(C_6H_{15}N)(H_2O)_2]$ **(c)** and $[Cu^{II}_2(C_{12}H_{12}N_6)_4(PMo^{VI}_9Mo^V_3O_{39})]$ **(d)**, where the two-dimensional compound **(c)** hosts free triethylamine and water molecules whereas the compound

(d) existed with a three-dimensional host–guest system, in which entrapment of one-dimensional POM chains has been observed (Chai, Xin, et al., 2019). Distinguished conclusions have been transpired where the composite (c) based electrode displayed higher order of specific capacitance that is not only superior to the generality of the reported SC POMOF materials but also superior than most advanced SC electrode materials that are based on POM and MOF individually. It is an overall indisputable point to culminate that the synergistic existence of POM-based MOF composites with the cooperative characteristics of water-soluble POM entities and insoluble crystalline MOF units, renders vast prospects to broaden the horizons of the POM applications and unfolds new research areas for the utilization of MOFs in the field of electrochemistry together with energy storage devices. Most recently, hierarchical mesoporous carbons (HMCs) were combined with phosphomolybdic acid, $H_3PMo_{12}O_{40}$ (PMo_{12}), with a facile synthetic route that can be tuned up to the industrial level, for its application as a hybrid supercapacitor electrode material (HMC–PMo_{12}) (Fuentes-Quezada et al., 2022). The results proposed the idea of the adsorption of PMo_{12} Keggins on mesopores can exhibit maximum incorporative as well as electrochemical effects than that of incorporation occurring in the micropores. This novel Keggin-Carbon material hybrid electrode improved the electrolyte diffusion owing to the relatively larger interconnected mesopores of dimensions about 25 nm, and a considerable enhancement (119%) was observed in the capacitance of this hybrid electrode along with improved rate capability performance and electron transport. To incapacitate the stability issues of polymers, mixed Keggin compound of vanadium and tungstate, $K_5H_2[PV_4W_8O_{40}] \cdot 11H_2O$ (PV_4W_8) was established into the matrix system of polypyrrol, yielding a conventional metal oxide–conducting polymer hybrid electrode material (PV_4W_8/PPy (symmetric)) for the energy storage applications (Vannathan et al., 2022). The improvement of ion transfer was reported due to this very incorporation of metal oxide on the conducting polymer system. Meanwhile, the reported asymmetric PV_4W_8-PPy/PPy hybrid composite displayed a profound cycle stability with excellent specific capacitance of 291 F g^{-1}, which was greater than that of conventional PPy and PV_4W_8 at a current density of 0.4 A g^{-1}.

Functionalization of the unique properties of the POMs, like the adaptable redox actions and tunable surface architecture along with the association of other POM entities, can result in a more versatile prospective in designing the future generation energy storage electrodes. So recognizing these facts for the upcoming endeavors of POM chemistry should be focused on examining novel ideas to optimize the functionalities of the composites as well as upgrade their design for electrode surfaces for high-throughput energy storage applications and other objectives.

3.4 CONCLUSION AND FUTURE PROSPECTS

It is a great challenge to develop portable, cost-effective, environmentally benign and easily modulable electronic materials consisting of relatively larger power density along with long cycling life as a charge storage device. In this context, POMs have emerged as potential material in energy storage applications due to their outstanding redox effects, readily at the molecular level; there is a lot of flexibility, unique structures and physicochemical effects. One of the major issues to direct the application of POMs as a charge storage device is that they suffer from limitations in terms of low conductivity, high solubility and performance issues. So, POM-based hybrid electrode materials are employed as effective energy storage systems. POM-based hybrid electrode materials can be modulated by fabricating compatible inorganic/organic molecules, ligand mediation, metal-doping, framework assembly and others.

This chapter illustrates the synthesis methodology of different types of POMs, the synthesis of nanocarbon particulate (mainly CNTs and graphene)–supported composites and their beneficial applications as viable electrode materials of the SCs. Specific attention needs to be carried out on the amended performance of these energy-harvesting systems related to literature-documented reference schemes. Furthermore, we have conferred the probable techniques for the optimized fabrication procedures along with covalent and noncovalent interactions between POMs and nanocarbon particulates. The nanocarbon particulates enhance the intrinsic redox activity of POM-based composites and facilitate fast reversible electrochemical redox reactions. These POM-based nanocomposite materials have the ability to boost the capability of discharge along with an upsurge the power density of electrochemical SCs. These types of systems make storage reservoirs for charge carriers like electrons and thus make ideal environments for charge storage devices. The fabrication of POM-based composite systems for the charge storage device is still an incipient field and considerable improvements are required for its upgradation to industrial applications.

ACKNOWLEDGMENTS

Mr. Phukan, Mr. Sah, Dr. Garai is thankful to DST IC#3 Projects (DST/TM/EWO/MI/CCUS/15 & DST/TMD/HFC/2K18/06) while Dr. Garai is also thankful to SERB Project (ECR/2018/000254), BHU-IOE-SEED Grant and SPARC Project (SPARC/2018–2019/P124/SL) for kind financial help. Mr. Sah is thankful to UGC-CSIR for the JRF fellowships. Dr. Sankaranarayanan thanks DST, Govt. of India for the Inspire grant (IFA13-PH-82) and SERB-SRG grant No. SRG/2020/001894. The authors are also thankful to their respective Institutions/Universities/Departments as affiliated herewith for basic research and infrastructural facilities.

REFERENCES

Akutagawa, T., Jin, R., Tunashima, R., Noro, S., Cronin, L., & Nakamura, T. (2008). Nanoscale assemblies of gigantic molecular $\{Mo_{154}\}$-rings: (dimethyldioctadecylammonium)$_{20}$[Mo$_{154}$O$_{462}$H$_8$ (H$_2$O) $_{70}$]. *Langmuir*, 24(1), 231–238.

Bajwa, G., Genovese, M., & Lian, K. (2013). Multilayer polyoxometalates-carbon nanotube composites for electrochemical capacitors. *ECS Journal of Solid-State Science and Technology*, 2(10), M3046.

Baker, L. C. W., & Glick, D. C. (1998). Present general status of understanding of heteropoly electrolytes and a tracing of some major highlights in the history of their elucidation. *Chemical Reviews*, 98(1), 3–50.

Becker, H. I. (1957). *Low voltage electrolytic capacitor*. Google Patents.

Blazevic, A., & Rompel, A. (2016). The Anderson—Evans polyoxometalate: From inorganic building blocks via hybrid organic—inorganic structures to tomorrows "Bio-POM." *Coordination Chemistry Reviews*, 307, 42–64.

Chai, D., Gómez-García, C. J., Li, B., Pang, H., Ma, H., Wang, X., & Tan, L. (2019). Polyoxometalate-based metal-organic frameworks for boosting electrochemical capacitor performance. *Chemical Engineering Journal*, 373, 587–597.

Chai, D., Xin, J., Li, B., Pang, H., Ma, H., Li, K., Xiao, B., Wang, X., & Tan, L. (2019). Mo-Based crystal POMOFs with a high electrochemical capacitor performance. *Dalton Transactions*, 48(34), 13026–13033.

Chen, S., Zhu, J., Wu, X., Han, Q., & Wang, X. (2010). Graphene oxide-MnO_2 nanocomposites for supercapacitors. *ACS Nano*, 4(5), 2822–2830.

Chen, X., Zhu, X., Xiao, Y., & Yang, X. (2015). PEDOT/g-C_3N_4 binary electrode material for supercapacitors. *Journal of Electroanalytical Chemistry*, 743, 99–104.

Contant, R., & Herveb, G. (2002). The heteropolyoxotungstates: Relationships between routes of formation and structures. *Reviews in Inorganic Chemistry*, 22(2), 63–112.

Contant, R., Klemperer, W. G., & Yaghi, O. (1990). Potassium octadecatungstodiphosphates (V) and related lacunary compounds. *Inorganic Syntheses*, 27, 104–111.

Cuentas-Gallegos, A. K., Martínez-Rosales, R., Rincón, M. E., Hirata, G. A., & Orozco, G. (2006). Design of hybrid materials based on carbon nanotubes and polyoxometalates. *Optical Materials*, 29(1), 126–133.

Cui, L., Yu, K., Lv, J., Guo, C., & Zhou, B. (2020). A 3D POMOF based on a {AsW_{12}} cluster and a Ag-MOF with interpenetrating channels for large-capacity aqueous asymmetric supercapacitors and highly selective biosensors for the detection of hydrogen peroxide. *Journal of Materials Chemistry A*, 8(43), 22918–22928.

Da Silva, L. M., Cesar, R., Moreira, C. M., Santos, J. H., De Souza, L. G., Pires, B. M., Vicentini, R., Nunes, W., & Zanin, H. (2020). Reviewing the fundamentals of supercapacitors and the difficulties involving the analysis of the electrochemical findings obtained for porous electrode materials. *Energy Storage Materials*, 27, 555–590.

Demarconnay, L., Raymundo-Piñero, E., & Béguin, F. (2011). Adjustment of electrodes potential window in an asymmetric carbon/MnO_2 supercapacitor. *Journal of Power Sources*, 196(1), 580–586.

Dubal, D. P., Ayyad, O., Ruiz, V., & Gomez-Romero, P. (2015). Hybrid energy storage: The merging of battery and supercapacitor chemistries. *Chemical Society Reviews*, 44(7), 1777–1790.

Du Pasquier, A., Plitz, I., Menocal, S., & Amatucci, G. (2003). A comparative study of Li-ion battery, supercapacitor and nonaqueous asymmetric hybrid devices for automotive applications. *Journal of Power Sources*, 115(1), 171–178.

Evans Jr, H. T. (1948). The crystal structures of ammonium and potassium molybdotellurates. *Journal of the American Chemical Society*, 70(3), 1291–1292.

Fan, D., Hao, J., & Wei, Q. (2012). Assembly of polyoxometalate-based composite materials. *Journal of Inorganic and Organometallic Polymers and Materials*, 22(2), 301–306.

Finke, R. G., Droege, M. W., & Domaille, P. J. (1987). Trivacant heteropolytungstate derivatives. 3. Rational syntheses, characterization, two-dimensional tungsten-183 NMR, and properties of tungstometallophosphates $P_2W_{18}M_4 (H_2O)_2O_{68}^{10-}$ and $P_4W_{30}M_4 (H_2O)_2O_{112}^{16-}$ (M= cobalt, copper, zinc). *Inorganic Chemistry*, 26(23), 3886–3896.

Fuentes-Quezada, E., Martínez-Casillas, D. C., Cuentas Gallegos, A. K., & de la Llave, E. (2022). Effect of hierarchical porosity on PMo_{12} adsorption and capacitance in hybrid carbon—PMo_{12} electrodes for supercapacitors. *Energy & Fuels*, 36(7), 3987–3996.

Genovese, M., & Lian, K. (2015). Polyoxometalate modified inorganic—organic nanocomposite materials for energy storage applications: A review. *Current Opinion in Solid State and Materials Science*, 19(2), 126–137.

Gómez-Romero, P., Chojak, M., Cuentas-Gallegos, K., Asensio, J. A., Kulesza, P. J., Casañ-Pastor, N., & Lira-Cantú, M. (2003). Hybrid organic—inorganic nanocomposite materials for application in solid state electrochemical supercapacitors. *Electrochemistry Communications*, 5(2), 149–153.

Gouzerh, P., & Che, M. (2006). From Scheele and Berzelius to Müller. *L'actualité Chimique*, 298, 1.

Gujar, T. P., Kim, W.-Y., Puspitasari, I., Jung, K.-D., & Joo, O.-S. (2007). Electrochemically deposited nanograin ruthenium oxide as a pseudocapacitive electrode. *International Journal of Electrochemical Science*, 2, 666–673.

Honma, N., Kusaka, K., & Ozeki, T. (2002). Self-assembly of a lacunary α -Keggin undecatungstophosphate into a three-dimensional network linked by s-block cations. *Chemical Communications*, 23, 2896–2897.

Hou, Y., Chai, D., Li, B., Pang, H., Ma, H., Wang, X., & Tan, L. (2019). Polyoxometalate-incorporated metallacalixarene@ graphene composite electrodes for high-performance supercapacitors. *ACS Applied Materials & Interfaces*, 11(23), 20845–20853.

Izumi, Y., Urabe, K., & Onaka, M. (1992). *Zeolite, clay, and heteropoly acid in organic reactions.* ISBN: 3-527-29011-7; 4-06-205752-2, reference number: DE-94-0G3240; EDB-94-046433.

Kampf, A. R., Hughes, J. M., Nash, B. P., & Marty, J. (2017). Kegginite, $Pb_3Ca_3[AsV_{12}O_{40}(VO)]·20H_2O$, a new mineral with a novel ε -isomer of the Keggin anion. *American Mineralogist*, 102(2), 461–465.

Kang, Z., Wang, Y., Wang, E., Lian, S., Gao, L., You, W., Hu, C., & Xu, L. (2004). Polyoxometalates nanoparticles: Synthesis, characterization and carbon nanotube modification. *Solid State Communications*, 129(9), 559–564.

Karthikeyan, K., Aravindan, V., Lee, S. B., Jang, I. C., Lim, H. H., Park, G. J., Yoshio, M., & Lee, Y. S. (2010). A novel asymmetric hybrid supercapacitor based on Li_2FeSiO_4 and activated carbon electrodes. *Journal of Alloys and Compounds*, 504(1), 224–227.

Keggin, J. F. (1934). The structure and formula of 12-phosphotungstic acid. *Proceedings of the Royal Society of London. Series A, Containing Papers of a Mathematical and Physical Character*, 144(851), 75–100.

Kemp, M. (1998). Kepler's cosmos. *Nature*, 393(6681), 123–123.

Kim, T. Y., Lee, H. W., Stoller, M., Dreyer, D. R., Bielawski, C. W., Ruoff, R. S., & Suh, K. S. (2011). High-performance supercapacitors based on poly (ionic liquid)-modified graphene electrodes. *ACS Nano*, 5(1), 436–442.

Kim, Y., & Shanmugam, S. (2013). Polyoxometalate—reduced graphene oxide hybrid catalyst: Synthesis, structure, and electrochemical properties. *ACS Applied Materials & Interfaces*, 5(22), 12197–12204.

Klemperer, W. G. (1990). Tetrabutylammonium isopolyoxometalates. *Inorganic Syntheses*, 27, 74–85.

Li, J., Xie, H., Li, Y., Liu, J., & Li, Z. (2011). Electrochemical properties of graphene nanosheets/polyaniline nanofibers composites as electrode for supercapacitors. *Journal of Power Sources*, 196(24), 10775–10781.

Lindqvist, I. (1953). The Structure of the hexaniobate ion in $7Na_2O.6Nb_2O_5.32H_2O$. *Arkiv for Kemi*, 5(3), 247–250.

Liu, S., Volkmer, D., & Kurth, D. G. (2003). Functional polyoxometalate thin films via electrostatic layer-by-layer self-assembly. *Journal of Cluster Science*, 14(3), 405–419.

Ma, W., Chen, S., Yang, S., Chen, W., Weng, W., Cheng, Y., & Zhu, M. (2017). Flexible all-solid-state asymmetric supercapacitor based on transition metal oxide nanorods/reduced graphene oxide hybrid fibers with high energy density. *Carbon*, 113, 151–158.

Miller, J. R., & Simon, P. (2008). The chalkboard: Fundamentals of electrochemical capacitor design and operation. *The Electrochemical Society Interface*, 17(1), 31.

Mu, A., Li, J., Chen, W., Sang, X., Su, Z., & Wang, E. (2015). The composite material based on Dawson-type polyoxometalate and activated carbon as the supercapacitor electrode. *Inorganic Chemistry Communications*, 55, 149–152.

Müller, A., Das, S. K., Bögge, H., Schmidtmann, M., Botar, A., & Patrut, A. (2001). Generation of cluster capsules (Ih) from decomposition products of a smaller cluster (Keggin-Td) while surviving ones get encapsulated: Species with core—shell topology formed by a fundamental symmetry-driven reaction. *Chemical Communications*, 7, 657–658.

Müller, A., & Gouzerh, P. (2012). From linking of metal-oxide building blocks in a dynamic library to giant clusters with unique properties and towards adaptive chemistry. *Chemical Society Reviews*, 41(22), 7431–7463.

Müller, A., & Serain, C. (2000). Soluble molybdenum blues "des pudels kern." *Accounts of Chemical Research*, 33(1), 2–10.

Nyman, M., Alam, T. M., Bonhomme, F., Rodriguez, M. A., Frazer, C. S., & Welk, M. E. (2006). Solid-state structures and solution behavior of alkali salts of the $\{Nb_6O_{19}\}^{8-}$ Lindqvist ion. *Journal of Cluster Science*, 17(2), 197–219.

Nyman, M. D., Anderson, T. M., Alam, T. M., Rodriguez, M. A., Bixler, J. N., & Bonhomme, F. (2007). *An aqueous route to [Ta$_6$O$_{19}$]$^{8-}$ and solid-state studies of isostructural niobium and tantalum oxide complexes.* Sandia National Lab (SNL-NM).

Pakulski, D., Gorczyński, A., Czepa, W., Liu, Z., Ortolani, L., Morandi, V., Patroniak, V., Ciesielski, A., & Samorì, P. (2019). Novel Keplerate type polyoxometalate-surfactant-graphene hybrids as advanced electrode materials for supercapacitors. *Energy Storage Materials*, 17, 186–193.

Piştea, I. C., Roba, C., & Roşu, C. (2017). Green syntheses of anderson-evans polyanions $\{XMo_6O_{24}\}$ in one-pot procedure. *International Multidisciplinary Scientific GeoConference: SGEM*, 17, 157–161.

Pope, M. T., & Müller, A. (1991). Polyoxometalate chemistry: An old field with new dimensions in several disciplines. *Angewandte Chemie International Edition in English*, 30(1), 34–48.

Rani, J., Thangavel, R., Oh, S.-I., Lee, Y. S., & Jang, J.-H. (2019). An ultra-high-energy density supercapacitor; fabrication based on thiol-functionalized graphene oxide scrolls. *Nanomaterials*, 9(2), 148.

Redondo, E., Le Fevre, L. W., Fields, R., Todd, R., Forsyth, A. J., & Dryfe, R. A. (2020). Enhancing supercapacitor energy density by mass-balancing of graphene composite electrodes. *Electrochimica Acta*, 360, 136957.

Roy, S., Vemuri, V., Maiti, S., Manoj, K. S., Subbarao, U., & Peter, S. C. (2018). Two Keggin-based isostructural POMOF hybrids: Synthesis, crystal structure, and catalytic properties. *Inorganic Chemistry*, 57(19), 12078–12092.

Sadakane, M., Yamagata, K., Kodato, K., Endo, K., Toriumi, K., Ozawa, Y., Ozeki, T., Nagai, T., Matsui, Y., & Sakaguchi, N. (2009). Synthesis of orthorhombic Mo-V-Sb oxide species by assembly of pentagonal Mo$_6$O$_{21}$ polyoxometalate building blocks. *Angewandte Chemie International Edition*, 48(21), 3782–3786.

Sagadevan, S., Marlinda, A. R., Chowdhury, Z. Z., Wahab, Y. B. A., Hamizi, N. A., Shahid, M. M., Mohammad, F., Podder, J., & Johan, M. R. (2021). Fundamental electrochemical energy storage systems. In *Advances in supercapacitor and Supercapattery* (pp. 27–43). Elsevier.

Sarwar, S., Oppel, I., & Kögerler, P. (2021). *Organic-inorganic hybrid materials based on phthalocyanine, calix [4] arene and polyoxometalate scaffolds.* Fachgruppe Chemie.

Saubanère, M., McCalla, E., Tarascon, J.-M., & Doublet, M.-L. (2016). The intriguing question of anionic redox in high-energy density cathodes for Li-ion batteries. *Energy & Environmental Science*, 9(3), 984–991.

Schwegler, M. A., Vinke, P., Van der Eijk, M., & Van Bekkum, H. (1992). Activated carbon as a support for heteropolyanion catalysts. *Applied Catalysis A: General*, 80(1), 41–57.

Sheberla, D., Bachman, J. C., Elias, J. S., Sun, C.-J., Shao-Horn, Y., & Dincă, M. (2017). Conductive MOF electrodes for stable supercapacitors with high areal capacitance. *Nature Materials*, 16(2), 220–224.

Shi, H. (1996). Activated carbons and double layer capacitance. *Electrochimica Acta*, 41(10), 1633–1639.

Souchay, P. (1945). Contribution a l'etude des heteropolyacides tungstiques. [PhD Thesis]. Univ. de Paris.

Suárez-Guevara, J., Ruiz, V., & Gomez-Romero, P. (2014a). Hybrid energy storage: High voltage aqueous supercapacitors based on activated carbon—phosphotungstate hybrid materials. *Journal of Materials Chemistry A*, 2(4), 1014–1021.

Suárez-Guevara, J., Ruiz, V., & Gómez-Romero, P. (2014b). Stable graphene—polyoxometalate nanomaterials for application in hybrid supercapacitors. *Physical Chemistry Chemical Physics*, 16(38), 20411–20414.

Teze, A., & Herve, G. (1990). Investigation on stability in emulsion stabilized with different surface. *Inorg. Synth*, 27, 85.

Ueda, T. (2018). Electrochemistry of polyoxometalates: From fundamental aspects to applications. *ChemElectroChem*, 5(6), 823–838.

Vannathan, A. A., Chandewar, P. R., Shee, D., & Mal, S. S. (2022). Asymmetric polyoxometalate-polypyrrole composite electrode material for electrochemical energy storage supercapacitors. *Journal of Electroanalytical Chemistry*, 904, 115856.

Volkmer, D., Du Chesne, A., Kurth, D. G., Schnablegger, H., Lehmann, P., Koop, M. J., & Müller, A. (2000). Toward nanodevices: Synthesis and characterization of the nanoporous surfactant-encapsulated keplerate (DODA) $_{40}(NH_4)_2[(H_2O)_n \subset Mo_{132}O_{372} (CH_3COO)_{30}(H_2O)_{72}]$. *Journal of the American Chemical Society*, 122(9), 1995–1998.

Wang, G., Chen, T., Gómez-García, C. J., Zhang, F., Zhang, M., Ma, H., Pang, H., Wang, X., & Tan, L. (2020). A high-capacity negative electrode for asymmetric supercapacitors based on a PMo_{12} coordination polymer with novel water-assisted proton channels. *Small*, 16(29), 2001626.

Wang, X., Lu, X., Liu, B., Chen, D., Tong, Y., & Shen, G. (2014). Flexible energy-storage devices: Design consideration and recent progress. *Advanced Materials*, 26(28), 4763–4782.

Wang, Y., Shen, C., Niu, L., Li, R., Guo, H., Shi, Y., Li, C., Liu, X., & Gong, Y. (2016). Hydrothermal synthesis of $CuCo_2O_4$/CuO nanowire arrays and RGO/Fe_2O_3 composites for high-performance aqueous asymmetric supercapacitors. *Journal of Materials Chemistry A*, 4(25), 9977–9985.

Wu, H.-Y., & Wang, H.-W. (2012). Electrochemical synthesis of nickel oxide nanoparticulate films on nickel foils for high-performance electrode materials of supercapacitors. *International Journal of Electrochemical Science*, 7, 4405–4417.

Xie, Z., Jin, X., Chen, G., Xu, J., Chen, D., & Shen, G. (2014). Integrated smart electrochromic windows for energy saving and storage applications. *Chemical Communications*, 50(5), 608–610.

Yan, X., Mo, S., Ju, Z., Wu, J., & Yao, K. (2008). Synthesis and structure of a series of Anderson-Evans type heteropolymolybdates: $[Ln(H_2O)_n]_2[TeMo_6O_{24}]\cdot 6H_2O$. *Synthesis and Reactivity in Inorganic, Metal-Organic, and Nano-Metal Chemistry*, 38(6), 529–533.

Yang, H., Kannappan, S., Pandian, A. S., Jang, J.-H., Lee, Y. S., & Lu, W. (2013). Achieving both high power and energy density in electrochemical supercapacitors with nanoporous graphene materials. *arXiv:1311.1413*. https://doi.org/10.48550/arXiv.1311.1413

Yang, Y., Li, S., Huang, W., Shangguan, H., Engelbrekt, C., Duan, S., Ci, L., & Si, P. (2019). Effective synthetic strategy for $Zn_{0.76}Co_{0.24}S$ encapsulated in stabilized N-doped carbon nanoarchitecture towards ultra-long-life hybrid supercapacitors. *Journal of Materials Chemistry A*, 7(24), 14670–14680.

Zhang, Y., Zeng, T., Huang, D., Yan, W., Zhang, Y., Wan, Q., & Yang, N. (2020). High-energy-density supercapacitors from dual pseudocapacitive nanoelectrodes. *ACS Applied Energy Materials*, 3(11), 10685–10694.

Zhou, G., Li, F., & Cheng, H.-M. (2014). Progress in flexible lithium batteries and future prospects. *Energy & Environmental Science*, 7(4), 1307–1338.

4 Supercapacitors
Carbon-Based Nanostructures for Supercapacitor Application

Shanu Mishra and Ashish Kumar Mishra

4.1 INTRODUCTION

The continuous consumption of existing fossil fuels, their soaring price and the associated alarming global warming situation have triggered the urge for developing sustainable and renewable sources of energy. Renewable energy sources (solar, wind, hydroelectric) are regarded as next-generation energy sources, but the sun does not shine all the time and wind does not always blow. So, due to their intermittent nature, imbalanced regional distribution and low energy density, these energy sources do not show a potential impact unless they are integrated with efficient energy storage technologies [1, 2]. To resolve issues related to energy demand, the research community needs to design clean, efficacious, economical, and eco-friendly energy storage and conversion systems. For these reasons, devices like capacitors, conventional batteries, fuel cells, and supercapacitors (SCs) are contemplated as electrical and electrochemical storage systems [3, 4]. The different structural/assembly and charge–discharge capacities of conventional capacitors, SCs, and batteries are displayed in Figure 4.1a. The efficiency of various energy conversion and storage devices is determined by their energy and power density. As shown in the Ragone plot (Figure 4.1b) [5], various batteries (Li-ion, etc.) are regarded as energy sources (~10^5 J kg^{-1}), but they suffer from slow power delivery, approximately <100 W kg^{-1}. Thus, the extensive use of batteries is explicitly limited in energy storage systems where high power is required. However, conventional capacitors can deliver high power density (10^6 W kg^{-1}) and can provide higher current than batteries, but their energy density (0.1 Wh kg^{-1}) is limited [6, 7]. A comparison of key properties (charging time, no of cycles, energy density, power density) of SCs with conventional capacitors and batteries is outlined in Table 4.1.

Recently, high energy and power density systems are desired for fast storage/delivery. For instance, in hybrid electric vehicles, a high energy-density device is required to travel a longer distance, while a high power-density device provides initial acceleration in a short time. Conventional batteries and capacitors are both incapable of delivering high energy and power density simultaneously. Nippon

DOI: 10.1201/9781003208709-4

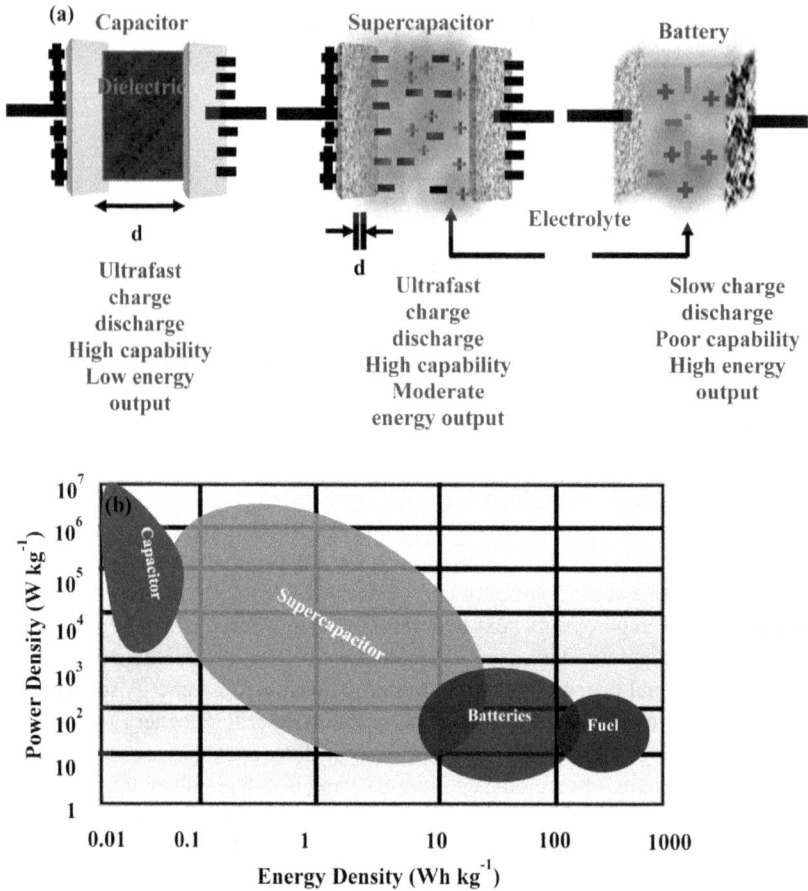

FIGURE 4.1 (a) Configurational outlines of conventional capacitors, supercapacitors, and batteries; (b) Ragone plot for different energy storage and conversion devices.

Electric Corporation was the first to commercialize an electrochemical capacitor called a "supercapacitor" in 1978. SC technology has the potential to provide a bridge between traditional capacitors and batteries as they possess higher energy density (1–10 Wh kg^{-1}) than conventional capacitors and higher power density (2–5 kW kg^{-1}) than conventional batteries, as well as prolonged cycle life, which makes them a suitable applicant for industrial uses, such as consumer electronics, memory backup and regenerative braking energy of electric vehicles, where high power and optimal energy capacity is required [8, 9]. A typical example is electric buses (traction vehicles) launched in Shanghai (China) in 2006 that uses an SC to store energy but can store only 5% of energy compared to batteries. This shortcoming can limit them to a couple of miles per charge. The key issue with SCs is producing

TABLE 4.1

Comparison of Key Properties of Capacitors, Supercapacitors and Batteries [4, 6]

Functions	Capacitor	Supercapacitor	Battery
Time of charging	10^{-6} to 10^{-3} s	0.3 to 30 s	1 to 5 h
Number of cycles	Almost infinite	1 million	500–1000
Energy density (Wh kg^{-1})	<0.1	1–10	10–100
Power density (W kg^{-1})	>100,000	10,000	<1000
Coulombic efficiency (%)	Almost 100	>100	70–80
Charge storage determinants	Electrode area and	Microstructure of	Thermodynamics and
Operating temperature (°C)	dielectric	electrode &	active mass
	−20 to 100	electrolyte	−20 to 65
		−40 to 85	

them with a high energy density comparable to that of present rechargeable batteries while keeping their intrinsic qualities (high power and prolonged cycling life). The development of SCs for industrial application rely on electrode materials and comprehending a unified model of electrochemical interfaces at the nanoscale. This inspires the researcher toward designing and preparing many novel electrode materials with various components and structures. Carbon-based nanostructures are immensely attractive as electrode materials for flexible and wearable energy storage applications due to their intriguing properties, such as the graphene, showing high electrical conductivity up to 10^8 S m^{-1}; good mechanical stability, approximately 240 to 280 GPa; and high theoretical surface area up to 2600 m^2 g^{-1} [4]. Besides the physicochemical stability, their unique pore structure, large-scale synthesis, and the presence of accessible electrocatalytic active sites make carbon materials suitable for energy storage applications [10–15]. In this chapter, a brief overview of the classification of SCs, an explanation of charge–storage mechanisms, and characterization techniques for studying charge-storage capacity are emphasized. In addition, the design and development of SCs with high energy and power density having different architectures of carbon materials are discussed in this chapter.

4.2 FUNDAMENTALS AND CHARGE STORAGE MECHANISM OF SCS

According to charge-storage mechanism, SCs are divided into two categories: electrochemical double-layer capacitors and pseudocapacitors [16], which are described in the following sections.

4.2.1 ELECTROCHEMICAL DOUBLE-LAYER CAPACITORS

The working principle of electrochemical double-layer capacitors (EDLCs) is similar to conventional capacitors. The energy is stored via the physical adsorption of electrolyte ions at the surface of a porous electrode during charging of the device. Owing to their high surface area and very short charge separation distances, EDLCs possess higher energy density than traditional electric capacitors. As shown in Figure 4.2a, EDLCs consist of two carbon electrodes dipped in a suitable electrolyte (KOH, H_2SO_4, Na_2SO_4, etc.) and an ion-exchange membrane situated between the electrode [17]. The concept of the EDLC model was first described by Hermann von Helmholtz in 1853. According to this model, when an electrical conductor is immersed in an electrolyte, the opposite charges are accumulated at the interface of the electrode/electrolyte and form the double layer separated by an atomic gap. This layer is known as a compact or Helmholtz layer [18]. The model is identical to traditional two-plate capacitors; these charges are separated electrostatically, and no charge transfer occurs during the formation of EDLCs. Furthermore, the Gouy-Chapman modified the oversimplified model of a rigid charged surface and contemplated the diffuse layer that arises due to the thermal motion of electrolyte ions. The capacitance value of two charged arrays separated by a certain distance should increase in proportion to the decrease in distance between them. In other words, as point charges get closer to the electrode surface, their capacitance increases dramatically. However, the model fails to explain the formation of highly charged double layers at the electrode/electrolyte interface. Otto Stern later integrated the essentials of both Helmholtz-Perrin and Guoy-Chapmann electrical double-layer theories. According to this theory, the double layer has two distinct ion distribution regions. The first region is approximately single ion thick, and these ions are firmly adsorbed to the solid surface and form the inner Helmholtz plane (Helmholtz view), as shown in Figure 4.2b [19, 20]. The remaining charges are distributed through the electrolyte next to the compact layer as a diffuse layer (Guoy's view) and are called the outer Helmholtz plane (OHP). So the Stern model suggests that as a result of thermal agitation, a part of ions in solution are 'fixed' to the electrode and the rest are 'scattered' in a cloud-like pattern (Figure 4.2b). In the case of SCs, the charge accumulation mechanism is based on the surfaces rather than the bulk volume, which is opposite to the charge accumulation in batteries. During charging electrons migrate from the

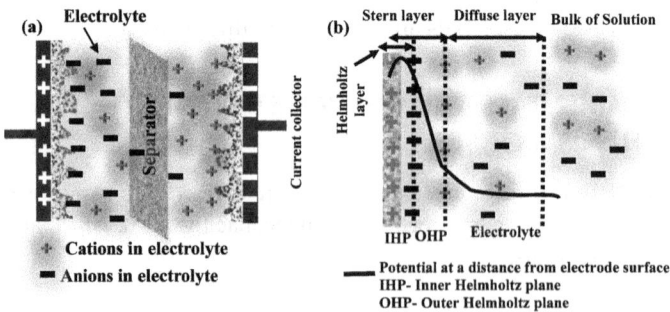

FIGURE 4.2　(a) Schematic illustration of an electrochemical double-layer capacitor; (b) the charge storage mechanism of the double-layer capacitor.

negative electrode to the positive electrode as a result of an applied external voltage. As a result, an anion of electrolyte (A^-) travels toward the positive electrode (E_1). The electrochemical charging process is expressed as $E_1 + A^- = E_1^+//A^- + e^-$. While discharging, cations (C^+) of electrolyte moves toward negative electrode (E_2). The electrochemical charging process is expressed as $E_2 + C^+ + e^- = E_2\_//C^+$, where $//$ represents the electrode–electrolyte interface [21].

EDLCs are the most commercialized SCs, in which the charge is physically held via electrostatic charge adsorption at the electrode–electrolyte interface. The notable characteristic of EDLCs is the absence of charge transfer between the electrode and electrolyte interfaces, that is, no Faradaic processes. The capacitance of an EDLC electrode can be predicted according to Equation 4.1 [5]:

$$\alpha \ H_4XM_{12}O_{40}{}^{n-} + 2 \ e^- + 2 \ H^+ \rightarrow \alpha \ H_6XM_{12}O_{40}{}^{n-} , \qquad (4.1)$$

where ε_r is the dielectric constant of the electrolyte; ε_o is the permittivity of free space; d is the effective charge separation, that is, Debye length; and A is the accessible surface area of the electrode material. In comparison to conventional capacitors, SCs have a three-dimensional (3D) porous electrode which maximizes the surface area. Also, the dielectric layer of SCs consists of a single or several molecular layers; that is, d is a short molecular distance (few to several Å) while in conventional capacitor distance, d reaches up to several microns. As a result, EDLCs may store significantly more electric energy than typical dielectric capacitors due to their larger effective surface area (A) and shorter charge separation distance (d). The electrochemical double layer's effective thickness varies between 5 to 10 Å, depending on the electrolyte ions' concentration and size, as well as the solvation shell. Based on the relative permittivity of the electrolyte medium, the specific capacitance of a carbon-based system is expected to be between 10 and 21 $\mu F \ cm^{-2}$. Carbon-based materials with large specific surface area (1000–3000 $m^2 \ g^{-1}$) can theoretically achieve double-layer capacitances of 300–550 F g^{-1} [5, 17]. Thus, the carbon-based materials (activated carbon, carbon aerogel, carbon nanotubes [CNTs], graphene, etc.) with specific surface area show good EDLC behavior with higher specific capacitance ideally, but experimentally achieved specific capacitance of these materials are lower than expected value due to limited accessibility of surface sites.

4.2.2 PSEUDOCAPACITORS AND HYBRID SCs

The word *pseudocapacitance* was coined by Conway to identify those materials that display electrochemical signatures, which are close to those of typical double-layer capacitors (EDLC) [22]. Unlike EDLCs, pseudocapacitors involve a different charge-storing mechanism. Pseudocapacitors store electric energy through reversible Faradic reaction or redox reaction, which occurs at the electrode/electrolyte interface involving the passage of charges between the electrode and the electrolyte (Figure 4.3a) [23]. During the oxidation process, the electrolyte ions move to the electrode, and during reduction, they move back to the electrolyte. The charge–discharge mechanism of a pseudocapacitor is similar to that of a battery. The electrochemical charging process is expressed as $E_1 + A^- = E_1^{\delta+}//A^- + \delta e^-$. However, the discharging

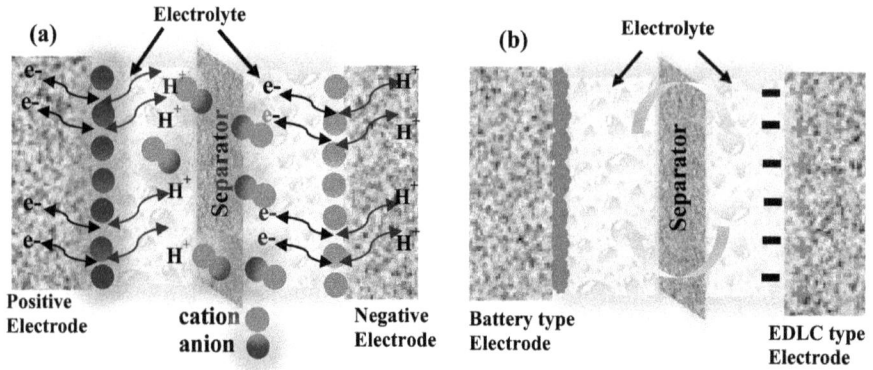

FIGURE 4.3 Schematic illustration of (a) pseudocapacitors and (b) hybrid SCs.

process is expressed as $E_2 + C^+ + \delta e^- = E_2^{\delta-} \,// \, C^+$, where E_1 and E_2 are positive and negative electrodes and the parameter δe^- represents electrosorption valence, related to the oxidation–reduction reactions [21]. Generally, metal oxides and conducting polymers are fundamental candidates explored as pseudocapacitive electrode material [10, 24]. These materials can provide multiple oxidation states for efficient redox reaction at the electrode surface, and this may allow the pseudocapacitor to achieve high energy density and capacitance. The selection of a suitable potential widow is another crucial factor for pseudocapacitor, beyond the appropriate potential the electrode will be degraded. The first electrode material to show pseudocapacitive behavior was ruthenium dioxide (RuO_2). In the case of RuO_2 thin film, the charge is stored via a Faradic reaction while cyclic voltammetry (CV) curve shows a typical capacitive behavior. Thus, the surface redox thermodynamics of pseudocapacitors is Faradic, but the capacitance is originated from a linear relationship between the amount of electric charge stored (ΔQ) and applied potential (ΔU). So, at the time of charging, for each addition of electric charges, electrical work must be done. The energy storing of the pseudocapacitor is midway between EDLCs and Faradaic reactions. The active centers of pseudocapacitive materials are located near the metal oxide surface at a distance $< (2Dt)^{1/2}$, where D is the diffusion coefficient ($cm^2 \, s^{-1}$) and t is the diffusion time (s). As reported by Dunn, pseudocapacitors have three different Faradaic mechanisms, including underpotential deposition, redox pseudocapacitance, and intercalation pseudocapacitance [25]. Langmuir-type electrosorption of hydrogen atoms on a noble metal substrate is a good example of underpotential deposition, where hydrogen atoms form an adsorbed monolayer at a distinct metal's surface (Pt, Rh, Ru, etc.) considerably above their redox potential in underpotential deposition, as shown in the equation 4.2 [26]:

$$M + xC^{z^+} + xze^- = C.M \,, \tag{4.2}$$

where M represents the noble metal (Pt, Ru), C represents the absorbed atoms, x denotes the number of absorbed atoms, z represents the absorbed valency, and so zx represents the number of electrons exchanged. However, in the case of the underpotential deposition process, the overpotential ranges are small ranging from 0.3 to 0.6 V, and the capacitance values are potential dependent [26]. As a result, the energy density of the system is limited compared to other pseudocapacitive systems. The redox pseudocapacitance occurs by transferring electrons between an oxidized (Ox) and a reduced species (Red), that is, electrochemical adsorption of cations on the surface of oxidized species, with quick and reversible electron transfer across the electrode–electrolyte interface per Equation 4.3 [26]:

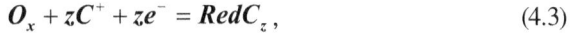

$$O_x + zC^+ + ze^- = RedC_z,$$ (4.3)

where C^+ is the adsorbed cation of electrolyte and z denotes the number of transferred electrons. The metal oxides RuO_2, NiO, and MnO_2 and conducting polymers (polypyrrole and polyaniline) are examples of redox pseudocapacitive materials [10]. In such redox pseudocapacitor devices, the capacitance is determined by the reactant ions and active site density and the maximum achievable capacitance value is around 5000 F cm^{-3} [27]. The intercalation pseudocapacitance is caused via ions intercalation into the layers of a redox-active material. The ion intercalation process involves no crystallographic phase shift per Equation 4.4 [26]:

$$MA_y + xLi^+ + xe^- = Li_xMA_y,$$ (4)

MA_y is the host material for lattice intercalation and x denotes the number of transferred electrons. In the case of cation, intercalation pseudocapacitors show an intermediate behavior between battery and SCs. All three pseudocapacitive mechanisms are outcome of various electrochemical interfacial processes, the electrochemical fingerprints are similar due to the relationship between potential and the amount of charge that emerged as a result of adsorption/desorption at the electrode/electrolyte contact. According to the thermodynamic considerations, the charge that forms at the interface and the potential (E) are related in Equation 4.5 [26, 28]:

$$E = E^o - \frac{RT}{nF} \ln \frac{X}{1-X},$$ (4.5)

where E denotes the potential (V), E^0 denotes standard potential of the redox couple, R stands for ideal gas constant (8.314 J mol^{-1} K^{-1}), n represents the number of electrons, T represents temperature (K), F represents the Faraday constant with usual value 96485 C mol^{-1}, and X denotes the fractional charge coverage. Using Equation 4.6, the capacitance (F g^{-1}) can be calculated in areas where the plot of E vs. X is linear:

$$C = \left(\frac{nF}{m}\right)\frac{X}{E},$$ (6)

where m represents the active material's molecular weight. The capacitance is not always constant because the graph of E vs. X is not completely linear, as it is in a capacitor, so it is called pseudocapacitance. The thermodynamic basis for pseudocapacitance is described by the preceding relationship that clearly depicts that the reactions near the electrode surface are not constrained by diffusion and so possesses a good rate of capability. EDLCs have the advantage of good cyclic stability and high power density, while pseudocapacitors offer high energy density. Another class of SC is hybrid SC (Figure 4.3b), which combines two different types of electrodes. The electrodes are made up of high-power EDLCs materials (carbon-based), on one side, and battery-like materials (metal oxides/conducting polymer), on the other side [29]. Hence, the device follows both capacitive and Faradaic charge storage mechanisms in order to obtain high energy density as well as power density.

4.3 MEASUREMENT TECHNIQUES

Generally, the electrochemical characteristics of a SCs device are evaluated using three main techniques: CV, galvanostatic charge/discharge (GCD), and electrochemical impedance spectroscopy (EIS). For instance, in CV, the current is estimated at a fixed scan rate in a particular potential window while GCD uses fixed current density and the potential is ramped from open circuit voltage to a fixed potential value, and EIS uses the capacitance or impedance. The other important parameters as high-specific capacitance, equivalent series resistance (ESR), time constant, energy, and power density can be subsequently derived from them. Two-electrode cells, three-electrode cells, and a fully packaged actual device could be used in performance characterization experiments. The three-cell electrode is composed of a working, a counter, and a reference electrode, and it emphasizes the electrochemical behavior of the working electrode and provides vital understandings of a variety of applications, including energy storage device design. The two-cell electrode layout closely resembles a real SC cell and makes it easy to optimize important features or parameters for future prototype design and manufacture.

4.3.1 CV

CV is an effective and broadly used operative technique in electrochemistry. It determines the current that emerges in an electrochemical cell when the voltage exceeds the value indicated by the Nernst equation. Using the CV technique, qualitative and pseudo-quantitative studies can be performed to determine the voltage range of the device or electrode over a wide range of scan rates. In essence, the working electrode's voltage is increased linearly over time. The electric potential is measured between the working and the reference electrode, while the current is estimated between the working and counter electrodes. The rate of the potential ramp is known as a scan or sweep rate. The voltage sweep rate also efficiently affects the device performance of SCs, the capacitance value increases at a slower scan rate because at a slower sweep rate, the ion diffusion has sufficient time to reach the inner pores of electrode material. The CV curve can easily differentiate the electrochemical performance of EDLCs, pseudocapacitors, and batteries, as shown in Figure 4.4. The CV curve of EDLCs shows

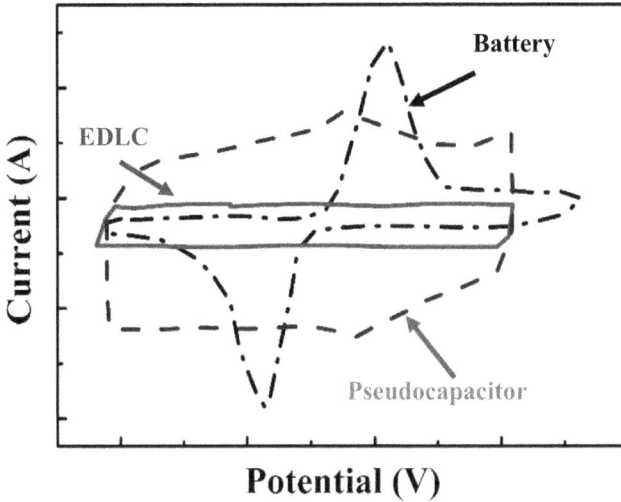

Potential (V)

FIGURE 4.4 Schematic illustration of typical CV curves of an EDLC, a pseudocapacitor, and a battery

Source: Adapted from Ref. 30.

a rectangular shape while pseudocapacitors and batteries have prominent redox peaks [30]. The redox nature of pseudocapacitors has symmetric and broad pairings, but batteries possess separated oxidation and reduction peaks that clearly differentiate the separate charge storage phenomena in batteries and pseudocapacitors. In order to distinguish the pseudocapacitive charge–storage process from batteries, a slow scan CV is utilized; that is, the electrode kinetics can be revealed by examining the relationships between peak (i_p) current and scan rate (v) per Equation 4.7 [26]:

$$i = av^{\frac{1}{b}}, \tag{4.7}$$

where i is the current acquired at a certain sweep rate, v, and a and b are variables that can be changed. In the case of batteries, $b = \frac{1}{2}$, and for SCs, $b = 1$, and hence, EDLCs and pseudocapacitors show similar CV behavior, which is distinct from batteries.

The instantaneous voltage and current response are recorded to examine the electrochemical response. The capacitance can be calculated using Equation 4.8 [20, 31]:

$$C = \frac{\Delta Q}{\Delta V}, \tag{4.8}$$

where ΔQ is the integral area of the voltammetry curve, which reflects the total charge, and ΔV is the working potential window. The specific capacitance of a two-cell SC device is evaluated using Equation 4.9:

$$C_s = \frac{2 \times \int_{V_i}^{V_f} IdV}{mv\Delta V} \, , \tag{4.9}$$

where $\int I \, dV$ represents an integral area of one CV curve, m represents the total mass loading of active material in prepared device, and v is the scan rate. There are some advantages and disadvantages of the CV technique. The CV test is used to interpret the type of electrochemical reaction and its contribution from an EDLC (capacitive) and a propylene carbonate (PC) (Faradic). The CV test is an important method to evaluate the specific capacitance value of the electrochemical cell, and it can easily detect the degradation process of the electrode. Still, it shows only kinetic aspects, and thermodynamic aspects are neglected.

4.3.2 GCD

The GCD is the most versatile and accurate technique for characterizing electro-chemical devices under direct current. The GCD curve can clearly distinguish bat-tery and capacitive performances. In GCD, the potential varies as a function of time with an applied constant current. In accordance with the energy storage mechanism, at a constant discharge current, voltage vs. discharge time has a linear relationship for EDLCs and pseudocapacitors as shown in Figure 4.5. Pseudocapacitors have lon-ger charge and discharge periods, indicating that more electrons and electrolyte ions are involved in the charge and discharge processes. EDLCs have a sharper slope compared to pseudocapacitors because of their lower capacitance and faster dis-charge. In the ideal capacitor, capacitance is a constant value, independent of the potentials. The linear voltage time response is described in Equation 4.10 [30]:

$$\Delta V = \Delta t \frac{I}{C} \, , \tag{10}$$

where the potential is linearly proportional to time and I and C are constant. However, the battery maintains a constant voltage during discharge or charge and shows a signifi-cant voltage plateau with various polarization zones due to its phase change mechanism.

An IR drop is common for EDLCs, pseudocapacitors, and batteries, and it is orig-inated due to the internal resistance of electrolytes, electrode material, and a current collector. The batteries show an additional sharp potential drop due to concentration polarization that is absent in pseudocapacitors and EDLCs as shown in Figure 4.5. The characteristic parameters, such as specific capacitance, life cycle, electrochemi-cal impedance, ESR, and energy and power density, can be evaluated using the GCD process. GCD follows a repetitive charging–discharging loop of a SC device. In the GCD process, a constant current is applied until a given potential is reached, and the output is plotted as potential vs. time. The specific capacitance of SC devices can be calculated using Equation 4.11 [31–33]:

$$C_s = \frac{2 \times I \times \Delta t}{mass \times \Delta V} \, , \tag{4.11}$$

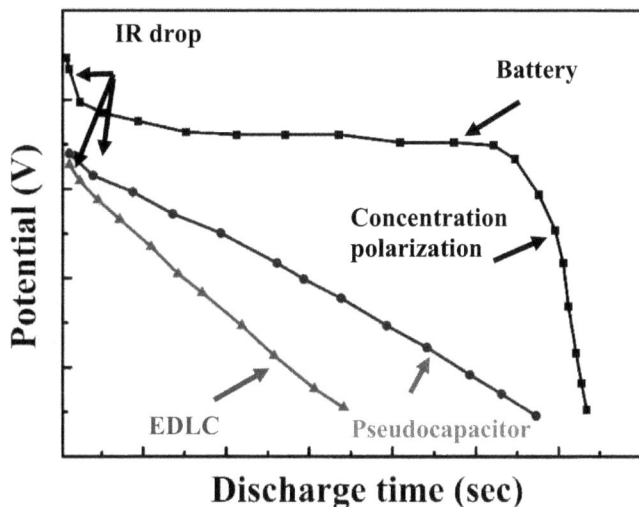

FIGURE 4.5 Schematic illustration of typical GCD curves of EDLC, pseudocapacitor, and battery.

Source: Adapted from Ref. 30.

where I denotes the discharge current density, Δt represents the discharge time, and m represents the active mass of the both electrodes. The energy density (E) and power density (P) of SC devices can be computed using the GCD curves per Equations 4.12 and 4.13:

$$E = \frac{1}{2}\frac{C_s(\Delta V)^2}{3.6},$$ (4.12)

$$P = \frac{E * 3600}{\Delta t},$$ (4.13)

where C_s is specific capacitance obtained from GCD, ΔV is the potential difference, and Δt is discharge time calculated from the GCD curve. Also, the performance of the supercapacitor device can be determined using volumetric-specific capacitance and areal capacitance. In order to estimate the volumetric-specific capacitance (F cm^{-3}), the density of the active material of the electrode is multiplied by the gravimetric-specific capacitance. Hence, the density of the active material and the volumetric capacitance are directly proportional. Similarly, the areal capacitance (F cm^{-2}) is calculated by dividing the electrode's capacitance with the area of the working electrode. The ESR value can be evaluated using IR drop. The ESR is computed by dividing the applied constant current by the IR drop (obtained from the GCD curve) as ESR= IR$_{drop}$/2I. The GCD test is the most widely used method for determining the stability of SCs. As observed from Equation 4.12, the high specific capacitance value and voltage window can greatly enhance the energy density value of SC devices. Using asymmetric or hybrid supercapacitors is one technique to dramatically enhance the voltage window.

4.3.3   EIS

EIS is an extremely sensitive and nondestructive characterization technique for subsuming small signals and evaluating the resulting response of physio-electrochemical characteristics. The EIS can be used to investigate chemical mechanisms at the electrode/electrolyte interface. The impedance data is measured by applying a low-amplitude alternative voltage (5–10 mV) over a wide frequency range (0.01–10^6 Hz), and the results are graphically described as a Nyquist plot. The Nyquist plot illustrates the imaginary (Z'') and real (Z') components of the cell impedances, composed of three separate zones: an intercept on the real axis (solution resistance, R_s), a semicircle in the high-frequency zone (charge transfer resistance, R_{ct}), and a linear component over low-frequency zones (Warburg resistance, W). The Nyquist plot can also be used to distinguish between EDLC, pseudocapacitor, and batteries, as shown in Figure 4.6. The EDLC shows a vertical line (90°) from the real axis since there is an absence of any Faradic process. EDLCs do not display any semicircle or ion diffusion, but practical EDLCs display a semicircle due the presence of functionality or doping. The ideal pseudocapacitor has a very small semicircle accompanied by a vertical line at 90°; the small semicircle depicts rapid charge transfer Faradic process. Compared to pseudocapacitors, batteries have a significantly bigger semicircle followed by a straight line, making a 45° angle from the real axis. The bigger semicircle illustrates the slower charge transfer, and the 45° straight line depicts the slow diffusion process.

The Bode plot (Figure 4.6b) is another representation of EIS, which shows relationship between phase angle and frequency of a cell [34]. Bode plot is an effective technique to reveal the capacitive and battery behavior of the device by simple observation of phase shift at specified frequency values. The gradual increase in the phase angle to reach a value of 90° indicate a capacitive behavior. The conventional method of calculating capacitance from Bode frequency response can be given by equation 14 [35, 36]:

$$C_s = \frac{1}{2\Pi f\,|Z|}, \tag{4.14}$$

where f is the frequency and Z is the impedance. The EIS technique is also used to estimate charge transfer or pseudo-charge transfer, capacitive behavior, and charge–storage mechanisms. The charge transfer resistance can be calculated from the Nyquist plot as the diameter of the semicircle on the real axis [37]. Although the EIS is a nondestructive method to evaluate specific capacitance, relaxation time for recharging along with detection of various resistances in a system, but it evaluates the system at low voltage only, and the system shows discrete behavior above 10^6 Hz.

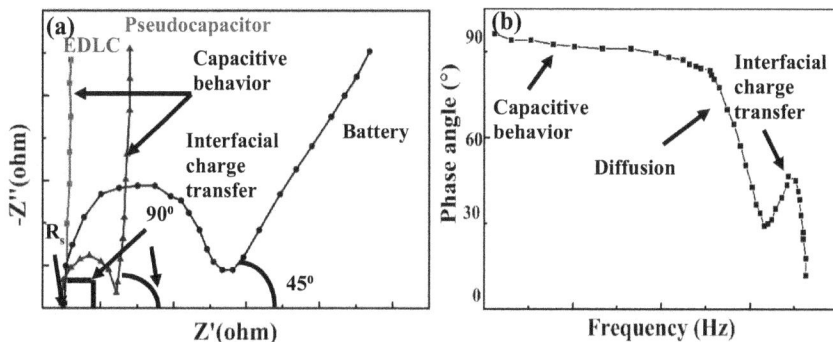

FIGURE 4.6 (a) Schematic illustration of typical EIS curves of an EDLC, a pseudocapacitor, and battery (adapted from ref. 30); (b) schematic illustration of Bode plot.

4.4 SC DEVICES

The supercapacitors are made in different styles for commercial applications, such as coin-type assembly, cylindrical assembly, and flexible assembly, as shown in Figure 4.7. Two electrodes and electrolyte-saturated separators are placed in a conductive metal case during coin cell manufacturing, and the assembly is sealed under high pressure to prevent leakage and sorting of cells (Figure 4.7a). The coin-cell design is ideal for powering compact devices. The cylindrical cell design of supercapacitors is also used in commercial purposes. The design consists of rolling layers of electrode and separator sheets packed into a metal casing to enhance mechanical durability, as shown in Figure 4.7b. Figure 4.7c depicts an alternative design known as the flexible SC assembly (pouch cell), which likewise uses layer-by-layer sheets of electrodes and separators. These parts are put together in polymer bags, which are flexible enough and reduce the extra weight of devices. So, flexible SCs could benefit from very flexible thin-film electrodes with soft substrates and good electrical conductivity and mechanical strength along with flexible electrolytes to provide the shape variable support [38]. Generally, three types of electrolytes- aqueous (H_2SO_4, KCl, KOH, etc.), organic (PC, tetrahydrofuran [THF], etc.), and ionic liquids (1-Ethyl-3-methylimidazolium tetrafluoroborate EMI-BF$_4$ etc.) are used in SCs. Among them, polymer- and ionic liquid (IL)–based electrolytes fulfilled the standard of commercial demand but the inadequate ionic conductivity of ILs hinder the capacitance of the cell. The current research on supercapacitors is focused on the miniaturization of mechanically flexible solid-state devices (SSDs) that follow the significant trend of portable and wearable electronics. Carbon nanomaterials, such as graphene, CNTs, activated carbon, and others, are extensively used in the construction of flexible SSDs because of their extraordinary mechanical strength (~200 GPa), which allows them to endure bending, folding, or rolling, as well as outstanding surface area (~2600 m^2g^{-1}), and superior electrical conductivity (~10^6 to 10^7 S m^{-1}) [39]. The development of practical methodologies for achieving high-performance flexible SCs with planar topologies based on CNTs, graphene, and hybrid electrodes are discussed in the following sections [40–42].

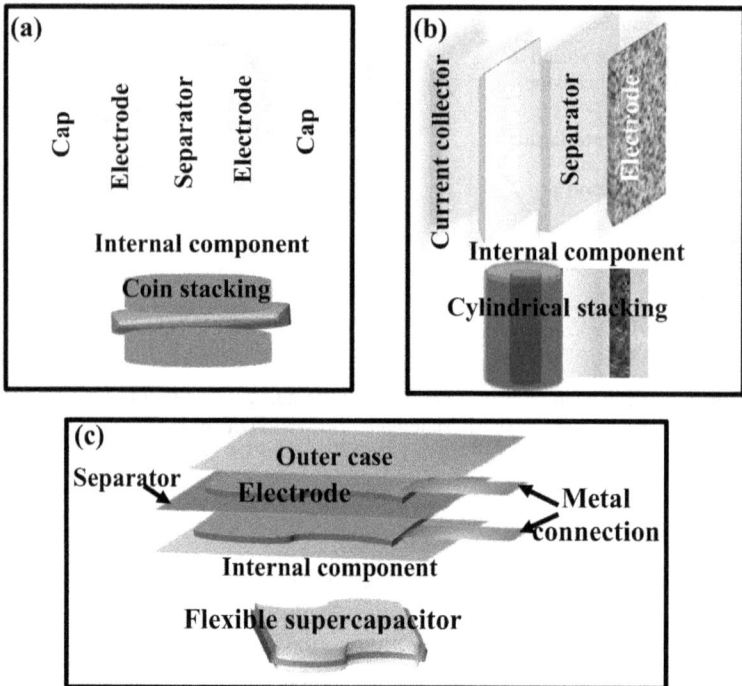

FIGURE 4.7 Schematic of different types of SCs: (a) coin-type assembly, (b) stack rectangular assembly, and (c) flexible assemble.

4.4.1 CNT-Based Flexible SCs

CNTs, as excellent polarizable electrodes, have received a lot of attention because of its outstanding electrical conductivity, high surface area, unique porous structure, and exceptional mechanical and thermal durability. Also, their high aspect ratio and open tubular network make them suitable candidates for flexible SCs. Thin-film SC configurations can be achieved by spray or brush coating, inkjet printing, and dip-drying a CNT suspension onto a highly flexible substrate. Due to the van der Waals interaction between 1D CNTs and substrates, CNTs can strongly attach to the substrate and provide a continuous conducting path along the length. For example, Kaempgen et al. [43] fabricated thin-film SCs using a spray coating of a single-walled carbon nanotube (SWCNT) suspension on a flexible polyethylene terephthalate (PET) substrate. The network of SWCNTs functions serve as both current collectors and electrodes. To achieve a fully flexible device, a printable polymer gel electrolyte (PVA-H_3PO_4) was used as both the separator and the electrolyte. The flexible SWCNT film coated on a PET substrate and a polymer gel electrolyte were clubbed to construct the flexible SC devices (Figure 4.8a). Figures 4.8b and 4.8c represent the CV and GCD curves of the device, showing the specific capacitance of 36 F g^{-1}. Various microfabrication techniques have been established to accurately regulate the interdigitated electrode's structure, made from freestanding CNT films and composites.

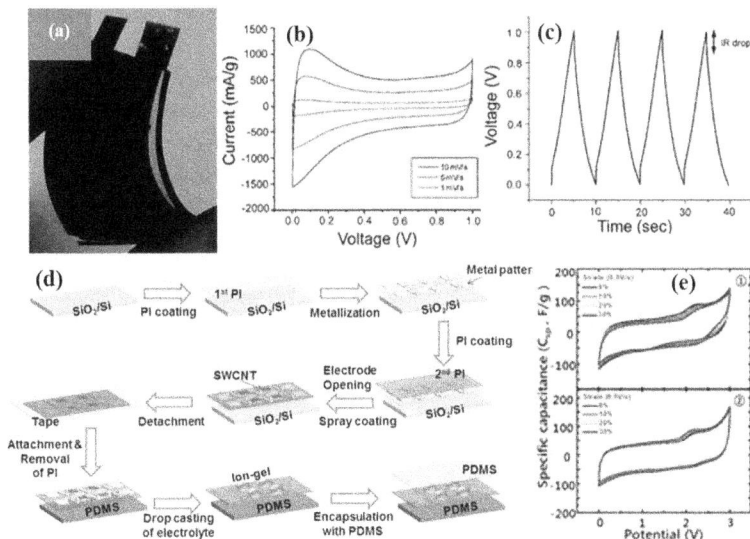

FIGURE 4.8 (a) Thin-film SC using sprayed SWCNT films on PET as the electrodes and a PVA/H$_3$PO$_4$ as polymer electrolyte and separator; (b) the cyclic voltammetry; and (c) the galvanostatic charge–discharge curves of the thin film supercapacitor using PVA/H$_3$PO$_4$ polymer electrolyte (Reproduced with permission from ref. 43. Copyright 2009 American Chemical Society); (d) schematics of cross-sectional view of a stretchable micro-SC array on a PDMS substrate; and (e) CV curves of micro-SC array obtained while increasing and decreasing the strain from 0% to 30% and 30% to 0%. (Reproduced with permission from ref. 44. Copyright 2013 American Chemical Society).

Kim *et al.* [44] used bottom-up technique to produce interdigitated electrodes for micro-SCs. They fabricated a stretchable micro-SC array by spray coating SWCNTs as electrodes and using an ion-gel electrolyte, 1-Ethyl-3-methylimidazolium bis (trifluoromethylsulfonyl) imide [EMIM][NTf$_2$] (Figure 4.8d). The fabricated flexible micro-SCs array exhibited a specific capacitance of 55.3 F g^{-1}, and the device was not deteriorated upon increasing and decreasing the strain up to 30% (Figure 4.8e). To fulfill the need for a fully stretchable device with 100% stretchability, Niu *et al.* [45] fabricated highly flexible and stretchable SCs using buckled SWCNT films. In order to synthesize SWCNT film electrodes having continuous reticulate architecture, they combined directly grown SWCNT film with polydimethylsiloxane (PDMS) substrate. The electrode performance remains nearly unchanged even under the strain of up to 140%. The device showed a high specific capacitance of 48 F g^{-1} and 53 F g^{-1} with no strain and 120% strain, respectively, using PVA-H$_2$SO$_4$ as both separator and electrolyte.

CNT composites with conductive polymers or transition metal oxides have also been examined to further enhance the performance of substrate-supported pristine CNT film electrodes. Based on the electron transfer process, the composite endows extra pseudocapacitance to carbon network and enhances the specific capacitance of the device [46, 47]. Meng *et al.* [48] fabricated highly flexible and paper-like CNT/polyaniline (PANI) thin-film electrodes using H_2SO_4-PVA gel as solid-state electrolytes (Figure 4.9a). In Figure 4.9b, three very flexible electronics are connected in series to light a red LED. The device had the thickness of commercial standard A4 print paper. The integrated device showed a discharge capacitance of 332 F g^{-1} for electrode material, and the device capacitance was 31.4 F g^{-1} using polymer-gel electrolyte (PVA-H_2SO_4). The specific capacitance of the device showed more sustainable performance in a polymer-gel electrolyte in comparison to an aqueous electrolyte; the inset shows the GCD curve at 1 A g^{-1} for both aqueous and gel electrolyte (Figure 4.9c). They showed good stability in the device over 1000 GCD cycles. Lin *et al.* [49] developed a flexible and transparent supercapacitor based on an aligned multiwalled carbon nanotube/polyaniline (MWCNT/PANI) composite. The maximum capacitance of 233 F g^{-1} was obtained from the GCD discharge curve for single electrode. The specific capacitances value of the system remains almost unchanged with the increasing bent angle from 0° to 180° due to the aligned structure. To further increase the integral electrochemical performances, yarn-based SCs were used. In this regard, Choi *et al.* [50] developed waivable super-elastic

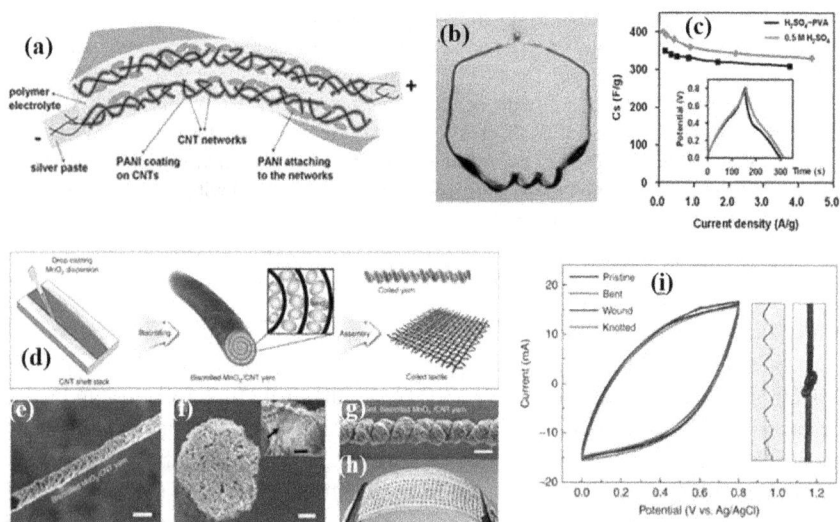

FIGURE 4.9 (a) Schematic illustration of the highly flexible and all-solid-state paper-like polymer supercapacitors. (b) A digital picture that shows three highly flexible devices in series to light a red LED (c) The sustainability of PANI/CNT nanocomposite thin-film device in aqueous and polymer-gel electrolyte with inset of GDD at 1 Ag^{-1} (Reproduced with permission from Ref. 48. Copyright 2010 American Chemical Society). (d) Schematic showing the preparation and structure of biscrolled CNT/MnO$_2$ hybrid fiber, SEM morphologies of (e) As-spun biscrolled hybrid fiber at scale bar of 100 μm. (f) The cross-sectional view of yarn at a scale bar of 15 μm. (g) The SEM image of coiled 5-ply, biscrolled yarn (70 wt% MnO$_2$/CNT). (h) Image of two-ply coiled yarn integrated in a fabric. (i) CV curves (at 50 mV s^{-1}) for nondeformed, bent, mandrel-wrapped, and knotted biscrolled MnO$_2$/CNT yarn.

Source: Reproduced with permission from Ref. 50. Copyright 2016 Nature communication

biscrolled yarn SCs. The biscrolled yarns were made from the MnO_2-coated CNT sheets as shown in Figure 4.9d. Biscrolling can boost the loading mass of active nanoparticles by up to 99% without compromising the fiber's mechanical properties. Scanning electron microscope (SEM) images of MnO_2-loaded biscrolled yarn electrode (Figure 4.9e), and its cross-sectional view (Figure 4.9f) shows the production of uniform and roughly circular yarn. The five yarns were piled together (5-ply electrode) for additional strength as shown in Figure 4.9g. The biscrolled yarn of MnO_2/CNT was strong and flexible enough to be woven into a textile integrated into a fabric (Figure 4.9h). The biscrolled MnO_2/CNT fiber showed a specific capacitance of 889 mF cm^{-2} and possessed an energy density of 35.8 μWh cm^{-2} using PVA-LiCl gel electrolyte. The deformed capacitor fibers' CV was nearly identical to that of the nondeformed fibers as showed in Figure 4.9i. No significant changes were found when strains of 10%, 20%, and 30% were applied to the CV curves of the stretchable supercapacitor. Hence, the biscrolling technique provides strength and electrical conductivity and enhances the performance of the capacitance.

4.4.2 GRAPHENE-BASED FLEXIBLE SCs

Graphene is made up of a hexagonally arranged single layer of sp^2 hybridized carbon atoms. It is considered one of the most enticing materials for flexible SCs due to unrivaled physical and chemical features such as unique structural properties, large surface area, and high electronic and thermal conductivities along with superior chemical stability [51–54]. Xie *et al.* measured the quantum capacitance of graphene, suggesting that single-layer graphene has an inherent capacitance of 21 μF cm^{-2}, which set up the highest limit of double-layer capacitance for all carbon-based materials [55].

In a traditional stacked design in supercapacitors, graphitic carbon-based materials are typically orientated arbitrarily with respect to the current collectors, which limits the access of electrolyte ions into the graphitic planes. As a result, the electrochemical surface area of graphene layers is not fully utilized. In this regard, Yoo *et al.* [56] demonstrated the in-plane fabrication of reduced multilayer graphene oxide (RMGO)–based ultrathin supercapacitors (Figure 4.10a, b). The CV and GCD curves of RMGO are shown in Figures 4.10c and 4.10d. The in-plane fabrication enables the maximum utilization of electrochemical surface area and a higher capacitance value of 390 μF cm^{-2} for the RMGO device using PVA-H$_2$SO$_4$ gel electrolyte. Few-layer graphene produced from graphite oxide (GO) is a potentially cost-effective material in the fabrication of SCs. Due to its strong sheet-to-sheet van der Waals interactions, restacking of graphene sheets reduces the surface area. To avoid the problem of restacking in graphene sheets, El-Kady *et al.* fabricated a conventional LightScribe DVD optical drive for laser-induced GO-layer reduction to a graphene film electrode without any additional binders. The device showed high power density (20 W cm^{-3}), and the system remains almost unchanged even under bending from 0° to 180° [57]. To further boost the performance of flexible devices, conducting polymer and metal-oxide composites of graphene has been extensively studied. The introduction of polymer offers good electrochemical and mechanical properties in flexible devices. In this regard, Chi *et al.* [58] fabricated a freestanding graphene

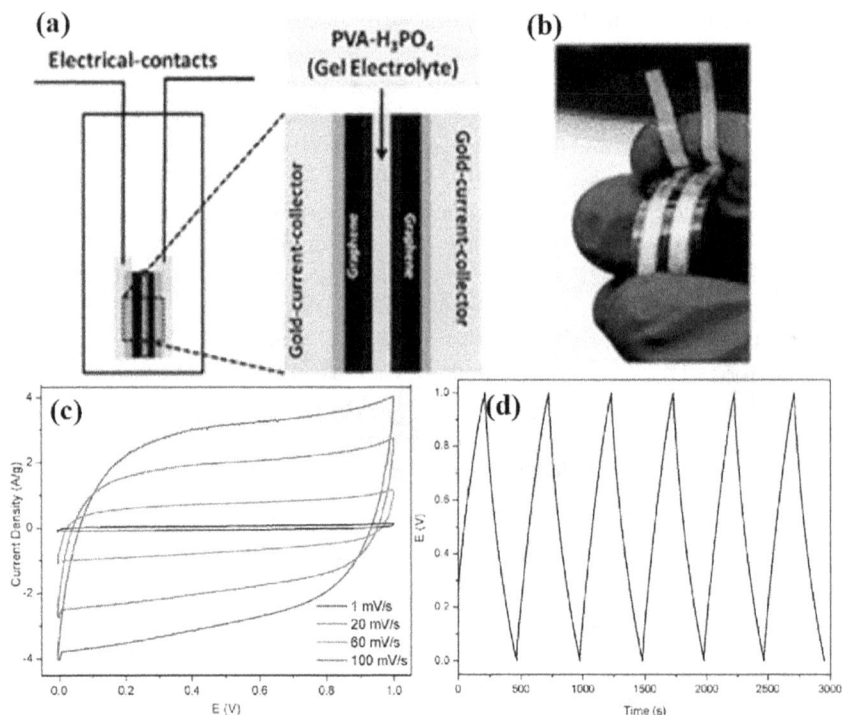

FIGURE 4.10 (a) Schematic illustration of the device fabricated using the concept of 2D in-plane SCs; (b) a prototype flexible supercapacitor device based on reduced multilayer graphene oxide (RMGO) developed using the new in-plane geometry; (c) CV curves obtained at different scan rates for RMGO; and (d) the GCD curve of an RMGO SC measured at a constant current density of 176 mA g^{-1}.

Source: Reproduced with permission from Ref. 56. Copyright 2011 American Chemical Society.

paper (GP)–based 3D porous graphene hydrogel-polyaniline (GH-PANI) electrode using an inkjet printing method. The schematics of ink-printed GH-PANI/GP electrode is shown in Figure 4.11a. Owing to interactions of the graphene nanosheets' stacking, the SEM image of the graphene hydrogel exhibits a 3D porous structure (Figure 4.11b). Figure 4.11c shows an SEM image of coral-like PANI loaded onto the pore walls of the graphene with numerous interconnected nanorods. The layered stack of GH-PANI nanocomposite on GP is shown in Figure 4.11d. This method proposed a low-cost and efficient synthesis to fabricate flexible graphene–PANI nanohybrid paper with a sizable area and light weight, which also uses the collective impact of graphene and PANI. The flexible GH-PANI/GP device showed a maximum energy density of 24.02 Wh kg^{-1} along with outstanding mechanical flexibility.

SCs constructed of robust and flexible yarn can be used as power sources for small electronic devices such as mobile phones, medical equipment, and wearable electronic fabrics, as they can be easily molded into various sizes and structures. Huang *et al.* [59] used the electrodeposition technique to fabricate high-tensile-strength stainless steel yarns coated with a PPy@MnO$_2$@rGO hierarchical structure as shown in Figure 4.12a. The cross-sectional view of the PPy@MnO$_2$@rGO-deposited conductive yarn (Figure 4.12b) depicts a thin layer of hierarchical structure that facilitates rapid ion

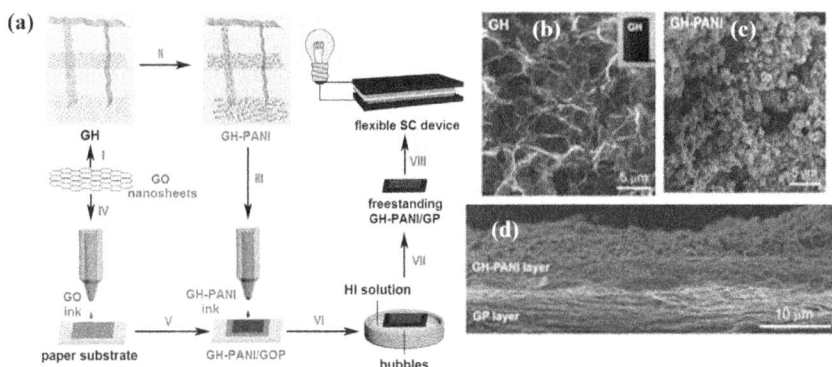

FIGURE 4.11 (a) Schematic illustration of the fabrication process of GH–PANI, SEM images of GH (b), GH-PANI (c), and (d) GH–PANI layer on GH–PANI/GP at high magnification.

Source: Reproduced with permission from ref. 58. Copyright 2014 American Chemical Society

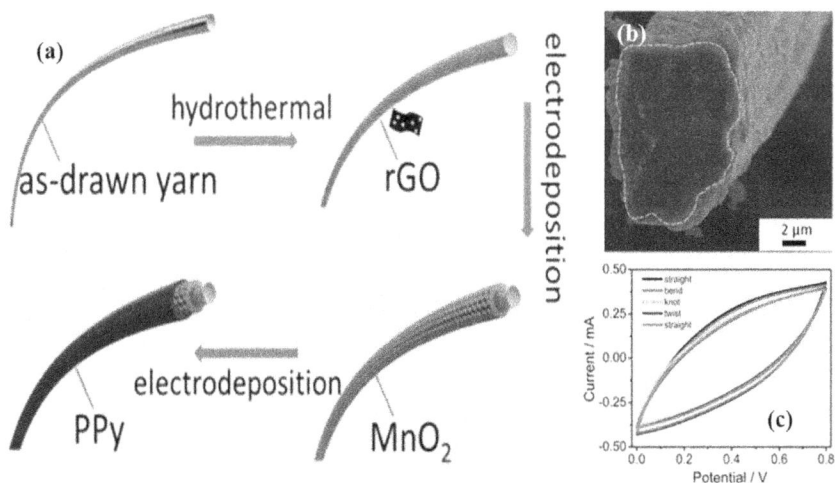

FIGURE 4.12 (a) Schematic illustration of the yarn modified by deposition of rGO, MnO$_2$ and PPy; (b) cross section of the PPy@MnO$_2$@rGO-deposited conductive yarn; and (c) CV curves of the all-solid-state yarn SC undergoing consecutive deformations at a scan rate of 100 mV s^{-1}.

Source: Reproduced with permission from ref. 59. Copyright 2015 American Chemical Society.

movement during charge–discharge. The resultant modified yarns exhibit an outstanding specific capacitance of 411 mF cm^{-2} in flexible electrodes using PVA-H$_3$PO$_4$ gel electrolyte. The yarn supercapacitor was then put through a series of deformation tests, in which it was straight, bent, knotted, twisted, and straight again, which shows a slight change in the voltammetry curve (Figure 4.12c). This provides clear evidence that yarn-based SCs possess excellent electrochemical and mechanical properties, and it can be used in the modern textile industry to create a variety of wearable electronics.

Furthermore, Kim *et al.* [60] developed highly conducting graphene/organic semiconducting polymer hybrid core/shell fibers (G@PFs) composed of Cu and N co-doped porous graphene fiber through a wet-spinning process (Figure 4.13a). Figure 4.13b shows an SEM image of G@PEDOT fiber core-shell morphology. The graphene core provides a high surface area of 1174.4 m^2 g^{-1} and open channels for the infiltration of electrolytes while Cu and N-doping improve the wettability of a graphene sheet. As a consequence, PEDOT-coated porous graphene fibers showed a high volumetric-specific capacitance of 263.1 F cm^{-3} and energy density of 7.0 mW h cm^{-3}, along with high cycle life (Figure 4.13c). G@PEDOT fibers show exceptional cyclic stability after 20,000 charge-discharge cycles along with excellent mechanical flexibility no cyclic deuteriation is observed on bending up to 180°. In addition to the previously mentioned studies, there have been few recent studies on biomass-derived porous carbon as electrode material for high energy density SCs. Ghosh and co-workers constructed a high-performance all-solid-state flexible asymmetric SC device using coconut fiber–derived porous carbon as the anode and CoFe$_2$O$_4$/porous carbon as the cathode, showing a high energy density of 50.34 Wh kg^{-1} and approximately 91% retention for 5000 cycles [61]. Xu *et al.* showed dung beetle forewing carbon (DBFC)–derived nitrogen and oxygen self-doped porous carbon nanostructure–based symmetric solid-state SC with a high specific capacitance of 260 F g^{-1} at 0.5 A g^{-1} and excellent capacitance retention (~90.3%) after 8000 consecutive cycles [62].

4.5 SUMMARY

The aforementioned study summarized the recent progress in SC devices, charge-storage mechanisms, characterization techniques, and electrode materials. The broad range of carbon, with various dimensions and unique morphology, has been extensively studied as electrode material for supercapacitor. Carbon-based nano-structures, such as CNTs and graphene, have been studied substantially as electrode materials for portable and wearable electronics power sources due to their phys-iochemical stability, exceptional conductivity, huge surface area, high mechanical strength, and low manufacturing cost. The main methods for fabricating flexible carbon electrodes are the deposition of carbon nanomaterials onto a flexible sub-strate, the production of a flexible, freestanding film and yarn of electrochemically active materials. Furthermore, the design and development of graphene and CNT composite with conducting polymers and metal oxides are effective approaches to high-performance carbon-based SCs, introducing pseudo-active sites to improve interfacial interactions. Although significant progress has been made, there are several advancements that are required, particularly for carbon-based nanomaterials

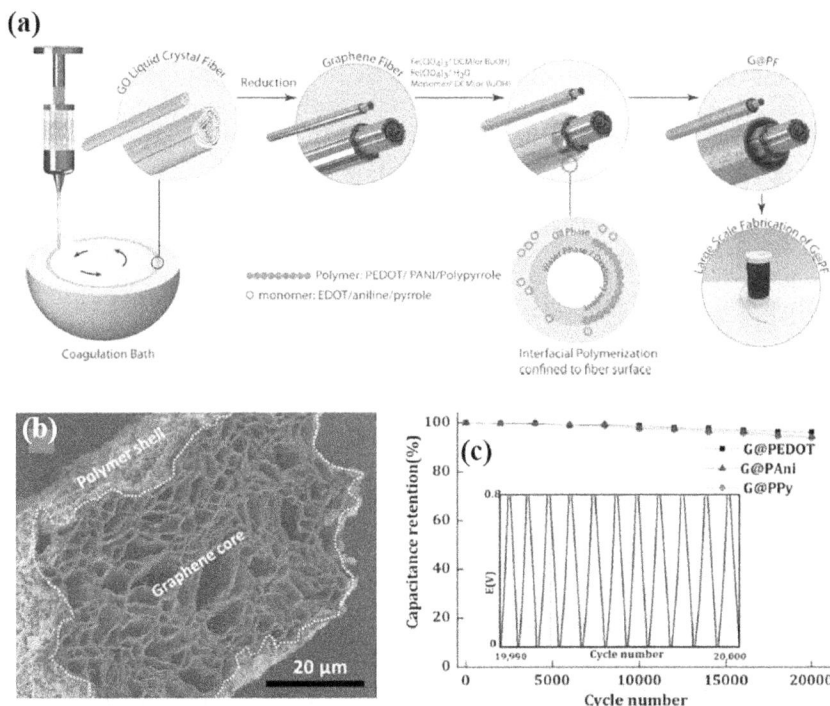

FIGURE 4.13 (a) Schematic illustration displaying the fabrication steps for G@PFs; (b) SEM images showing the cross sections of G@PEDOT fiber; and (c) cycle life of G@PF SC devices with inset is GCD profiles of G@PEDOT after 19899 cycles up to 20,000 cycles at 1.05 A cm^{-3}.

Source: Reproduced with permission from ref. 60. Copyright 2017 American Chemical Society

such as lowering of operating cost, carbon/electrolyte synergy, and new proposals to try different synthesis methods for higher specific capacitance and energy density, before these can become commercially available products.

REFERENCES

1. Liu, C., Li, F., Ma, L. P., & Cheng, H. M. (2010). Advanced materials for energy storage. *Adv Mater*, 22, 28–62.
2. Simon, P., & Gogotsi, Y. (2008). Materials for electrochemical capacitors. *Nat Mater*, 7, 845–854.
3. Zhang, H., Cao, G., & Yang, Y. (2009). Carbon nanotube arrays and their composites for electrochemical capacitors and lithium-ion batteries. *Energy Environ Sci*, 2, 932–943.
4. Zhai, Y., Dou, Y., Zhao, D., Fulvio, P. F., Mayes, R. T., & Dai, S. (2011). Carbon materials for chemical capacitive energy storage. *Adv Mater*, 23, 4828–4850.
5. Lokhande, P. E., Chavan, U. S., & Pandey, A. (2020). Materials and fabrication methods for electrochemical supercapacitors: Overview. *Electrochem Ener Rev*, 3, 155–186.

6. Hu, B., Wang, K., Wu, L., Yu, S. H., Antonietti, M., & Titirici, M. M. (2010). Engineering carbon materials from the hydrothermal carbonization process of biomass. *Adv Mater*, 22, 813–828.

7. Meng, C., Gall, O. Z., & Irazoqui, P. P. (2013). A flexible super-capacitive solid-state power supply for miniature implantable medical devices. *Biomed Microdevices*, 15, 973–983.

8. Jiang, H., Lee, P. S., & Li, C. (2013). 3D carbon based nanostructures for advanced super-capacitors. *Energy Environ Sci*, 6, 41–53.

9. Zhang, L. L., & Zhao, X. S. (2009). Carbon-based materials as supercapacitor electrodes. *Chem Soc Rev*, 38, 2520–2531.

10. Mishra, A. K., & Ramaprabhu, S. (2011). Functionalized graphene-based nanocomposites for supercapacitor application. *J Phys Chem C*, 115, 14006–14013.

11. Mishra, S., Majee, B. P., Maurya, P. K., & Mishra, A. K. (2019). Multifunctional low temperature reduced graphite oxides for high performance supercapacitors and SERS application. *Mater Res Express*, 6, 85527.

12. Paquin, F., Rivnay, J., Salleo, A., Stingelin, N., & Silva, C. (2013). Multi-phase semicrystalline microstructures drive exciton dissociation in neat plastic semiconductors. *J Mater Chem C*, 3, 10715–10722.

13. Pumera, M. (2011). Graphene-based nanomaterials for energy storage. *Energy Environ Sci*, 4, 668–674.

14. Sun, Y., Wu, Q., & Shi, G. (2011). Graphene based new energy materials. *Energy Environ Sci*, 4, 1113–1132.

15. Zhang, B., Liang, J., Xu, C. L., Wei, B. Q., Ruan, D. B., & Wu, D. H. (2001). Electric double-layer capacitors using carbon nanotube electrodes and organic electrolyte. *Mater Lett*, 51, 539–542.

16. Guan, L., Yu, L., & Chen, G. Z. (2016). Capacitive and non-capacitive faradaic charge storage. *Electrochim Acta*, 206, 464–478.

17. Bokhari, S. W., Siddique, A. H., Sherrell, P. C., Yue, X., Karumbaiah, K. M., Wei, S., Ellis, A.V., & Gao, W. (2020). Advances in graphene-based supercapacitor electrodes. *Ener Rep*, 6, 2768–2784.

18. Wang, Y., Zhang, L., Hou, H., Xu, W., Duan, G., He, S., Liu, K., & Jiang, S. (2021). Recent progress in carbon-based materials for supercapacitor electrodes: A review. *J Mater Sci*, 56, 173–200.

19. Wang, Y., Song, Y., & Xia, Y. (2016). Electrochemical capacitors: Mechanism, materials, systems, characterization and applications. *Chem Soc Rev*, 45, 5925–5950.

20. Endo, M., Takeda, T., Kim, Y. J., Koshiba, K., & Ishii, K. (2001). High power electric double layer capacitor (EDLC's); from operating principle to pore size control in advanced activated carbons. *Carbon Sci*, 1, 117–128.

21. Sarno, M. (2020). Nanotechnology in energy storage: The supercapacitors. *Stud Surf Sci Catal*, 179, 431–458.

22. Chodankar, N. R., Pham, H. D., Nanjundan, A. K., Fernando, J. F., Jayaramulu, K., Golberg, D., Han, Y. K., & Dubal, D. P. (2020). True meaning of pseudocapacitors and their performance metrics: Asymmetric versus hybrid supercapacitors. *Small*, 16, 2002806.

23. Chen, X., Paul, R., & Dai, L. (2017). Carbon-based supercapacitors for efficient energy storage. *Natl Sci Rev*, 4, 453–489.

24. Jiang, Y., & Liu, J. (2019). Definitions of pseudocapacitive materials: A brief review. *Energy Environ Mater*, 2, 30–37.

25. Augustyn, V., Simon, P., & Dunn, B. (2014). Pseudocapacitive oxide materials for high-rate electrochemical energy storage. *Energy Environ Sci*, 7, 1597–1614.

26. Shao, Y., El-Kady, M.F., Sun, J., Li, Y., Zhang, Q., Zhu, M., Wang, H., Dunn, B., & Kaner, R. B. (2018). Design and mechanisms of asymmetric supercapacitors. *Chem Rev*, 118, 9233–9280.

27. Costentin, C., & Savéant, J. M. (2019). Energy storage: Pseudocapacitance in prospect. *Chem Sci*, 10, 5656–5666.
28. Winter, M., & Brodd, R. J. (2004).What are batteries, fuel cells, and supercapacitors? *Chem Rev*, 104, 4245–4269.
29. Zuo, W., Li, R., Zhou, C., Li, Y., Xia, J., & Liu, J. (2017). Battery-supercapacitor hybrid devices: Recent progress and future prospects. *Adv Sci*, 4, 1–21.
30. Xie, J., Yang, P., Wang, Y., Qi, T., Lei, Y., & Li, C. M. (2018). Puzzles and confusions in supercapacitor and battery: Theory and solutions. *J Power Sources*, 401, 213–223.
31. Zhang, R., Xu, Y., Harrison, D., Fyson, J., Southee, D., & Tanwilaisiri, A. (2014). Fabrication and characterization of energy storage fibres. In *2014 20th International Conference on Automation and Computing*, Cranfield, UK, IEEE, 228–230.
32. Zhang, R. R., Xu, Y. M., Harrison, D., Fyson, J., Qiu, F. L., & Southee, D. (2015). Flexible strip supercapacitors for future energy storage. *Int J Autom Comput*, 12, 43–49.
33. Zhang, L., & Gong, H. (2015). Improvement in flexibility and volumetric performance for supercapacitor application and the effect of Ni-Fe ratio on electrode behaviour. *J Mater Chem A*, 3, 7607–7615.
34. Zhang, J., & Zhao, X. S. (2012). On the configuration of supercapacitors for maximizing electrochemical performance. *ChemSusChem*, 5, 818–841.
35. Kampouris, D. K., Ji, X., Randviir, E. P., & Banks, C. E. (2015). A new approach for the improved interpretation of capacitance measurements for materials utilised in energy storage. *RSC Adv*, 5, 12782–12791.
36. Subramani, K., Sudhan, N., Divya, R., & Sathish, M. (2017). All-solid-state asymmetric supercapacitors based on cobalt hexacyanoferrate-derived CoS and activated carbon. *RSC Adv*, 7, 6648–6659.
37. Fletcher, S., Black, V. J., & Kirkpatrick, I. (2014). A universal equivalent circuit for carbon-based supercapacitors. *J Solid State Electrochem*, 18, 1377–1387.
38. Lu, X., Yu, M., Wang, G., Tong, Y., & Li, Y. (2014). Flexible solid-state supercapacitors: Design, fabrication and applications. *Energy Environ Sci*, 7, 2160–2181.
39. Yang, P., & Mai, W. (2014). Flexible solid-state electrochemical supercapacitors. *Nano Ener*, 8, 274–290.
40. Wang, G., Lu, X., Ling, Y., Zhai, T., Wang, H., Tong, Y., & Li, Y. (2012). LiCl/PVA gel electrolyte stabilizes vanadium oxide nanowire electrodes for pseudocapacitors. *ACS Nano*, 6, 10296–10302.
41. Sawangphruk, M., Srimuk, P., Chiochan, P., Krittayavathananon, A., Luanwuthi, S., & Limtrakul, J. (2013). High-performance supercapacitor of manganese oxide/reduced graphene oxide nanocomposite coated on flexible carbon fiber paper. *Carbon N Y*, 60, 109–116.
42. Panero, S., Clemente, A., & Spila, E. (1996). Solid state supercapacitors. *Solid State Ion*, 86, 1285–1289.
43. Kaempgen, M., Chan, C. K., Ma, J., Cui, Y., & Gruner, G. (2009). Printable thin film supercapacitors using single-walled carbon nanotubes. *Nano Lett*, 9, 1872–1876.
44. Kim, D., Shin, G., Kang, Y. J., Kim, W., & Ha, J. S. (2013). Fabrication of a stretchable solid-state micro-supercapacitor array. *ACS Nano*, 7, 7975–7982.
45. Niu, Z., Dong, H., Zhu, B., Li, J., Hng, H.H., Zhou, W., Chen, X., & Xie, S. (2013). Highly stretchable, integrated supercapacitors based on single-walled carbon nanotube films with continuous reticulate architecture. *Adv Mat*, 25, 1058–1064.
46. Yu, G., Xie, X., Pan, L., Bao, Z., & Cui, Y. (2013). Hybrid nanostructured materials for high-performance electrochemical capacitors. *Nano Ener*, 2, 213–234.
47. Hu, L., Chen, W., Xie, X., Liu, N., Yang, Y., Wu, H., Yao, Y., Pasta, M., Alshareef, H. N., & Cui, Y. (2011). Symmetrical MnO_2—carbon nanotube—textile nanostructures for wearable pseudocapacitors with high mass loading. *ACS Nano*, 5, 8904–8913.
48. Meng, C., Liu, C., Chen, L., Hu, C., & Fan, S. (2010). Highly flexible and all-solid-state paperlike polymer supercapacitors. *Nano Lett*, 10, 4025–4031.

49. Lin, H., Li, L., Ren, J., Cai, Z., Qiu, L., Yang, Z., & Peng, H. (2013). Conducting polymer composite film incorporated with aligned carbon nanotubes for transparent, flexible and efficient supercapacitor. *Sci Rep*, 3, 1–6.
50. Choi, C., Kim, K. M., Kim, K. J., Lepró, X., Spinks, G. M., Baughman, R. H., & Kim, S. J. (2016). Improvement of system capacitance via weavable superelastic biscrolled yarn supercapacitors. *Nat Commun*, 7, 1–8.
51. Niu, Z., Zhang, L., Liu, L., Zhu, B., Dong, H., & Chen, X. (2013). All-solid-state flexible ultrathin micro-supercapacitors based on graphene. *Adv Mater*, 25, 4035–4042.
52. Chang, C. W., & Liao, Y. C. (2016). Accelerated sedimentation velocity assessment for nanowires stabilized in a non-newtonian fluid. *Langmuir*, 32, 13620–13626.
53. Pumera, M. (2010). Graphene-based nanomaterials and their electrochemistry. *Chem Soc Rev*, 39, 4146–4157.
54. Rao, C. E. E., Sood, A. E., Subrahmanyam, K. E., & Govindaraj, A. (2009). Graphene: The new two-dimensional nanomaterial. *Angew Chemie Int Ed*, 48, 7752–7777.
55. Xia, J., Chen, F., Li, J., & Tao, N. (2009). Measurement of the quantum capacitance of graphene. *Nat Nanotechnol*, 4, 505–509.
56. Yoo, J. J., Balakrishnan, K., Huang, J., Meunier, V., Sumpter, B. G., Srivastava, A., Conway, M., Mohana Reddy, A. L., Yu, J., Vajtai, R., & Ajayan, P. M. (2011). Ultrathin planar graphene supercapacitors. *Nano Lett*, 11, 1423–1427.
57. El-Kady, M. F., Strong, V., Dubin, S., & Kaner, R. B. (2012). Laser scribing of high-performance and flexible graphene-based electrochemical capacitors. *Science*, 335, 1326–1330.
58. Chi, K., Zhang, Z., Xi, J., Huang, Y., Xiao, F., Wang, S., & Liu, Y. (2014). Freestanding graphene paper supported three-dimensional porous graphene-polyaniline nanocomposite synthesized by inkjet printing and in flexible all-solid-state supercapacitor. *ACS Appl Mater Interfaces*, 6, 16312–16319.
59. Huang, Y., Hu, H., Huang, Y., Zhu, M., Meng, W., Liu, C., Pei, Z., Hao, C., Wang, Z., & Zhi, C. (2015). From industrially weavable and knittable highly conductive yarns to large wearable energy storage textiles. *ACS Nano*, 9, 4766–4775.
60. Padmajan Sasikala, S., Lee, K. E., Lim, J., Lee, H. J., Koo, S. H., Kim, I. H., Jung, H. J., & Kim, S. O. (2017). Interface-confined high crystalline growth of semiconducting polymers at graphene fibers for high-performance wearable supercapacitors. *ACS Nano*, 11, 9424–9434.
61. Gogoi, D., Makkar, P., Das, M. R., & Ghosh, N. N. (2022). $CoFe_2O_4$ nanoparticle decorated hierarchical biomass derived porous carbon based nanocomposites for high-performance all-solid-state flexible asymmetric supercapacitor devices. *ACS Appl Electron Mater*, 4, 795–806.
62. Xu, P., Tong, J., Zhang, L., Yang, Y., Chen, X., Wang, J., & Zhang, S. (2022). Dung beetle forewing-derived nitrogen and oxygen self-doped porous carbon for high performance solid-state supercapacitors. *J Alloys Compd*, 892, 162129.

5 Designing Hollow Structured Materials for Sustainable Electrochemical Energy Conversion

Baghendra Singh and Arindam Indra

5.1 INTRODUCTION

Sustainable energy conversion technologies based on electrocatalysis are considered as the most promising clean and renewable pathways to solve energy and environmental problems (Yi Li et al., 2021). Electrocatalytic energy conversion involves various processes like water splitting, reduction of O_2, N_2, and CO_2, or oxidation of H_2. Generally, electrocatalysis is kinetically sluggish and proceeds through the input of extra energy known as overpotential (Singh et al., 2021). The requirement of high overpotential hinders the practical applications of the electrocatalysts in electrochemical water splitting as well as in fuel cells.

Therefore, the development of highly efficient and durable electrocatalysts with excellent electrochemical performance is essential for the development of eco-friendly energy conversion technologies (Fan et al., 2022; J. Wang et al., 2019). Designing nanostructured materials with desired compositions, highly porous structures, large surface areas, and desired morphology can improve the electrochemical performance (S. Kim et al., 2022; Xue et al., 2022; Yasin et al., 2022). In this respect, hollow-structured materials have attained immense attention in the field of electrochemical energy conversion (J. Wang et al., 2019). Hollow-structured nanomaterials are defined as the nano- or microstructure having well-defined boundaries and interior cavities. The unique structural features like low density, high surface-to-volume ratio, large pore volume, and small mass- and charge-transport pathways make them a potential candidate for various applications in energy conversion (L. Zhou et al., 2017). Hollow-structured materials are widely employed in batteries, supercapacitors, fuel cells, and electrochemical energy conversion owing to their unique structural properties (J. Wang et al., 2019). Hence, efforts have been undertaken to design hollow-structured materials with different compositions and morphology for excellent electrochemical performance.

DOI: 10.1201/9781003208709-5

5.2   IMPORTANCE OF HOLLOW-STRUCTURED MATERIALS

Hollow-structured materials have lower density attributed to their extra void spaces than their solid counterparts (X. W. Lou et al., 2008, L. Zhou et al., 2017). In addition, hollow-structured materials provide a large surface area with rich electrochemical active sites and a large contact area between electrode and electrolyte enhancing the mass diffusion (Zhao & Jiang, 2009, J. Wang et al., 2019).

Besides, the interior cavity of the hollow-structured materials results in the exposure of the active sites improving the adsorption properties (X. H. Liu et al., 2020, Yu et al., 2016). The open-pore channels in hollow-structured materials lead to very low diffusion blockage and sufficient transportation of electrolytes into the inner region of the hollow structure (Yu et al., 2016).

5.3   STRUCTURAL DIVERSITY OF HOLLOW-STRUCTURED MATERIALS

The hollow materials with diverse structures and morphologies have been designed by optimizing synthetic methodologies and reaction conditions (Jun Liu et al., 2008). Simple hollow structures with polyhedral, cubic, or bowl-like shapes have been developed for the various electrochemical energy conversion processes (J. Wang et al., 2019). Complex hollow structures having multiple shells or multiple internal chambers or cores have been demonstrated for the various energy conversion processes. The hollow-structured materials with closed single-shelled, open hollow, complex hollow, bubble-within-bubble, tube-in-tube, and wire-in-tube hollow structures have been synthesized for the electrochemical energy conversion process (Figure 5.1; X. W. Lou et al., 2008). The hollow spheres, core-shell spheres, yolk-shell spheres, double-shell spheres, and multi-shell spheres have been designed with tunable compositions (Figure 5.1). The open and closed single-shell structures are the simplest form of hollow-structured nanomaterials, which can be synthesized by the soft and hard templating methods. Generally, oil-in-water or water-in-oil emulsions are utilized to form single-shell hollow materials (L. Zhou et al., 2017).

The open hollow structures have an interior cavity with a polyhedral cage or frame-like morphology (Nai et al., 2018). The cage structure has multiple openings on its wall while the frame structure involves the complete removal of the walls preserving the edges and corners. Furthermore, the single-shelled hollow structures can be assembled into secondary spheres, hollow spheres, and fibers to form bubble-within-bubble, bubble-within-sphere, and bubble-within-fiber structures, respectively, and other different types of hollow structures (J. Wang et al., 2019).

The spherical hollow-structured materials have an exterior hollow shell and interior core, which can move inside the shell (Gong et al., 2019). The spherical structures have an extra void space, which can provide total exposure of the active center inside the shell. Generally, selective etching, Kirkendall effect, Ostwald ripening, mass relocation, and ion-exchange strategies have been explored to design hollow-structured nanomaterials with various structures and morphologies (J. Wang et al., 2019).

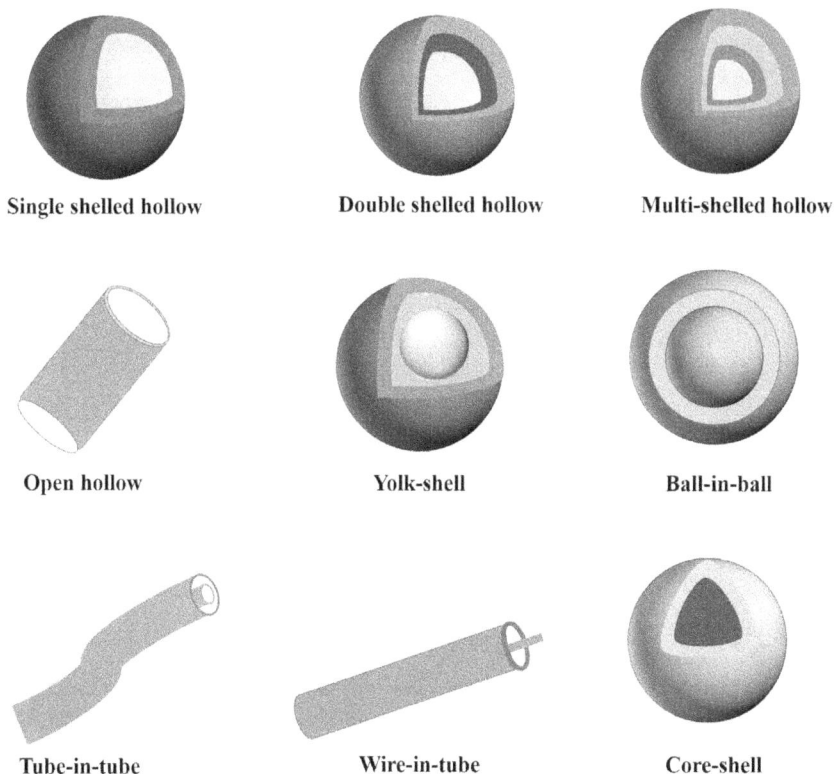

Single shelled hollow   Double shelled hollow   Multi-shelled hollow

Open hollow   Yolk-shell   Ball-in-ball

Tube-in-tube   Wire-in-tube   Core-shell

FIGURE 5.1 Schematic illustration of the various hollow structured nanomaterials employed in the electrochemical energy conversion processes.

5.4 PRINCIPLE OF DESIGNING THE HOLLOW STRUCTURE

Hollow-structured nanomaterials have been designed by using different templates or template-free methods. The templated methods involve hard and soft templating strategies whereas the template-free methods include Ostwald ripening, Kirkendall effect, ion exchange, and selective etching strategy.

5.4.1 SOFT TEMPLATING METHOD

The amphiphilic molecules such as surfactants and block copolymers are self-assembled into micelles with different structures when their concentration is higher than the critical micelle concentration (CMC; M. Zhu et al., 2020). Such micelles can be widely utilized as soft templates to form various hollow nanostructures. The micelles are thermodynamically meta-stable and observed to be highly sensitive to pH value; the concentration of the amphiphilic molecules, solvents, and organic/inorganic additives; and the ionic strength of the solution. Furthermore, their sensitivity can be utilized for effective control over the morphology and structure of the micelle templates (Y. Liu et al., 2013). Hence, micelles are used as the soft templates for the construction of hollow-structured nanomaterials with various compositions and morphologies (Figure 5.2).

Soft templates, such as gas bubbles, supramolecular micelles, polymer vesicles, and so on, have been employed to fabricate the hollow nanostructures (Yongjun Li et al., 2006; Jian Liu et al., 2010; Xu & Wang, 2007). First, the shells are grown along the interfacial region, and further the template is removed to form hollow nanostructures (X. W. Lou et al., 2008). For example, Pinnavaia and co-workers synthesized mesoporous molecular sieves with vesicle-like hierarchical structures using a supramolecular assembly pathway (S. S. Kim et al., 1998). The vesicle shells were constructed of one or more silica sheets having a 3-nanometer thickness.

Zhang et al. also synthesized organosiliceous multilamellar vesicles utilizing a single-surfactant vesicle template with Pluronic P85 as the structure-directing agent and 1,2-bis(triethoxysilyl)ethane as the organosilica precursor (Y. Zhang et al., 2008). Similarly, hollow nanostructures of multi-shelled Cu_2O (H. Xu & Wang, 2007), SiO_2 (Gu et al., 2010), and double-shelled ferrihydrite hollow spheres (Z. Wu et al., 2008) were designed using cetyltrimethylammonium bromide (CTAB), carbonaceous silica, and gas-bubble soft templates, respectively.

5.4.2 HARD TEMPLATING METHOD

The designing of hollow structured materials by hard-templating method consists of four main steps: (1) template preparation, (2) surface modification of the hard template, (3) coating/deposition of the desired material, and (4) removal of the template (Figure 5.2; Gong et al., 2019). The monodispersity, control over the size and shape, and easy synthesis are the main criteria for selecting the templates. The monodispersed polymer, polystyrene spheres, silica, carbonaceous microsphere, metal, and metal-oxide colloids are the commonly utilized hard templates for the synthesis of hollow-structured materials (Pérez-Page et al., 2016; M. Zhu et al., 2021). For example, the Si microparticles encapsulated in graphene cages (Si-MP@Gr) were synthesized with yolk-shell structure (Yuzhang Li et al., 2016). Furthermore, the Si-MP@Gr were coated with Ni, and the thickness of the coating was tuned to provide the void space. In addition to the hollow carbon nanostructures, nanofiber arrays and nanospheres have been successfully designed using anodic aluminum oxide or polystyrene sphere as templates.

5.4.3 SACRIFICIAL TEMPLATING METHOD

The sacrificial template method has been explored for the development of hollow structured materials (X. W. Lou et al., 2008). In this method, the shells are formed around the surface of the sacrificial template, and furthermore, these shells are observed to adopt the shape of the template (Miao et al., 2007). Simultaneously, the hollow interior cavity is obtained by destroying the sacrificial template. This method avoids the template removal process and generates a controllable morphology. The template removal is observed to prevent possible damage to the produced shell (Figure 5.2; Miao et al., 2007).

Soft-template method

Hard-template method

Semi-sacrificial template method

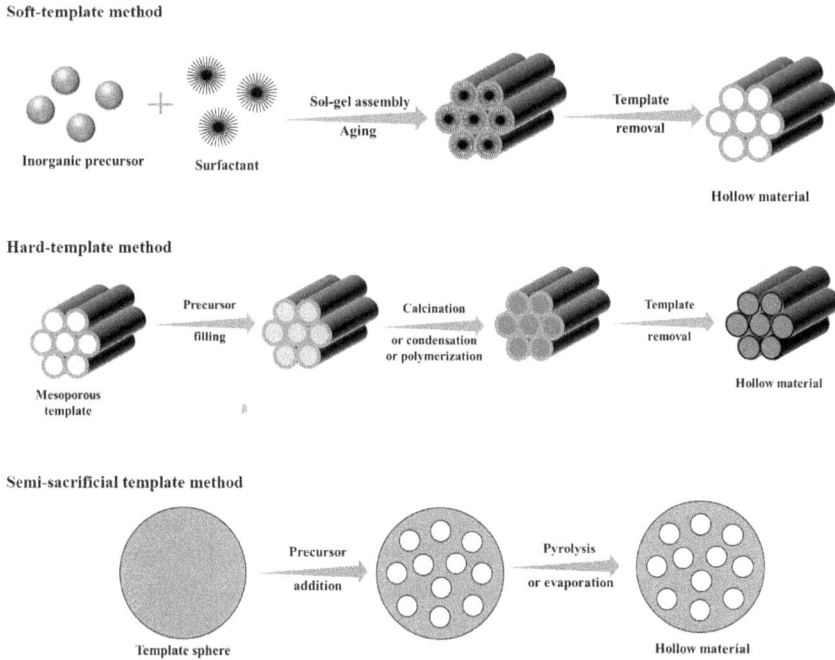

FIGURE 5.2 Schematic representation for the designing of the various hollow-structured nanomaterials using soft-template, hard-template, and semi-sacrificial template methods.

This method is facile as it can be performed in a stirring aqueous solution or a hydrothermal system. Several hollow nanotubes and hollow spheres have been designed using the sacrificial template method. For example, semiconductor CdX (X = Te, Se, S) hollow structures were successfully synthesized utilizing Cd(OH)Cl as a sacrificial template by simple stirring method (Miao et al., 2007). The hollow structures with hollow spheres or hollow tubes were formed by controlling the shape of the sacrificial template. The CdTe hollow spheres and hollow tubes were synthesized from Cd(OH)Cl precursors. A double-shelled poly(3,4-ethylenedioxythiophene) (PEDOT) and polypyrrole (PPy) hollow spheres were also synthesized using Fe_3O_4 hollow spheres as sacrificial templates (Pang et al., 2014).

5.4.4 SELECTIVE ETCHING

Selective etching is known as the selective removal of a part of single-component material to generate a hollow cavity by chemical treatment or calcination (Figure 5.3; Y. Liu et al., 2013). Recently, the formation of hollow structures has also been demonstrated through the selective etching process of a single-component material, such as Prussian blue (PB), Prussian blue analogue (PBA), $ZnSn(OH)_6$, and $CoSn(OH)_6$ (L. Zhou et al., 2017). For example, the uniform PB hollow particles were developed using an acid etching strategy. The PB cubes were treated with 1.0 M HCl solution at 140 °C for 4 h to produce a hollow interior (Nai et al., 2018). Similarly, Lou group synthesized Ni–Co PBA nanocages through a chemical etching approach (Nai et al., 2018). The Ni–Co PBA nanocubes were treated with ammonia. As a result, the etching of the Ni–Co PBA nanocubes occurred selectively at the corners and along the diagonal direction to form the Ni–Co PBA nanocages.

The selective etching strategy has also been employed to design the intricate hollow structures with various compositions. For example, highly porous and single-crystalline $ZnSn(OH)_6$ and $CoSn(OH)_6$ nanoboxes were developed by the selective etching of the core of $ZnSn(OH)_6$ (L. Wang et al., 2011), and $CoSn(OH)_6$ (Z. Wang et al., 2013) nanocubes by NaOH, respectively. Furthermore, a series of $CoSn(OH)_6$ complex hollow structures are designed including yolk-shell cubes and double-, triple-, and quadruple-shelled nanoboxes by the repeated deposition of $CoSn(OH)_6$ onto pre-grown nanocubes and nanoboxes followed by the etching in the alkaline solution. The $ZnSn(OH)_6$ multi-shelled hollow structure was also fabricated using the same method and further converted into $ZnSnO_3$ (Y. Ma et al., 2015) and Zn_2SnO_4/SnO_2 multi-shelled hollow structure (L. Sun et al., 2016).

The SiO_2 spheres and their derivatives having yolk-shell and hollow structures were also prepared by the selective etching method (D. Chen et al., 2009; Teng et al., 2014; X.-J. Wu & Xu, 2010; Q. Zhang et al., 2008). For example, Wong et al. demonstrated the silica shell on Au-nanoparticles (AuNP@silica) by a typical Stöber method. Furthermore, AuNP@silica resulted in the formation of yolk-shell structured nanoparticles when etched with hot water for 30 min (Wong et al., 2011).

5.4.5 Ion Exchange

The ion-exchange method involves the exchange of ions (either cations or anions) between the solution and the solid particles (Figure 5.3; Yan & Rosei, 2014). For example, Shen et al. developed a nickel–cobalt sulfide ($NiCo_2S_4$) hollow structure by the controlled ion exchange during the sulfidation process of nickel cobalt glycerate (NiCo-glycerate) spheres (Shen et al., 2015). First, sulfide (S^{2-}) ions were observed to react with the surface metal ions of NiCo-glycerate to form NiCo-glycerate@$NiCo_2S_4$ core-shell structure. Furthermore, the inside diffusion of S^{2-} ions and outside diffusion of metal ions resulted in the continuous growth of the $NiCo_2S_4$ shell and generated the well-defined gap between the shell and the inner NiCo-glycerate core. The diffusion process became slow with the progress of the reaction and resulted in the formation of a secondary $NiCo_2S_4$ shell. This ultimately formed a yolk-shell or ball-in-ball hollow sphere structure.

5.4.6 Ostwald Ripening

The interface formation between the smaller particles and larger particles results in a large difference in the interfacial energy (X. W. Lou et al., 2008). As a result, smaller particles are dissolved and redeposited on the larger particles during the crystal growth (Figure 5.3). Generally, the dissolution of small crystals or particles and the redeposition of the dissolved species on the surfaces of larger crystals or

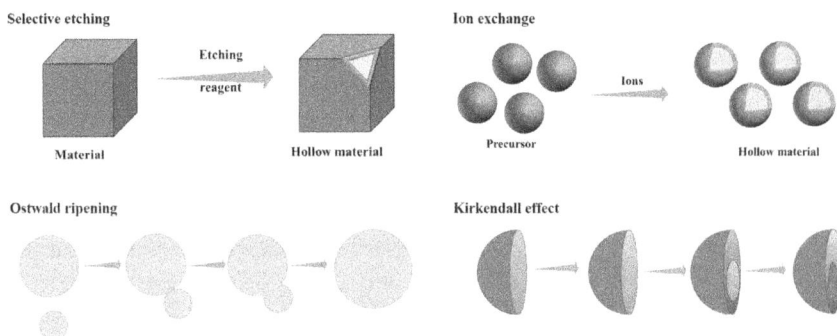

FIGURE 5.3 Schematic representation for the designing of the various hollow-structured nanomaterials using selective etching, ion exchange, Ostwald ripening, and Kirkendall effect.

particles are referred to as Ostwald ripening (Yec & Zeng, 2014). Yang and Zeng developed the Ostwald ripening mechanism as a template-free route for the designing of hollow structures (H. G. Yang & Zeng, 2004). In the first step, primary TiO_2 crystallites are synthesized through the hydrolysis of TiF_4 under hydrothermal conditions. Furthermore, the TiO_2 crystallites aggregate into spherical assemblies for the minimization of the total interfacial energy. The inner crystallites are relatively small and observed to possess high surface energies and the outer surface crystallites are relatively large. As a result, a hollow cavity is formed through dissolution due to the removal of the central region of the spheres.

Recently, Ostwald ripening has been employed to design various hollow nanostructures, such as Cu_2O (L. Zhang & Wang, 2011), TiO_2 (H. G. Yang & Zeng, 2004), SnO_2 (X. W. Lou et al., 2006), and $Ni(OH)_2$ (Y. Wang et al., 2005). For example, a variety of Cu_2O hollow nanostructures were synthesized using the Ostwald ripening approach (L. Zhang & Wang, 2011). The controlled Ostwald ripening resulted in the formation of Cu_2O hollow structures with single-layer, double-layer, triple-layer, and quadruple-layer nanoshells.

5.4.7 THE KIRKENDALL EFFECT

The Kirkendall effect refers to the dissimilar mutual diffusion rates of two metals leading to the formation of voids in the zone of the more rapidly diffusing material to compensate for the unequal material flow (Figure 5.3; W. Wang et al., 2013). The small voids are formed at the interface, and the voids are enlarged due to the surface diffusion of atoms. Furthermore, a hollow cavity is generated inside the shell to form a hollow structure (W. Wang et al., 2013, Yin et al., 2004).

The Kirkendall effect has been widely utilized to design various hollow nanostructures like Cu_2O (Gou & Murphy, 2003), Fe_2O_3 (Cabot et al., 2007), ZnO (B. Liu & Zeng, 2004), ZnS (Shao et al., 2005), $CoSe_2$ (J. Gao et al., 2006), and Ag_2Se (Bernard Ng et al., 2006). For example, Chen et al. synthesized Pt_3Ni hollow nanoframes through the Kirkendall effect (C. Chen et al., 2014). It was observed that the surface Ni atoms were more easily oxidized and dissolved than Pt atoms after the dispersion of oleylamine-capped PtNi$_3$ polyhedra in nonpolar solvents. This resulted in the surface compositional change of PtNi$_3$ from Ni-rich to Pt-rich, leading to the formation and enlargement of the void cavity to give a nanoframe-like structure.

5.5   APPLICATION IN ELECTROCHEMICAL ENERGY CONVERSION

The sudden global energy demand, continuous industrial development, and population growth give rise to the excessive combustion of fossil fuels (Singh et al., 2021; Singh & Indra, 2020b). Electrochemical energy conversion is regarded as the most promising pathway for the production of renewable energy (Indra et al., 2018; Singh & Indra, 2020b). As high thermodynamic efficiency is observed for the electrochemical energy conversion processes, it has been widely employed as an eco-friendly technique. In this regard, electrochemical water splitting has been explored to produce renewable H$_2$ fuel (Singh & Indra, 2020a). Electrochemical water splitting consists of two half-cell reactions: oxygen evolution reaction (OER) in the anode and hydrogen evolution reaction (HER) in the cathode.

5.5.1   OER

OER is a bottleneck of the electrochemical water splitting and a thermodynamically unfavorable reaction (Indra et al., 2015). The OER is accompanied by the transfer of four electrons and four protons involving high-energy reaction intermediates. As a result, extra potential in terms of overpotential is required to facilitate the overall reaction (Menezes et al., 2015, 2016). In this respect, hollow-structured nanomaterials have been explored with superior catalytic performance for OER owing to their unique structural and morphological features (Indra et al., 2018). Hollow-structured nanomaterials provide several key advantages: (1) individual catalytic functionality of species are placed in sequentially localized compartments, enabling the combination of different incompatible catalytic functions; (2) charge transport, as well as electrolyte diffusion, can be improved by controlling the shell pore structure; (3) the large specific surface area increases the contact area between catalyst and reactant; and (4) the thin shell facilitates charge transfer process (J. Wang et al., 2019).

In this respect, the α -Ni(OH)$_2$ hollow microspheres were developed as oxygen evolution catalyst (M. Gao et al., 2014). The hollow sphere α -Ni(OH)$_2$ delivered the current density of 10 mA cm^{-2} at an overpotential of 331 mV, whereas the Tafel slope was recorded to be 42 mV decade^{-1}. The three-dimensional (3D) hollow Co(OH)$_2$ nanoflowers were developed by using a Cu$_2$O template (H. Liu et al., 2018). The hollow Co(OH)$_2$ nanoflowers reached an OER current density of 10 mA cm^{-2} at an overpotential of 310 mV.

Furthermore, hollow NiCo$_2$O$_4$ octahedral nanocages were synthesized using a Cu$_2$O-template (Lv et al., 2015). The overpotential of 340 mV was observed for the hollow NiCo$_2$O$_4$ octahedral nanocages to reach the current density of 10 mA cm^{-2}. The improved OER activity was attributed to the hollow cavity, large roughness, and high porosity of the NiCo$_2$O$_4$ nanocages. Co$_3$O$_4$/NiCo$_2$O$_4$ double-shelled box-in-box nanocages were also designed with different shell compositions (Figure 5.4a; Hu et al., 2015). The transmission electron microscopy (TEM) images clearly showed the hollow structure with box-in-box morphology (Figure 5.4b-c). The overpotential of $Co_3O_4/NiCo_2O_4$ double-shelled nanocages was reduced by 70 mV compared to the Co_3O_4 nanocages to deliver the same current density (Figure 5.4d). The Tafel slope of 88 mV dec^{-1} was calculated for the double-shelled $Co_3O_4/NiCo_2O_4$ nanocages, lower than that of Co_3O_4 nanocages (110 mV dec^{-1}; Figure 5.4e). The lower Tafel slope revealed the faster OER kinetics for the double-shelled $Co_3O_4/NiCo_2O_4$ nanocages. This result indicated the importance of rational design of hollow complex structures. The $NiCo_2O_4$ with a hollow microcuboid structure was also synthesized and employed for the overall water splitting (X. Gao et al., 2016). The hierarchical $NiCo_2O_4$ hollow microcuboids showed an anodic current density of 10 mA cm^{-2} at an overpotential of 290 mV.

5.5.2 HER

HER is a complementary reaction of OER and involves the reduction of water to form H_2 (Long et al., 2016; K. Zhu et al., 2020). The hydrogen evolution requires the input of 0 V vs. reversible hydrogen electrode (RHE) at pH 0. The reaction is pH-dependent and takes place in both acidic and alkaline media. It is facile in the acidic medium due to the presence of a sufficient amount of protons. Although several strategies have been utilized to improve the catalytic performance of the electrocatalysts, the increment in the number of available active sites and surface area can be an effective method to improve the HER activity. Therefore, porous and hollow structures are particularly important for the electrochemical HER. For example, hollow carbon nanobox–encapsulating monodispersed FeP nanoparticles (denoted FeP/HCNB) were developed from Prussian blue precursors (Peng et al., 2019). During the synthesis, the evaporation of zinc species led to the formation of hollow-core whereas the decomposition of polydopamine resulted in the formation of a porous carbon shell. The resulting FeP/HCNB yielded a current density of 10 mA cm^{-2} at the overpotentials of 88 and 180 mV for HER in both acidic and alkaline media, respectively. The excellent electrochemical stability of FeP/HCNB was demonstrated for 15 h. The hollow mesoporous carbon-coated FeP microcubes were also designed for the hydrogen evolution reaction (X. Zhu et al., 2016). The Prussian blue was synthesized by solvothermal method and further phosphorization under a nitrogen atmosphere produced the hollow structured FeP. The catalyst required an overpotential of 115 mV to achieve a current density of 10 mA cm^{-2} for HER.

The hollow NiCoFeP nanocubes were demonstrated for the excellent HER activity in alkaline medium (Guo et al., 2018b). The NiCoFeP nanocubes were developed by the phosphidation of NiCoFe-PBA under the nitrogen atmosphere. The catalyst NiCoFeP showed an overpotential of 131 mV for HER at a current density of 10 mA cm^{-2}. The Tafel slope 56 mV dec^{-1} suggested improved HER kinetics for hollow NiCoFeP nanocubes. A hollow $Co_3S_4@MoS_2$ structure was also prepared by the ion-exchange method. The CoFe-PBA was treated with Na_2S and $Na_2MoO_4 \cdot 2H_2O$ at 120 °C using a hydrothermal approach, resulting in the formation of a hollow $Co_3S_4@MoS_2$ core-shell structure (Guo et al., 2018a). Scanning electron microscopy (SEM) images revealed the formation of a hollow core-shell structure of $Co_3S_4@MoS_2$ (Figure 5.5a–b). The hollow $Co_3S_4@MoS_2$ structure exhibited an overpotential of 136 mV for HER to achieve a current density of 10 mA cm^{-2} in a 1.0 M KOH solution (Figure 5.5b). The lowest Tafel slope value of $Co_3S_4@MoS_2$ also indicated faster HER kinetics compared to the synthesized catalysts (Figure 5.5c). Moreover, the lowest charge transfer resistance and

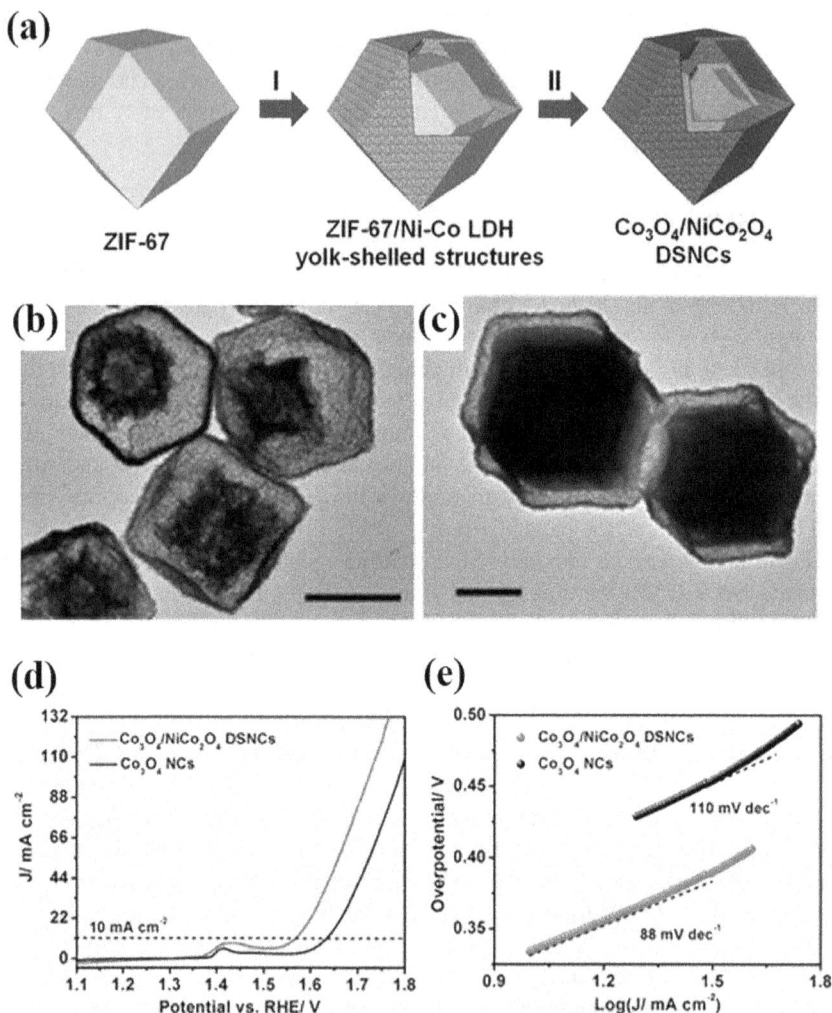

FIGURE 5.4   (a) Schematic representation for the synthesis of the $Co_3O_4/NiCo_2O_4$ double-shelled box-in-box nanocages; (b, c) TEM images of $Co_3O_4/NiCo_2O_4$ double-shelled box-in-box nanocages; (d) Linear sweep voltammetric profiles for the OER activity of the $Co_3O_4/NiCo_2O_4$ double-shelled box-in-box nanocages and Co_3O_4 nanocages indicating the improved OER activity of $Co_3O_4/NiCo_2O_4$; and (e) Tafel plots of $Co_3O_4/NiCo_2O_4$ double-shelled box-in-box nanocages and Co_3O_4 nanocages, showing the lowest Tafel slope value of $Co_3O_4/NiCo_2O_4$. Reproduced from Hu et al. (2015) with permission from the American Chemical Society.

enhanced stability of the core-shell Co_3S_4@MoS_2 were also established by electrochemical impedance spectroscopy (EIS; Figure 5.5d) and chronoamperometric test (Figure 5.5e), respectively. Other hollow nanostructured electrocatalysts, including Mo_2C, WC, IrNiN, Ni_2P, Co_2P, and $CoMoS_3$ have also been reported with improved hydrogen evolution activity.

5.6 ELECTROCHEMICAL CARBON DIOXIDE REDUCTION REACTION

The electrochemical carbon dioxide (CO_2) reduction is a cathodic reaction in which CO_2 is reduced to various carbon species (Kas et al., 2016). In a typical CO_2 electrolyzer, the anode and the cathode are separated into two chambers by an ion-conducting membrane (Da Silva Freitas et al., 2021). The thermodynamic potential to drive one-electron CO_2 reduction to $CO_2^{\cdot -}$ is determined to be 1.90 V vs. SHE (standard hydrogen electrode), making the reaction unfavorable. The proton-assisted reduction of CO_2 under the applied potential leads to a wide range of $CO_2^{\cdot -}$-derived products. It has been observed that the selective production of desired chemicals is challenging due to the similar redox potentials for all the reaction pathways (Da Silva Freitas et al., 2021). To reduce the extra input of potential and get the selective product, hollow-structured nanomaterials have been employed.

For example, a porous hollow fiber copper electrode with a compact 3D geometry was designed with a large surface area and three-phase boundary for the gas-liquid reactions. SEM studies revealed the formation of hollow fiber copper electrodes (Figure 5.6a–c; Kas et al., 2016). The linear sweep voltammetry (LSV) was performed in Ar or CO_2 saturated electrolyte. The increment in the current density during Ar purge was observed due to the evolution of hydrogen, which showed an onset potential of about 0.25 V vs. RHE. In contrast, the purging of CO_2 resulted in a twofold increment in the cathodic current density at potentials between 0.2 and 0.4 V vs. RHE (Figure 5.6d). The Faradaic efficiency (FE) of the major product was determined by varying the applied potential between 0.15 and 0.55 V vs. RHE. The onset potential for CO formation was recorded at 0.15 V vs. RHE. A maximum FE of 72% was determined toward CO at a potential of 0.4 V vs. RHE (Figure 5.6e). The Tafel slope of 255 mV dec^{-1} was assigned for the rate-determining step of the generation of COOH intermediate during the CO production (Figure 5.6f). Moreover, the lower Tafel slope of 93 mV dec^{-1} was observed due to a nonuniform potential or current distribution in the porous hollow fibers. The stability of the Cu hollow fibers was determined for 24 h at an applied potential of 0.4 V vs. RHE (Figure 5.6g). After a 10% drop in the activity in the first 7 h, the catalyst showed stability for 17 h.

Similarly, hollow cubic Cu_2O@Au nanocomposites were developed for electrochemical CO_2 reduction (Tan et al., 2019). The catalyst showed superior catalytic activity for CO_2 reduction reaction than Cu_2O cubes and hollow cubic Au catalysts. The Cu_2O@Au catalyst converted CO_2 to CO with a maximum (FE) of approximately 30.1% at −1.0 V vs. RHE. Thin-walled hollow Au–Cu nanoparticles were synthesized using the Kirkendall effect for the efficient CO_2 reduction reaction (J. H. Zhou et al., 2018). The hollow Au–Cu nanoparticles exhibited a maximum

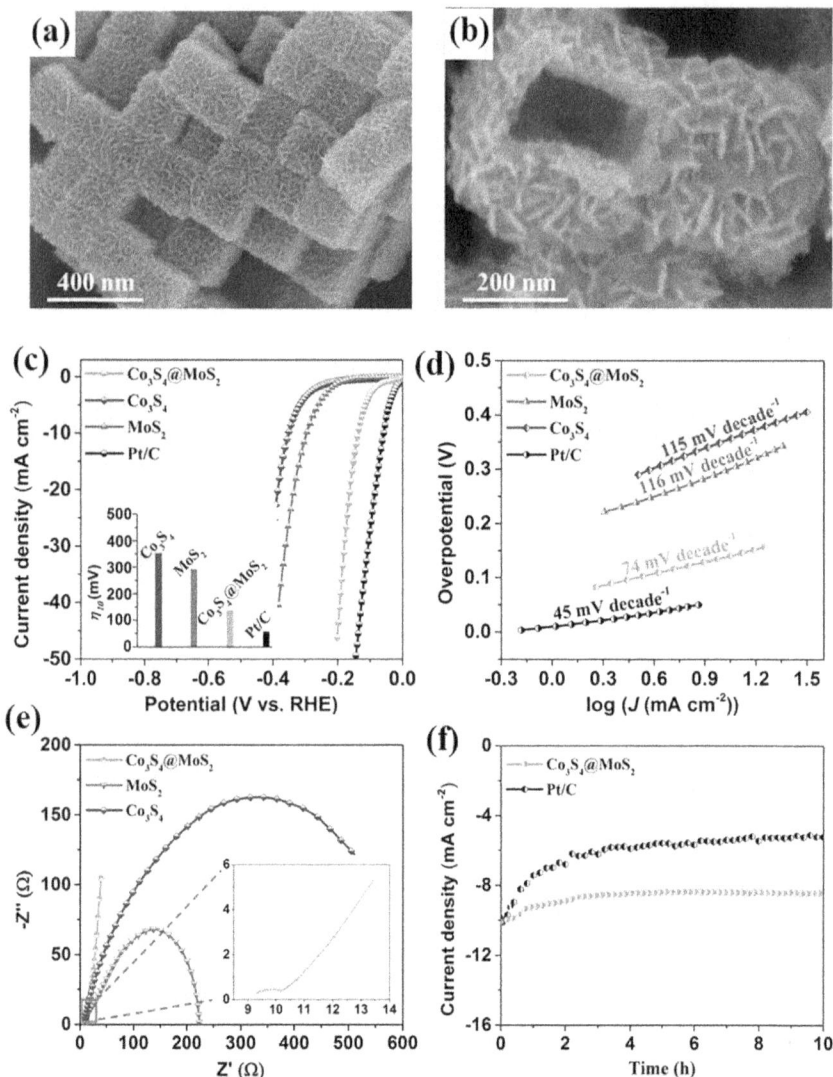

FIGURE 5.5 (a–b) SEM images of the $Co_3S_4@MoS_2$ showing the hollow structure morphology; (b) LSV curves of $Co_3S_4@MoS_2$ compared with Co_3S_4 and MoS_2 showing the improved HER activity of $Co_3S_4@MoS_2$; (c) Tafel plots of the $Co_3S_4@MoS_2$ compared with Co_3S_4 and MoS_2 showing the fastest HER kinetics for $Co_3S_4@MoS_2$; (c) EIS spectra of the $Co_3S_4@MoS_2$ compared with the Co_3S_4 and MoS_2 indicating the lowest charge transfer resistance for $Co_3S_4@MoS_2$; and (e) chronoamperometric stability test of the $Co_3S_4@MoS_2$ showing the enhanced stability of $Co_3S_4@MoS_2$ for 10 h.

Source: Reproduced from Guo et al. (2018a) with permission from Elsevier.

FE of 53.3% at −0.7 V (vs. RHE) for the conversion of CO_2 to CO. The hollow Au–Cu-NP/C showed a much higher current density than the Au-NP/C when normalized against the Au mass loading.

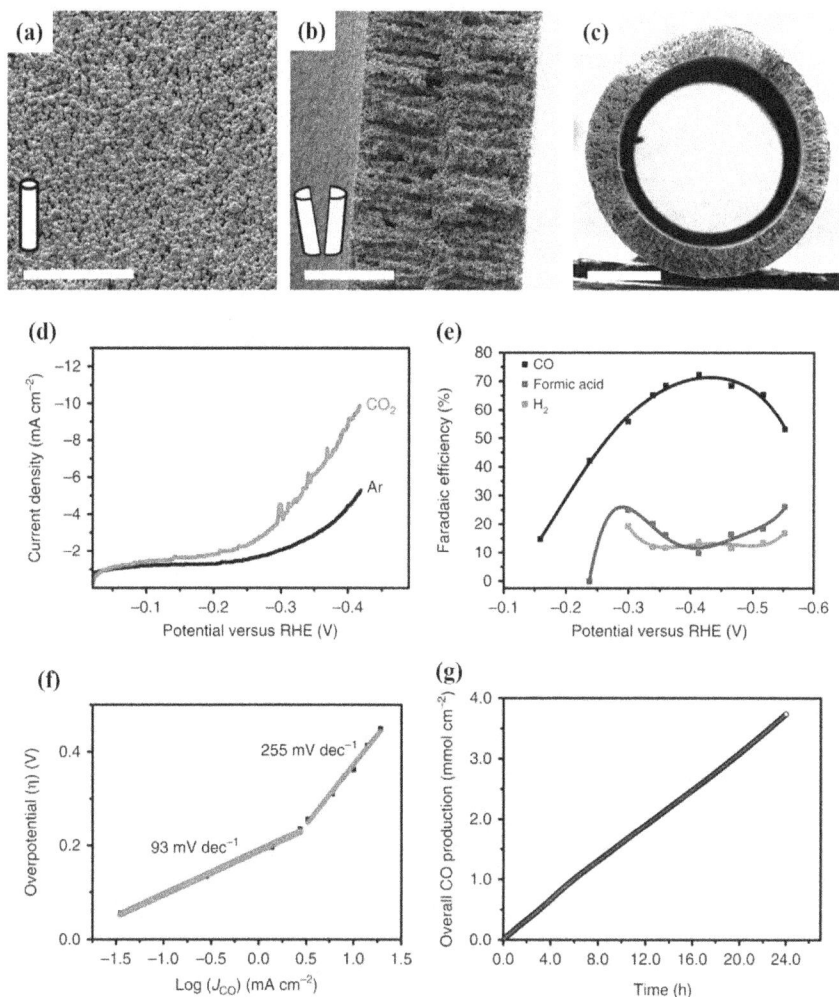

FIGURE 5.6 (a) SEM images of the Cu hollow fiber and (b) outer surface of the Cu hollow fiber; (c) cross-sectional image of a perpendicularly broken Cu hollow fiber; (d) linear polarization curves obtained for Cu hollow fibers when CO_2 or Ar was purged in 0.3 M $KHCO_3$ electrolyte; (e) FE of CO, formic acid and H_2 as a function of applied potential; (f) overpotential vs. the log of partial current density for the production of CO using Cu hollow fibers; and (g) production of CO at a constant applied potential of −0.4 V for 24 h.

Source: Reproduced from Kas et al. (2016) with permission from Springer Nature

5.7 ELECTROCHEMICAL NITROGEN REDUCTION REACTION

The production of ammonia is essential to improve the food supply for the increasing global population (Nazemi et al., 2018). To meet the global demand for ammonia, it is important to develop a sustainable and eco-friendly synthetic method that requires a smaller amount of energy than the current methods. Recently, nitrogen (N_2) has been considered as the potential source for the generation of ammonia (Nazemi et al., 2018). As the bond energy of diatomic nitrogen is observed to be high ($N \equiv N$ bond energy of 940.95 kJ mol^{-1}), the fixation of nitrogen into ammonia becomes difficult. As a result, a complex multistep process is followed to convert diatomic nitrogen into ammonia.

The electrochemical nitrogen reduction reaction (ENRR) has been considered to be one of the most effective pathways for the production of ammonia under ambient conditions (Shi et al., 2020). Generally, the electrochemical N_2 reduction is a potential-driven multistep process, and several proton-coupled electron transfer (PCET) processes are involved to convert N_2 into NH_3. It is a six-electron transfer reaction that requires +0.55 V vs. RHE at pH = 0. At pH = 14, the required potential is observed to be −1.48 V vs. RHE (Shi et al., 2020). The nitrogen reduction reaction mainly involves two pathways (i.e., dissociative pathway and associative pathway). In the dissociative pathway, the N–N bond breaking is observed to happen before the addition of the H atom followed by the conversion of the individual N atom into NH_3. However, in the associative pathway, the transformation of the $N \equiv N$ to the N–N bond and the generation of NH_3 occur simultaneously.

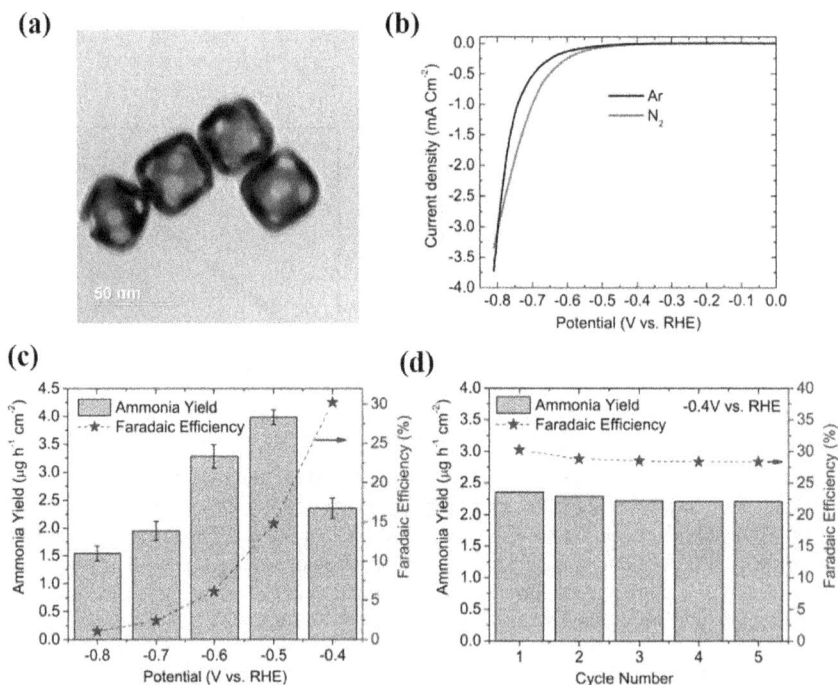

FIGURE 5.7   (a) TEM images of AuHNCs; (b) LSV profiles of AuHNCs for the electrochemical NRR in an Ar- and N_2-saturated environment in 0.5 M LiClO$_4$ aqueous solution under ambient conditions; (c) ammonia yield rate and FE at various potentials in 0.5 M LiClO$_4$ at 20 °C; (d) plot for the ammonia yield vs. cycle number showing a minor decrease in ammonia yield and FE in the fifth cycle.

Source: Reproduced from Nazemi et al. (2018) with permission from Elsevier.

In recent years, continuous efforts have been provided for the development of N_2 reduction reaction catalysts, improving catalytic selectivity and efficiency. In this respect, hollow structured materials have been explored recently with excellent catalytic activity for electrochemical N_2 reduction reaction (Shi et al., 2020). The hollow gold nanocages (AuHNCs) were developed for the electrochemical N_2 reduction reaction under ambient conditions (Nazemi et al., 2018). The formation of hollow gold nanocages was revealed by TEM studies (Figure 5.7a). LSV was performed in an Ar and N_2 saturated environment to qualitatively distinguish between HER and NRR (Figure 5.7b). As the potential was increased from 0.4 V vs. RHE, the enhancement in the current density was observed in the N_2-saturated environment. This was attributed to the formation of NH_3 from N_2 by the AuHNC catalyst.

The chronoamperometry (CA) tests were conducted to determine the ammonia yield and FE. The ENRR attained higher selectivity within the potential range of 0.4 V to 0.6 V vs. RHE. As a result, the highest ammonia yield rate of 3.98 μg cm^{-2} h^{-1} at 0.5 V and an FE of 30.2% at 0.4 V were recorded (Figure 5.7c). The FE was increased from 30.2% to 40.5% by increasing the operating temperature from 20 °C to 50 °C at −0.4 V vs. RHE. The cyclic stability for ammonia production was also well established for AuHNCs (Figure 5.7d). The AuHNCs demonstrated excellent stability with a minor decrease in ammonia yield rate and an FE (93.8% performance retention) in the fifth cycle.

Furthermore, hollow CuAu nanoboxes with Cu-rich inner walls were also synthesized to improve the FE in the ENRR (Talukdar et al., 2021). Density functional theory (DFT) calculations revealed that the Cu-rich inner walls in the hollow CuAu nanoboxes imparted a major role to improve their catalytic performance by reducing the free energy ΔG^*_{NNH} for the potential-determining step. The highest efficiency of 7.39% was observed at −0.2 V for hollow CuAu nanoboxes.

5.8   CONCLUSION

In this chapter, the importance of designing multi-shell hollow-structure electrocatalysts has been explained for electrochemical energy conversion. Complex hollow-structured materials produce better electrochemical performance compared to the hierarchical structure and simple hollow-structured materials. The different strategies adopted to attain hollow-structured materials with various morphology and structure have been explained. A series of materials like alloy, metal oxides, sulfides, phosphides, hydroxides, and others with multi-shell hollow structures have been explored for electrochemical energy conversion like oxygen evolution, hydrogen evolution, CO_2 reduction, and N_2 reduction.

ACKNOWLEDGMENT

The financial support from the CSIR [Grant no. 01(2977)/19/EMR-II], Govt. of India is gratefully acknowledged. Baghendra Singh acknowledges the DST-INSPIRE (IF180147) for the research fellowship.

REFERENCES

Bernard Ng, C. H., Tan, H., & Fan, W. Y. (2006). Formation of Ag_2Se nanotubes and dendrite-like structures from UV irradiation of a CSe_2/Ag colloidal solution. *Langmuir, 22*(23), 9712–9717. https://doi.org/10.1021/la061253u

Cabot, A., Puntes, V. F., Shevchenko, E., Yin, Y., Balcells, L., Marcus, M. A., Hughes, S. M., & Alivisatos, A. P. (2007). Vacancy coalescence during oxidation of iron nanoparticles. *Journal of the American Chemical Society, 129*(34), 10358–10360. https://doi.org/10.1021/ja072574a

Chen, C., Kang, Y., Huo, Z., Zhu, Z., Huang, W., Xin, H. L., Snyder, J. D., Li, D., Herron, J. A., Mavrikakis, M., Chi, M., More, K. L., Li, Y., Markovic, N. M., Somorjai, G. A., Yang, P., & Stamenkovic, V. R. (2014). Highly crystalline multimetallic nanoframes with three-dimensional electrocatalytic surfaces. *Science, 343*(6177), 1339–1343. https://doi.org/10.1126/science.1249061

Chen, D., Li, L., Tang, F., & Qi, S. (2009). Facile and scalable synthesis of tailored silica "nanorattle" structures. *Advanced Materials, 21*(37), 3804–3807. https://doi.org/10.1002/adma.200900599

Da Silva Freitas, W., D'Epifanio, A., & Mecheri, B. (2021). Electrocatalytic CO_2 reduction on nanostructured metal-based materials: Challenges and constraints for a sustainable pathway to decarbonization. *Journal of CO2 Utilization, 50*, 101579. https://doi.org/10.1016/j.jcou.2021.101579

Fan, C., Wu, X., Li, M., Wang, X., Zhu, Y., Fu, G., Ma, T., & Tang, Y. (2022). Surface chemical reconstruction of hierarchical hollow inverse-spinel manganese cobalt oxide boosting oxygen evolution reaction. *Chemical Engineering Journal, 431*, 133829. https://doi.org/10.1016/j.cej.2021.133829

Gao, J., Zhang, B., Zhang, X., & Xu, B. (2006). Magnetic-dipolar-interaction-induced self-assembly affords wires of hollow nanocrystals of cobalt selenide. *Angewandte Chemie International Edition, 45*(8), 1220–1223. https://doi.org/10.1002/anie.200503486

Gao, M., Sheng, W., Zhuang, Z., Fang, Q., Gu, S., Jiang, J., & Yan, Y. (2014). Efficient water oxidation using nanostructured α-nickel-hydroxide as an electrocatalyst. *Journal of the American Chemical Society, 136*(19), 7077–7084. https://doi.org/10.1021/ja502128j

Gao, X., Zhang, H., Li, Q., Yu, X., Hong, Z., Zhang, X., Liang, C., & Lin, Z. (2016). Hierarchical $NiCo_2O_4$ hollow microcuboids as bifunctional electrocatalysts for overall water-splitting. *Angewandte Chemie International Edition, 55*(21), 6290–6294. https://doi.org/10.1002/anie.201600525

Gong, Q., Gao, T., Hu, T., & Zhou, G. (2019). Synthesis and electrochemical energy storage applications of micro/nanostructured spherical materials. *Nanomaterials, 9*(9). https://doi.org/10.3390/nano9091207

Gou, L., & Murphy, C. J. (2003). Solution-phase synthesis of Cu_2O nanocubes. *Nano Letters, 3*(2), 231–234. https://doi.org/10.1021/nl0258776

Gu, D., Bongard, H., Deng, Y., Feng, D., Wu, Z., Fang, Y., Mao, J., Tu, B., Schüth, F., & Zhao, D. (2010). An aqueous emulsion route to synthesize mesoporous carbon vesicles and their nanocomposites. *Advanced Materials, 22*(7), 833–837. https://doi.org/10.1002/adma.200902550

Guo, Y., Tang, J., Wang, Z., Kang, Y. M., Bando, Y., & Yamauchi, Y. (2018a). Elaborately assembled core-shell structured metal sulfides as a bifunctional catalyst for highly efficient electrochemical overall water splitting. *Nano Energy, 47*, 494–502. https://doi.org/10.1016/J.NANOEN.2018.03.012

Guo, Y., Tang, J., Wang, Z., Sugahara, Y., & Yamauchi, Y. (2018b). Hollow porous heterometallic phosphide nanocubes for enhanced electrochemical water splitting. *Small, 14*(44), 1802442. https://doi.org/10.1002/smll.201802442

Hu, H., Guan, B., Xia, B., & Lou, X. W. (David). (2015). Designed formation of Co_3O_4/$NiCo_2O_4$ double-shelled nanocages with enhanced pseudocapacitive and electrocatalytic properties. *Journal of the American Chemical Society*, *137*(16), 5590–5595. https://doi.org/10.1021/jacs.5b02465

Indra, A., Menezes, P. W., & Driess, M. (2015). Uncovering structure—activity relationships in manganese-oxide-based heterogeneous catalysts for efficient water oxidation. *ChemSusChem*, *8*(5), 776–785. https://doi.org/10.1002/cssc.201402812

Indra, A., Song, T., & Paik, U. (2018). Metal organic framework derived materials: Progress and prospects for the energy conversion and storage. *Advanced Materials*, *30*(39), 1705146. https://doi.org/10.1002/adma.201705146

Kas, R., Hummadi, K. K., Kortlever, R., De Wit, P., Milbrat, A., Luiten-Olieman, M. W. J., Benes, N. E., Koper, M. T. M., & Mul, G. (2016). Three-dimensional porous hollow fibre copper electrodes for efficient and high-rate electrochemical carbon dioxide reduction. *Nature Communications*, *7*. https://doi.org/10.1038/ncomms10748

Kim, S., Min, K., Kim, H., Yoo, R., Shim, S. E., Lim, D., & Baeck, S.-H. (2022). Bimetallic-metal organic framework-derived Ni_3S_2/MoS_2 hollow spheres as bifunctional electrocatalyst for highly efficient and stable overall water splitting. *International Journal of Hydrogen Energy*, *47*(13), 8165–8176. https://doi.org/10.1016/j.ijhydene.2021.12.208

Kim, S. S., Zhang, W., & Pinnavaia, T. J. (1998). Ultrastable mesostructured silica vesicles. *Science*, *282*(5392), 1302–1305. https://doi.org/10.1126/science.282.5392.1302

Li, Y, Li, X., Li, Y., Liu, H., Wang, S., Gan, H., Li, J., Wang, N., He, X., & Zhu, D. (2006). Controlled self-assembly behavior of an amphiphilic bisporphyrin—bipyridinium—palladium complex: From multibilayer vesicles to hollow capsules. *Angewandte Chemie International Edition*, *45*(22), 3639–3643. https://doi.org/10.1002/anie.200600554

Li, Y, & Shi, J. (2014). Hollow-structured mesoporous materials: Chemical synthesis, functionalization and applications. *Advanced Materials*, *26*(20), 3176–3205. https://doi.org/10.1002/adma.201305319

Li, Y., Wang, H., Priest, C., Li, S., Xu, P., & Wu, G. (2021). Advanced electrocatalysis for energy and environmental sustainability via water and nitrogen reactions. *Advanced Materials*, *33*(6), 2000381. https://doi.org/10.1002/adma.202000381

Li, Y., Yan, K., Lee, H.-W., Lu, Z., Liu, N., & Cui, Y. (2016). Growth of conformal graphene cages on micrometre-sized silicon particles as stable battery anodes. *Nature Energy*, *1*(2), 15029. https://doi.org/10.1038/nenergy.2015.29

Liu, B., & Zeng, H. C. (2004). Fabrication of ZnO "Dandelions" via a modified kirkendall process. *Journal of the American Chemical Society*, *126*(51), 16744–16746. https://doi.org/10.1021/ja044825a

Liu, H., Guo, D., Zhang, W., & Cao, R. (2018). $Co(OH)_2$ hollow nanoflowers as highly efficient electrocatalysts for oxygen evolution reaction. *Journal of Materials Research*, *33*(5), 568–580. https://doi.org/10.1557/jmr.2017.390

Liu, J., Cao, G., Yang, Z., Wang, D., Dubois, D., Zhou, X., Graff, G. L., Pederson, L. R., & Zhang, J.-G. (2008). Oriented nanostructures for energy conversion and storage. *ChemSusChem*, *1*(8–9), 676–697. https://doi.org/10.1002/cssc.200800087

Liu, J., Hartono, S. B., Jin, Y. G., Li, Z., Lu, G. Q. (Max), & Qiao, S. Z. (2010). A facile vesicle template route to multi-shelled mesoporous silica hollow nanospheres. *Journal of Materials Chemistry*, *20*(22), 4595–4601. https://doi.org/10.1039/B925201K

Liu, X. H., Lai, W. H., & Chou, S. L. (2020). The application of hollow micro-/nanostructured cathodes for sodium-ion batteries. *Materials Chemistry Frontiers*, *4*(5), 1289–1303. https://doi.org/10.1039/c9qm00674e

Liu, Y., Goebl, J., & Yin, Y. (2013). Templated synthesis of nanostructured materials. *Chemical Society Reviews*, *42*(7), 2610–2653. https://doi.org/10.1039/c2cs35369e

Long, X., Wang, Z., Xiao, S., An, Y., & Yang, S. (2016). Transition metal based layered double hydroxides tailored for energy conversion and storage. *Materials Today*, *19*(4), 213–226. https://doi.org/10.1016/J.MATTOD.2015.10.006

Lou, X. W. (David), Archer, L. A., & Yang, Z. (2008). Hollow micro-/nanostructures: Synthesis and applications. *Advanced Materials*, *20*(21), 3987–4019. https://doi.org/10.1002/adma.200800854

Lou, X. W., Wang, Y., Yuan, C., Lee, J. Y., & Archer, L. A. (2006). Template-free synthesis of SnO_2 hollow nanostructures with high lithium storage capacity. *Advanced Materials*, *18*(17), 2325–2329. https://doi.org/10.1002/adma.200600733

Lv, X., Zhu, Y., Jiang, H., Yang, X., Liu, Y., Su, Y., Huang, J., Yao, Y., & Li, C. (2015). Hollow mesoporous $NiCo_2O_4$ nanocages as efficient electrocatalysts for oxygen evolution reaction. *Dalton Transactions*, *44*(9), 4148–4154. https://doi.org/10.1039/C4DT03803G

Ma, Y., Xie, Q., Liu, X., Zhao, Y., Zeng, D., Wang, L., Zheng, Y., & Peng, D. L. (2015). Synthesis of amorphous $ZnSnO_3$ double-shell hollow microcubes as advanced anode materials for lithium ion batteries. *Electrochimica Acta*, *182*, 327–333. https://doi.org/10.1016/J.ELECTACTA.2015.09.102

Menezes, P. W., Indra, A., Das, C., Walter, C., Göbel, C., Gutkin, V., Schmeißer, D., & Driess, M. (2016). Uncovering the nature of active species of nickel phosphide catalysts in high-performance electrochemical overall water splitting. *ACS Catalysis*, *7*(1), 103–109. https://doi.org/10.1021/acscatal.6b02666

Menezes, P. W., Indra, A., Sahraie, N. R., Bergmann, A., Strasser, P., & Driess, M. (2015). Cobalt—manganese-based spinels as multifunctional materials that unify catalytic water oxidation and oxygen reduction reactions. *ChemSusChem*, *8*(1), 164–171. https://doi.org/10.1002/cssc.201402699

Miao, J.-J., Jiang, L.-P., Liu, C., Zhu, J.-M., & Zhu, J.-J. (2007). General sacrificial template method for the synthesis of cadmium chalcogenide hollow structures. *Inorganic Chemistry*, *46*(14), 5673–5677. https://doi.org/10.1021/ic700404n

Nai, J., Zhang, J., & Lou, X. W. (David). (2018). Construction of single-crystalline Prussian blue analog hollow nanostructures with tailorable topologies. *Chem*, *4*(8), 1967–1982. https://doi.org/10.1016/J.CHEMPR.2018.07.001

Nazemi, M., Panikkanvalappil, S. R., & El-Sayed, M. A. (2018). Enhancing the rate of electrochemical nitrogen reduction reaction for ammonia synthesis under ambient conditions using hollow gold nanocages. *Nano Energy*, *49*, 316–323. https://doi.org/10.1016/j.nanoen.2018.04.039

Pang, R., Hu, X., Zhou, S., Sun, C., Yan, J., Sun, X., Xiao, S., & Chen, P. (2014). Preparation of multi-shelled conductive polymer hollow microspheres by using Fe_3O_4 hollow spheres as sacrificial templates. *Chemical Communications*, *50*(83), 12493–12496. https://doi.org/10.1039/C4CC05469E

Peng, Z., Qiu, X., Yu, Y., Jiang, D., Wang, H., Cai, G., Zhang, X., & Dong, Z. (2019). Polydopamine coated prussian blue analogue derived hollow carbon nanoboxes with FeP encapsulated for hydrogen evolution. *Carbon*, *152*, 16–23. https://doi.org/10.1016/J.CARBON.2019.05.073

Pérez-Page, M., Yu, E., Li, J., Rahman, M., Dryden, D. M., Vidu, R., & Stroeve, P. (2016). Template-based syntheses for shape controlled nanostructures. *Advances in Colloid and Interface Science*, *234*, 51–79. https://doi.org/10.1016/J.CIS.2016.04.001

Shao, H. F., Qian, X. F., & Zhu, Z. K. (2005). The synthesis of ZnS hollow nanospheres with nanoporous shell. *Journal of Solid State Chemistry*, *178*(11), 3522–3528. https://doi.org/10.1016/J.JSSC.2005.09.007

Shen, L., Yu, L., Wu, H. Bin, Yu, X.-Y., Zhang, X., & Lou, X. W. (David). (2015). Formation of nickel cobalt sulfide ball-in-ball hollow spheres with enhanced electrochemical pseudocapacitive properties. *Nature Communications*, *6*(1), 6694. https://doi.org/10.1038/ncomms7694

Shi, L., Yin, Y., Wang, S., & Sun, H. (2020). Rational catalyst design for N_2 Reduction under ambient conditions: Strategies toward enhanced conversion efficiency. *ACS Catalysis*, *10*(12), 6870–6899. https://doi.org/10.1021/acscatal.0c01081

Singh, B., & Indra, A. (2020a). Designing self-supported metal-organic framework derived catalysts for electrochemical water splitting. *Chemistry—An Asian Journal*, *15*(6), 607–623. https://doi.org/10.1002/asia.201901810

Singh, B., & Indra, A. (2020b). Prussian blue- and Prussian blue analogue-derived materials: progress and prospects for electrochemical energy conversion. In *Materials Today Energy* (Vol. 16, p. 100404). Elsevier Ltd. https://doi.org/10.1016/j.mtener.2020.100404

Singh, B., Singh, A., Yadav, A., & Indra, A. (2021). Modulating electronic structure of metal-organic framework derived catalysts for electrochemical water oxidation. *Coordination Chemistry Reviews*, *447*, 214144. https://doi.org/10.1016/J.CCR.2021.214144

Sun, L., Han, X., Jiang, Z., Ye, T., Li, R., Zhao, X., & Han, X. (2016). Fabrication of cubic Zn_2SnO_4/SnO_2 complex hollow structures and their sunlight-driven photocatalytic activity. *Nanoscale*, *8*(26), 12858–12862. https://doi.org/10.1039/C5NR08004E

Talukdar, B., Kuo, T. C., Sneed, B. T., Lyu, L. M., Lin, H. M., Chuang, Y. C., Cheng, M. J., & Kuo, C. H. (2021). Enhancement of NH_3 Production in electrochemical N_2 Reduction by the Cu-Rich inner surfaces of beveled CuAu nanoboxes. *ACS Applied Materials and Interfaces*. https://doi.org/10.1021/acsami.1c03454

Tan, W., Cao, B., Xiao, W., Zhang, M., Wang, S., Xie, S., Xie, D., Cheng, F., Guo, Q., & Liu, P. (2019). Electrochemical reduction of CO_2 on hollow cubic $Cu_2O@Au$ nanocomposites. *Nanoscale Research Letters*, *14*(1), 63. https://doi.org/10.1186/s11671-019-2892-3

Teng, Z., Wang, S., Su, X., Chen, G., Liu, Y., Luo, Z., Luo, W., Tang, Y., Ju, H., Zhao, D., & Lu, G. (2014). Facile synthesis of yolk—shell structured inorganic—organic hybrid spheres with ordered radial mesochannels. *Advanced Materials*, *26*(22), 3741–3747. https://doi.org/https://doi.org/10.1002/adma.201400136

Wang, J., Cui, Y., & Wang, D. (2019). Design of hollow nanostructures for energy storage, conversion and production. *Advanced Materials*, *31*(38), 1–24. https://doi.org/10.1002/adma.201801993

Wang, L., Tang, K., Liu, Z., Wang, D., Sheng, J., & Cheng, W. (2011). Single-crystalline $ZnSn(OH)_6$ hollow cubes via self-templated synthesis at room temperature and their photocatalytic properties. *Journal of Materials Chemistry*, *21*(12), 4352–4357. https://doi.org/10.1039/C0JM03734F

Wang, W., Dahl, M., & Yin, Y. (2013). Hollow nanocrystals through the nanoscale kirkendall effect. *Chemistry of Materials*, *25*(8), 1179–1189. https://doi.org/10.1021/cm3030928

Wang, Y., Zhu, Q., & Zhang, H. (2005). Fabrication of β-$Ni(OH)_2$ and NiO hollow spheres by a facile template-free process. *Chemical Communications*, *41*, 5231–5233. https://doi.org/10.1039/B508807K

Wang, Z., Wang, Z., Wu, H., & Lou, X. W. (David). (2013). Mesoporous single-crystal $CoSn(OH)_6$ hollow structures with multilevel interiors. *Scientific Reports*, *3*(1), 1391. https://doi.org/10.1038/srep01391

Wong, Y. J., Zhu, L., Teo, W. S., Tan, Y. W., Yang, Y., Wang, C., & Chen, H. (2011). Revisiting the Stöber method: Inhomogeneity in silica shells. *Journal of the American Chemical Society*, *133*(30), 11422–11425. https://doi.org/10.1021/ja203316q

Wu, X.-J., & Xu, D. (2010). Soft template synthesis of yolk/silica shell particles. *Advanced Materials*, *22*(13), 1516–1520. https://doi.org/https://doi.org/10.1002/adma.200903879

Wu, Z., Zhang, M., Yu, K., Zhang, S., & Xie, Y. (2008). Self-assembled double-shelled ferrihydrite hollow spheres with a tunable aperture. *Chemistry—A European Journal*, *14*(17), 5346–5352. https://doi.org/https://doi.org/10.1002/chem.200701945

Xu, H., & Wang, W. (2007). Template synthesis of multishelled Cu_2O hollow spheres with a single-crystalline shell wall. *Angewandte Chemie International Edition*, *46*(9), 1489–1492. https://doi.org/10.1002/anie.200603895

Xue, Y., Yan, Q., Bai, X., Xu, Y., Zhang, X., Li, Y., Zhu, K., Ye, K., Yan, J., Cao, D., & Wang, G. (2022). Ruthenium-nickel-cobalt alloy nanoparticles embedded in hollow carbon microtubes as a bifunctional mosaic catalyst for overall water splitting. *Journal of Colloid and Interface Science*, *612*, 710–721. https://doi.org/10.1016/j.jcis.2022.01.001

Yan, C., & Rosei, F. (2014). Hollow micro/nanostructured materials prepared by ion exchange synthesis and their potential applications. *New Journal of Chemistry*, *38*(5), 1883–1904. https://doi.org/10.1039/C3NJ00888F

Yang, H. G., & Zeng, H. C. (2004). Preparation of hollow anatase TiO_2 nanospheres via ostwald ripening. *The Journal of Physical Chemistry B, 108*(11), 3492–3495. https://doi.org/10.1021/jp0377782

Yasin, G., Ibraheem, S., Ali, S., Arif, M., Ibrahim, S., Iqbal, R., Kumar, A., Tabish, M., Mushtaq, M. A., Saad, A., Xu, H., & Zhao, W. (2022). Defects-engineered tailoring of tri-doped interlinked metal-free bifunctional catalyst with lower gibbs free energy of OER/HER intermediates for overall water splitting. *Materials Today Chemistry, 23*, 100634. https://doi.org/10.1016/j.mtchem.2021.100634

Yec, C. C., & Zeng, H. C. (2014). Synthesis of complex nanomaterials via Ostwald ripening. *Journal of Materials Chemistry A, 2*(14), 4843–4851. https://doi.org/10.1039/C3TA14203E

Yin, Y., Rioux, R. M., Erdonmez, C. K., Hughes, S., Gabor, A., & Alivisatos, A. P. (2004). Formation of hollow nanocrystals through the nanoscale Kirkendall effect. *Science, 304*(5671), 711–714. www.jstor.org/stable/3836820

Yu, X. Y., Yu, L., & Lou, X. W. (2016). Metal sulfide hollow nanostructures for electro-chemical energy storage. *Advanced Energy Materials, 6*(3), 1–14. https://doi.org/10.1002/aenm.201501333

Zhang, L., & Wang, H. (2011). Interior structural tailoring of Cu_2O shell-in-shell nanostructures through multistep Ostwald ripening. *The Journal of Physical Chemistry C, 115*(38), 18479–18485. https://doi.org/10.1021/jp2059613

Zhang, Q., Zhang, T., Ge, J., & Yin, Y. (2008). Permeable silica shell through surface-protected etching. *Nano Letters, 8*(9), 2867–2871. https://doi.org/10.1021/nl8016187

Zhang, Y., Yu, M., Zhou, L., Zhou, X., Zhao, Q., Li, H., & Yu, C. (2008). Organosilica multilamellar vesicles with tunable number of layers and sponge-like walls via one surfactant templating. *Chemistry of Materials, 20*(19), 6238–6243. https://doi.org/10.1021/cm8011815

Zhao, Y., & Jiang, L. (2009). Hollow micro/nanomaterials with multilevel interior structures. *Advanced Materials, 21*(36), 3621–3638. https://doi.org/10.1002/adma.200803645

Zhou, J. H., Lan, D. W., Yang, S. S., Guo, Y., Yuan, K., Dai, L. X., & Zhang, Y. W. (2018). Thin-walled hollow Au-Cu nanostructures with high efficiency in electrochemical reduction of CO_2 to CO. *Inorganic Chemistry Frontiers, 5*(7), 1524–1532. https://doi.org/10.1039/c8qi00297e

Zhou, L., Zhuang, Z., Zhao, H., Lin, M., Zhao, D., & Mai, L. (2017). Intricate hollow structures: Controlled synthesis and applications in energy storage and conversion. *Advanced Materials, 29*(20). https://doi.org/10.1002/adma.201602914

Zhu, K., Shi, F., Zhu, X., & Yang, W. (2020). The roles of oxygen vacancies in electro-catalytic oxygen evolution reaction. *Nano Energy, 73*, 104761. https://doi.org/10.1016/j.nanoen.2020.104761

Zhu, M., Cheng, Y., Luo, Q., El-Khateeb, M., & Zhang, Q. (2021). A review of synthetic approaches to hollow nanostructures. *Materials Chemistry Frontiers, 5*(6), 2552–2587. https://doi.org/10.1039/D0QM00879F

Zhu, M., Tang, J., Wei, W., & Li, S. (2020). Recent progress in the syntheses and applications of multishelled hollow nanostructures. *Materials Chemistry Frontiers, 4*(4), 1105–1149. https://doi.org/10.1039/c9qm00700h

Zhu, X., Liu, M., Liu, Y., Chen, R., Nie, Z., Li, J., & Yao, S. (2016). Carbon-coated hollow mesoporous FeP microcubes: An efficient and stable electrocatalyst for hydrogen evolution. *Journal of Materials Chemistry A, 4*(23), 8974–8977. https://doi.org/10.1039/C6TA01923D

6 Polymer-Based Nanocomposites in Energy Storage Applications

Biplab Kumar Kuila

6.1 INTRODUCTION

Electric energy storage systems (ESSs) can store electrical energy from intermittent renewable sources and deliver timely and adequate electric power for portable electrical and electronic devices, electric vehicles and many other applications. Therefore, highly efficient and environmentally friendly energy storage devices with low cost are highly demandable in our modern life in order to combat the challenging issues of depletion of fossil fuels and climate change. Electrochemical energy storage systems (EESSs) comprise supercapacitors, batteries and fuel cells (Beaudin et al. 2010). There are three major components common in all three ESSs: the electrodes, the electrolyte and the current collector. Generally, the electrodes are made of carbon materials, metal oxide and other conducting materials, which are still not considered ideal materials. Liquid electrolytes containing lithium salts (e.g., $LiPF_6$) in organic solvents (e.g., propylene carbonate) are mostly used in commercial battery. However, issues like achieving high energy density, performance decay and safety related to the leakage of inflammable organic solvents, catching fire and explosion related to liquid electrolytes still remain unaddressed. However, polymers are used in almost every application and field of our day-to-day life. However, it is necessary to modify its properties or upgrade its performance to meet particular processing or application requirements. The preparation of polymer nanocomposites in which nanomaterials are integrated with polymer is an elegant way to achieve such modification or upgradations of polymer properties. Recently, polymer-based nanocomposite materials have drawn significant research attention in energy storage applications due to their excellent electrical, electrochemical and mechanical properties arising from the synergistic effect of both the components. Modern electronics demand flexible energy storage devices that can only be realized by using polymer or its composite as structural materials. In this chapter, we focus on the recent advances in developing different types of polymer nanocomposites and their application as components of electrochemical energy storage materials. Major challenges to be addressed and future perspectives are also highlighted.

DOI: 10.1201/9781003208709-6

6.2   POLYMER NANOCOMPOSITES

Polymer nanocomposites are hybrid materials of polymer and nanomaterial having at least one dimension between 1 to 100 nm (Paul & Robeson 2008). In polymer nanocomposites, polymer generally serves as the matrix where the nanomaterials with a small amount are dispersed and called nanofillers. The nature and properties of both the polymer matrix and the nanofiller have a great effect on the overall properties like electric conductivity, ionic conductivity and ease of processing, tensile strength and chemical, thermal and mechanical stability. For energy storage applications, two types of polymers are mainly considered due to their electro-active properties: (a) electron conducting polymers (CPs) and (b) ion CPs. Electron CPs are composed of organic monomers having conjugated double bonds and exhibit significant conductivity compared to conventional polymers (Das & Prusty 2012). Polyaniline (PANI), polythiophene (PTh) and polypyrrole (PPy) are examples of such polymers that are explored most for various applications together with energy storage (Otero & Cantero 1999, Abdelhamid et al. 2015, Han et al. 2007, Shown et al. 2015, Wang et al. 2016, Singh et al. 2022, Shah et al. 2022), whereas ion CPs or polyelectrolytes generally do not have conjugated structure but show conductivity due to presence of ions in the pristine form in its ionic structure or ions are externally added (Aziz et al. 2018). The good electrical properties of these polymeric materials in combination with other properties such as cheap, mechanical flexibility, thermal stability, light weight and good processability possessed by conventional polymers make them attractive candidates for energy storage applications (Armand 1990). Nanomaterials are integrated into the polymer matrix for further improvement of their properties to make them more suitable, with the potential as material for energy storage devices. Per the literature, different nanomaterials like metal, transition metal oxides, carbon, carbon nanotubes (CNTs), graphene (GN), ferromagnetic materials, transition-metal chalcogenides and dendrimers have been used as fillers (Yang et al. 2017, Abdelhamid et al. 2015, Shown et al. 2015, Wang et al. 2016, Bendrea et al. 2011, Snook et al. 2011, Yin et al. 2012, Zhang & Samorì 2017, Kuila 2020). Carbon nanomaterials due to their unique structural, chemical, mechanical, thermal, optical and electronic properties have attracted tremendous research attention as nanofillers for the preparation of polymer nanocomposites for a wide range of applications (Liu & Kumar 2014, Zhang & Samorì 2017, Oueiny et al. 2014). In the last decade, significant progress was made, resulting in the opening of new possibilities in the use of these nanocomposites for a variety of applications. Specially, graphene has been considered one of the most exciting materials as a rising star on the horizon of materials science (Geim & Novoselov 2007). GN, which is a single layer of carbon atoms with two-dimensional atomic crystal structure, possesses very high mechanical strength (Young's modulus: 1 TPa; intrinsic strength: 130 GPa), high electronic conductivity (room temperature electron mobility: 2.5×105 cm^2 V^{-1}s^{-1}) and thermal conductivity (>3,000 W/mK), impermeability to gases and many other superior properties, all of which make it extremely attractive for making novel polymer nanocomposites (Melechko et al. 2005). As the overall performance of the composite materials is highly dependent on the dispersion of nanomaterials in a polymer matrix, therefore, a homogeneous dispersion is an important issue and must be taken care of during polymer nanocomposite preparation (Mittal et al. 2015, Gupta & Price 2016).

6.3   PREPARATION OF CONDUCTIVE POLYMER NANOCOMPOSITES

Hybridization of nanomaterials (NM) with polymer requires prior knowledge of the surface chemistry of the nanomaterials, functionality of the polymer and methods that will promote polymer–NM interaction. Most generally, two strategies are adopted to prepare conductive polymer nanocomposites (CPNCs), either by attaching CPs to the surface of NMs through covalent bonding or noncovalent adsorption of the polymer to the NMs surface through different interactions like π–π, hydrophobic or electrostatic interaction. In the case of a noncovalent strategy, polymer nanocomposites may be prepared using different ways: (1) Direct mixing/blending: the polymer is dissolved in a suitable solvent and then nanofiller dispersion or solution is added to the polymer solution under constant stirring (Li et al. 2011). (2) In situ polymerization: in this method, surface-modified nanofillers are added to the monomer solution and polymerization is completed by adding an oxidant or an initiator under a stirring condition. The nanocomposite prepared using this strategy results better nanofiller–polymer interaction, better dispersion of the nanofiller in polymer matrix and thus enhancing properties (Mallick et al. 2005, Wei et al. 2013, Wu et al. 2013a). (3) In situ nanoparticle formation in the presence of polymer: In this strategy, the nanofiller precursor is added to a polymer solution, followed by electrochemical or thermal treatment to prepare CPNC. Using this strategy one can generate uniform nanostructures of various morphologies inside the polymer matrix (Rastogi et al. 2014). Composite materials of CPs with various nanomaterials like CNTs (Oueiny et al. 2014), GN (Zhang & Samorì 2017, Hong et al. 2019), metal nanoparticles (Rhazi et al. 2018, Sih & Wolf 2005) and metal oxide ((Zarrintaj et al. 2019) have been prepared via noncovalent approach with an aim of their applications for energy storage. In a covalent approach, the first method includes polymer chains with suitable functionality that are reacted with surface functionalized nanomaterials via various chemical reactions like click reaction (Arslan & Tasdelen 2017) and esterification reaction (Kuila et al. 2010, Yang et al. 2014; grafting to polymers with reacting groups are chemically attached to nanomaterials). In the second method, initiator or monomer is covalently attached on the nanomaterial surface from which polymer chains are initiated through polymerization method like oxidative polymerization (grafting from approach; Kumar et al. 2012). Covalent modification yields more stable and robust composite materials compared to noncovalent approach (Kumar et al. 2012, Zhang et al. 2011, Quintana et al. 2013). Up to date, several reviews have been published describing detailed synthesis, properties and applications of the polymer composites (Rhazi et al. 2018, Kumar et al. 2018, Oueiny et al. 2014, Utracki et al. 2007, Yang et al. 2015), which may be worth reading.

6.4   APPLICATIONS OF POLYMER NANOCOMPOSITES FOR ENERGY STORAGE DEVICES

6.4.1   SUPERCAPACITORS

Supercapacitors are one type of energy storage device that provides higher power output (~10 Kw/kg) at the cost of low energy density (~5–10 Wh kg^{-1}) (Donne 2013). They are mostly suitable for applications where a large amount of electric energy is required in a very short span of time like military applications, powering remote areas, satellite launches and hybrid electric cars (Conway 1999). Supercapacitors also show very low charging and discharging times due to their storage mechanism, which is physical in nature and does not involve any slow chemical reaction. Compared to batteries, which can deliver power for a maximum of 1000 cycles, supercapacitors can keep up its performance for millions of charging/discharging cycles because of reversible charging and discharging process without any irreversible chemical reaction (Faradaic process) at the electrodes (Yan et al. 2014). A supercapacitor comprises two electrodes separated by a dielectric electrolyte. Depending on charge storage mechanism, supercapacitors can be classified as (a) electrochemical double-layer (EDL) supercapacitors and (b) pseudosupercapacitors. The working principle of two types of supercapacitors is shown in Figure 6.1. In EDL supercapacitor, the charge is stored electrostatically at the electrode–electrolyte interface through a reversible physical adsorption/desorption process of ions from the electrolyte. Since the charge is stored at the interface, so electrode with high surface area per unit mass and high conductivity is highly desirable for achieving high capacitance and energy density value (Simon & Gogotsi 2008). Other criteria are porous structures of the electrode materials with high porosity, high surface area and pore size distribution between 0.5 and 2 nm. Activated carbon due to its feasibility of large-scale production, low cost, high conductivity and mechanical and thermal stability, has been commercially used as electrode materials for EDL supercapacitors (Lu et al. 2013). Due to certain limitations of activated carbon like large pore-size distribution, higher percentage of micropores and few mesopores, the capacitance (100–120 Fg^{-1}) and energy density value (4–5 Wh kg^{-1}) remain low (Chen et al. 2017). So other nanostructured carbon materials like CNTs, GN, GO and carbon dots have been widely explored in the past decade as electrode materials for supercapacitor applications. These nanomaterials show exciting results and possibilities for their use as electrode materials along with their certain advantages and disadvantages as summarized in recent review articles (Shown et al. 2015, Lu et al. 2013, Chen & Dai 2014, Meng et al. 2017, Zhang & Zhao 2009). CPs are very exciting candidates for pseudocapacitors due to their high

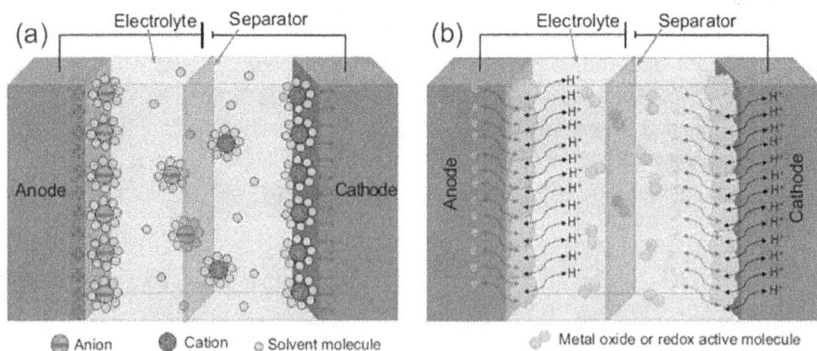

FIGURE 6.1   Schematic representation of (a) an EDLC and (b) a pseudocapacitor.

Source: From Chen, X. et al. *Natl. Sci. Rev. 4*, 453–489, 2017 with permission

theoretical capacitance value, good electrical conductivity, easy production at a large scale, low cost and usable for flexible device (Shown et al. 2015). But these polymers in pristine form show low cycle life because of the instability arises from expansion/contraction during the charging/discharging process (Shown et al. 2015) and cannot be utilized as electrode active material alone. The fabrication of nanocomposite is an elegant way to improve the electrochemical performance and stabilities of these CPs via synergistic effect. Researchers have explored a wide range of composite materials of CPs with other active nanomaterials, including carbon materials and metal oxides, and the progress of which is reviewed here.

6.4.1.1 CP/Carbon Nanostructured Composite Materials

Various CPs like PANI, PTh and PPy are integrated with carbon nanostructured materials with the aim of supercapacitor applications and significant progress has been made in designing new materials, preparation and their performance study. Among the CPs, PANI nanocomposite materials are one of the most extensively studied materials for supercapacitor applications (Meng et al. 2017). Different types of nanostructured materials like GN (Gomez et al. 2011, Wang et al. 2015), GO or reduced graphene oxide (rGO; Zhang & Samorì 2017, Hong et al. 2019, Huang et al. 2018) CNTs (Oueiny et al. 2014), carbon nanofibers (Yanilmaz et al. 2019), carbon spheres (Shen et al. 2015) and carbon nanoparticles (Khosrozadeh et al. 2015) are hybridized with PANI and supercapacitor performance of these hybrid materials have been studied. Gupta et al. first proposed that PANI/single-walled CNT (SWCNT) composite material can be successfully used as electrode material for supercapacitor application, and the highest capacitance value reported was $463Fg^{-1}$ was for 27wt% CNT (Gupta & Miura 2006). Mi et al. prepared PANI/MWCNT composite through a simple microwave-assisted polymerization method, and the resultant composite showed a capacitance value of 322 F/g and a high specific energy density of 22 W h/kg, which is approximately 12 times higher than pristine multiwalled CNTs (MWCNTs; Mi et al. 2007). Another CP that can be a composite with CNTs for supercapacitor applications that has been extensively studied is PPy (Kandasamy & Kandasamy 2018). Jurewicz et al. synthesized PPy/CNT nanocomposite by electrochemical polymerization of pyrrole, which showed a specific capacitance of 163 Fg^{-1} higher than pristine CNT (50 Fg^{-1}) (Jurewicz et al. 2001). The nanocomposite exhibited an open entangled network structure that favors the formation of a three-dimensional electrical double layer, allowing a more effective pseudocapacitance contribution from PPy.

GN or its derivatives have been widely explored for developing CP-based composite materials with superior electrochemical performance (like high capacitance value, long cycle life, higher electrical conductivity and better electrocatalytic activity) and processability (Zhang & Samorì 2017, Hong et al. 2019, Huang et al. 2018, Kumar & Baek 2014). For example, porous polyaniline/reduced graphene oxide composite materials with molecular-level uniformity were fabricated by a simple self-assembly approach (Figure 6.2) that exhibit a high specific capacitance of 808 Fg^{-1} (5717 $mFcm^{-2}$) at a current density of 53.33 Ag^{-1} (377.4 mA cm^{-2}) with an excellent rate of performance (Wu et al. 2018). Cyclic voltammetric (CV) curves of hybrid material (Figure 6.3A) exhibited a rectangular shape superimposed with a pair of redox peaks, suggesting the coexistence of both the electrical double-layer capacitance and pseudocapacitance. The galvanostatic charge–discharge plots reveal a high specific capacitance of 824 Fg^{-1} (5830 mFcm^{-2}) is achieved at 2.22 Ag^{-1} (15.72 mAcm^{-2}; Figure 6.3B). PANI/RGO composite gel (PGG) with 81.3% PANI content

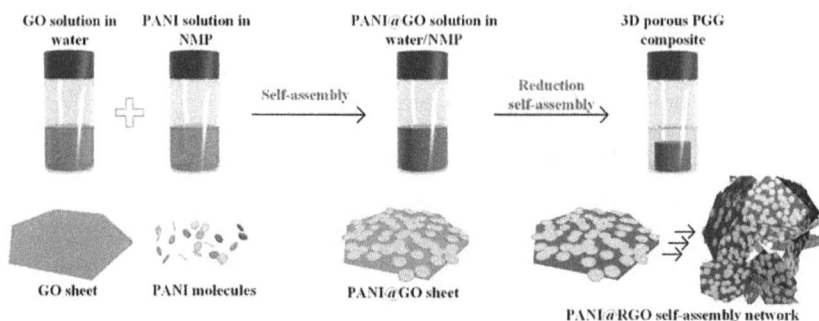

FIGURE 6.2 Schematic illustration of solution-based self-assembly method for preparation of PANI/RGO composite gel (PGG).

Source: From Wu et al. Energy Environ. Sci. *2018*, 11, 1280–1286 with permission

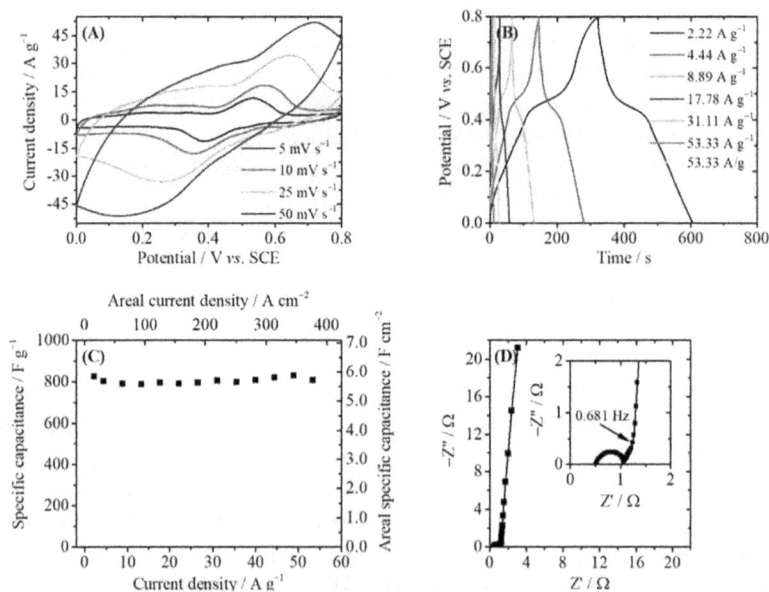

FIGURE 6.3 Capacitive performance of PGG. (A) CV curves of PGG-4 at the scan rate of 5 mV s^{-1}, 10ms^{-1}, 25 mVs^{-1} and 50 mVs^{-1}. (B) GCD curves of PGG-4 at different current densities. (C) Specific capacitances of PGG-4 versus current densities. (D) Nyquist plots of PGG-4.

Source: From Wu et al. *Energy Environ. Sci.* 2018, *11*, 1280–1286 with permission

(PGG-4) has an excellent rate of performance as shown in Figure 6.3C, even at a very high current density of 53.33 Ag^{-1} (377.4 $mAcm^{-2}$), the specific capacitance remains almost same, 808 Fg^{-1} (5717 $mFcm^{-2}$). In the low-frequency part of the Nyquist plot (Figure 6.3D), the curve tends toward a vertical line, implying that the electrode shows almost ideal capacitive behavior. Recently, Kumari et al. (2016) have described the fabrication of polyaniline graphene nanocomposite with 3D pillar structures using a simple supramolecular approach, which shows a high capacitance value of 630 Fg^{-1} at a current density 0.5 A/g with good electrochemical stability. Mini et al. (2011) fabricated PPy/GN nanocomposite by depositing nanolayers of GN acting as a template for the electropolymerization of PPy and the nanocomposite exhibited a porous 3D structure. The composite material showed very good electrochemical performance with an areal capacitance of 151 $mFcm^{-2}$ (a volume capacitance of 151 Fcm^{-3} and a specific capacitance of 1510 F g^{-1} at a scan rate of 10 mVs^{-1}) higher than the capacitance value of the individual components. A significant amount of research has also been devoted on composite material of PTh and carbon nanomaterials aiming superior supercapacitor performance (Fu et al. 2012, Zhang et al. 2014, Wang et al. 2013, Kim et al. 2015b). For example, Fu et al. (2012) prepared a PTh/MWCNT composite through electrochemical polymerization in an ionic liquid solution, and the composite exhibited a specific capacitance of the 110 F/g at the scan rate of 60 mV/s, which is higher than the individual components.

6.4.1.2   CP/Metal-Oxide Composite Materials

Metal oxides, especially transition metal oxides (such as RuO_2, NiO_2, MnO_2, V_2O_5, etc.), have been extensively studied in the last decade as rechargeable batteries and pseudocapacitor materials because of their excellent stability and redox properties. Despite such intriguing properties, poor conductivity and instability in acid electrolyte limit their rate capabilities for high power performances and a wide range of applications in energy storage and conversion (Belanger et al. 2008). To address such issues, considerable research efforts have been made to improve the electrochemical performance of metal oxide–based electrode materials by exploring composite structures with conducting materials like CPs (Liu & Lee 2008). Among them, MnO_2 is one of the mostly studied metal oxides due to its favorable characteristic properties like excellent charge storage capacity, longer cycle stability, low cost and environmental friendliness (Huang et al. 2015a, Wei et al. 2011). By hybridizing PANI into MnO_2 nanostructures, both conductivity and pseudocapacitive performance can be improved (Wang et al. 2012, Han et al. 2012). For example, Sun et al. (2015) demonstrated the synthesis of PANI-coated honeycomb-like MnO_2 nanosphere to take advantage of the synergistic effect from both PANI and MnO_2, by a two-step synthesis procedure (Figure 6.4A). Field emission scanning electron microscopy (FESEM) image (Figure 6.4B) revealed the homogeneous covering of MnO_2 nanosphere 90 nm in size with PANI. The capacitive performance of the binary composite is highly dependent on MnO_2/aniline mass ratio and composite from 1:1 ratio showed a high specific capacitance of 565 F g^{-1} at a discharge current density of 0.8 Ag^{-1} in a 0.5-M Na_2SO_4_0.5M H_2SO_4 electrolyte solution (Figure 6.4C & D). The nanocomposite retained 77% its initial capacitance after 1000 cycles at 8 A g^{-1} discharge rate (Figure 6.4E). MnO_2 nanostructures were also incorporated in other CPs like poly(3,4-ethylenedioxythiophene) (PEDOT; Tang

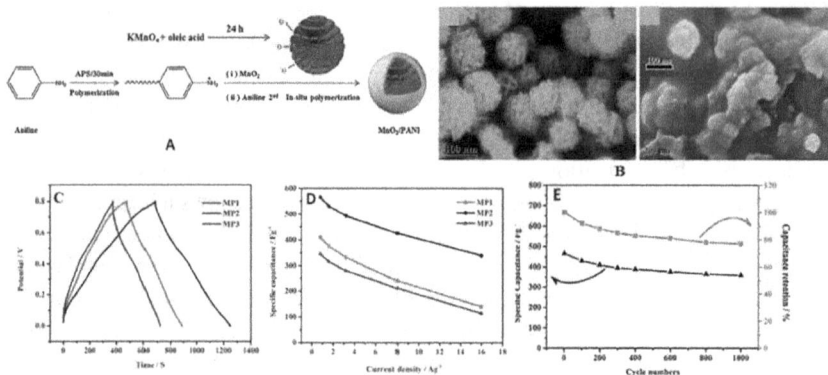

FIGURE 6.4 (A) Synthesis procedure of PANI@MnO$_2$; (B) FESEM image of the composite; (C) charge–discharge curves of different composite at current density of 0.8 A g^{-1}; (D) specific capacitances of composites at different current density (0.8–16 A g^{-1}); (E) cyclic stability of the nanocomposite.

Source: From Sun et al. *Electrochim. Acta* 2015, *180*: 977–82 with permission

et al. 2015) and PPy (Sharma et al. 2008, Huang et al. 2015a) for supercapacitor applications. Tang et al. (2015) demonstrated a simple but efficient method for the synthesis of a core–shell–branch hierarchical MnO$_2$@PEDOT@MnO$_2$ composite via an electrochemical deposition strategy. The composite material showed the maximum specific capacitance of 449 F g^{-1} at a current density of 0.5 A g^{-1} measured through galvanostatic charge/discharge test. Using in-situ polymerization technique, Zhang et al. prepared MnO$_2$-PPy composites and electrochemical study revealed better supercapacitor performance of the composite materials with a specific capacitance of 352.8 Fg^{-1} at a current density of 8 mAcm^{-2} (Zhang et al. 2013). CPs were also integrated with other nanomaterials like MoS$_2$ for supercapacitor applications. PPy/MoS$_2$ composite was prepared by growing ultra-thin layers of PPy on MoS$_2$ monolayers through in situ polymerization and utilized as efficient electrode material for a supercapacitor (Tang et al. 2015). PANI/MoS$_2$ composite materials also show very promising results as supercapacitor electrode material with a very high capacitance value of 575 Fg^{-1} at 1 Ag^{-1} (Huang et al. 2013).

6.4.1.3 CP-Based Ternary Composite

Despite tremendous progress in the area of CP-based composites for supercapacitor applications, the electrochemical performance of these materials is still below the maximum possible levels. Recently, research is more focused on ternary composites to achieve even better electrochemical performance (Chen et al. 2015, Sk et al. 2015). For achieving high electrochemical performance, the most critical issue is controlling the microstructure of the composite and the interaction between each component to maximize the synergistic effect of different types of materials. For example Mondal et al. (2017) have synthesized rGO/Fe$_3$O$_4$/PANI nanocomposite using a soft-template-based method and fabricated all solid-state supercapacitor device that exhibited capacitance of approximately 283.4 F/g at 1.0 A/g current density and a maximum energy density 47.7 Wh/kg at the power density of 550 W/kg. Recently, Khawas et al. (2017) demonstrated a facile method for preparing a PANI/MnO_2/rGO ternary hybrid with a surface decorated by ordered PANI whiskers for supercapacitor applications. The synthetic procedure involves two steps: the preparation of MnO_2-decorated rGO sheets followed by in situ polymerization of aniline. The composite materials show a well-defined morphology like tubular fiber, sphere resembling structures like natural tubular wiregrass sedge and spherical cactus. The hybrid materials show the best capacitance value of 762 F/g at current density 1.4 A/g with good electrochemical stability.

6.4.1.4 CP Composite Materials for Flexible Supercapacitors

Flexible supercapacitors, a state-of-the-art material, have come out with great potential for their use in portable and wearable electronic equipment. They follow the same fundamental standards as in conventional capacitors but give high flexibility, high charge storage capacity and low conductivity to electroactive materials to obtain high capacitance performance. CPs, which own some unique properties like high flexibility, high theoretical capacitance value, good electrical conductivity, easy production at large scale and low cost, are considered one of the most attractive candidates for flexible supercapacitor applications. There are some excellent reviews describing in detail the smart use of CPs or their composite for flexible supercapacitor applications (Shown et al. 2015, Meng et al. 2017, Han & Dai 2019, Lu et al. 2013). Cong et al. developed flexible GN–PANI composite paper by growing a PANI nanorod array on GN paper electrochemically for flexible supercapacitor application (Figure 6.5; Cong et al. 2013). The freestanding and highly conducting graphene paper was prepared by simply coating a thin layer of the mixture of GO suspensions and hydroiodic acid (HI) in a Teflon substrate followed by reduction through thermally heating. The GN–PANI composite electrodes show an increased capacitive performance of 763 Fg^{-1} at a discharge current of 1 Ag^{-1} compared to individual components (Figure 6.5b–c). Furthermore, the GN–PANI electrode exhibited good cycling stability, with 82% of capacity retention after 1000 cycles (Figure 6.5d).

6.4.2 LITHIUM-ION BATTERIES

Lithium (Li)-ion batteries are now the most popular storage devices and are widely used for power supplies in mobile phones, tablets, laptops, other portable electronic devices and hybrid vehicles. This is due to their extra advantages like high energy density (120–170 $Whkg^{-1}$), operating voltage of up to 4 V, high open-circuit potentials, minimal memory effects, fast charging, low self-discharge rates and environmental-friendliness compared with other conventional batteries (Scrosati & Garche 2010, Long et al. 2016). Figure 6.6 shows the major components of a basic Li-ion cell, which consists of a positive electrode (cathode) and a negative electrode (anode) separated by an electrolyte and a separator. The separator is generally made of a microporous polymer electrolyte, which allows only the transport of Li ions and prevents electron flow. However, the solid-state Li-ion battery uses a solid electrolyte, not a liquid

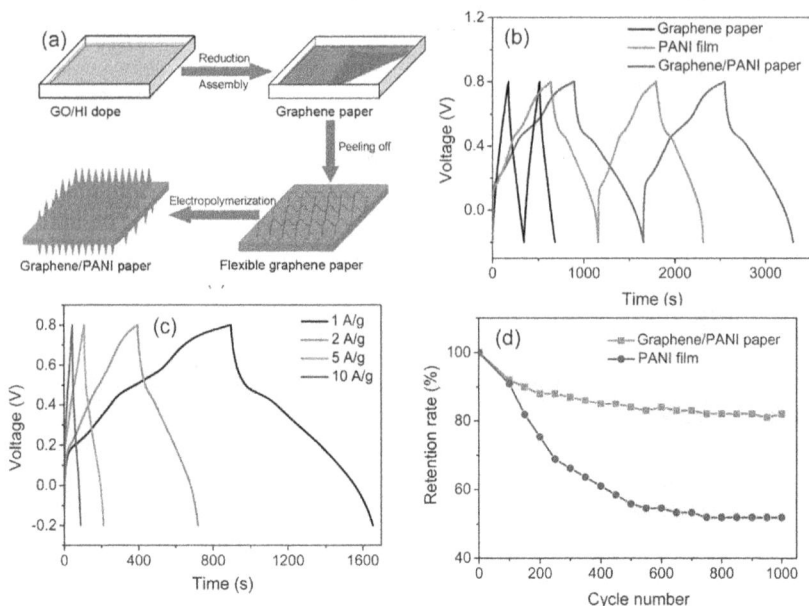

FIGURE 6.5 (a) Schematic illustrations of the formation process of GN–PANI paper; (b) GCD curves of graphene paper, PANI film on the Pt electrode and GN–PANI paper at a current density of 1 Ag^{-1}; (c) GCD curves of GN–PANI paper at different current densities; (d) cycle stability of PANI film on the Pt electrode and GN–PANI paper during the long-term charge–discharge process at a current density of 5 Ag^{-1}.

Source: From Cong et al. Energy Environ. Sci. *2013*, 6, 1185–1191 with permission

electrolyte solution, and the solid electrolyte plays a role of a separator as well. The common anode materials are carbon, silicon and metal oxides, whereas the Li salts, such as $LiCoO_2$, $LiFePO_4$ and $LiMn_2O_4$, are used as cathode material. Generally, Li salts, such as $LiClO_4$, $LiPF_6$, $LiAsF_6$ and $LiCF_3SO_3$, dissolved in nonaqueous organic solvents, such as ethylene carbonate or polyethylene carbonate, are used as the electrolyte (Arya & Sharma 2017). In the charging process, the electrodes are connected to an external electric supply. An electrochemical reaction occurs where Li$^+$ ions are generated (Figure 6.7) and further transported internally to the anode through the electrolyte. In the external circuit, electrons are forced to be transferred from the cathode to the anode. The reversible reaction continues until the full intercalation of layered anode materials with the Li$^+$ ions (Li$_x$C) complete. In the discharging process, electrons are spontaneously released from the anode to the cathode externally and the Li$^+$ ions move from the anode to the cathode internally through electrolyte. This process continues until all the Li$_x$C is fully transformed into carbon.

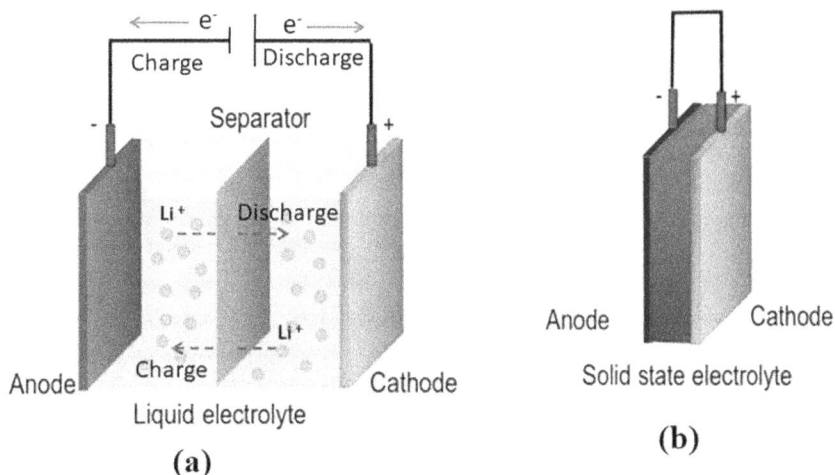

FIGURE 6.6 (a) Working principle of a Li-ion battery; (b) schematic diagram of a solid-state Li-ion battery.

Positive electrode   **M: transition metal**

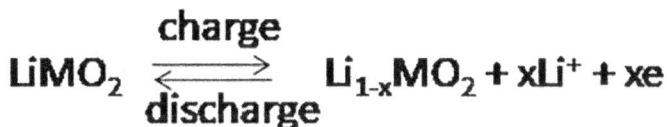

$$LiMO_2 \underset{\text{discharge}}{\overset{\text{charge}}{\rightleftharpoons}} Li_{1-x}MO_2 + xLi^+ + xe$$

Negative electrode

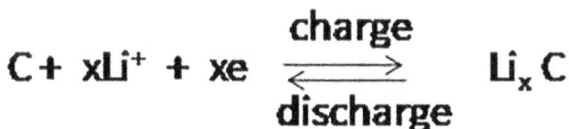

$$C + xLi^+ + xe \underset{\text{discharge}}{\overset{\text{charge}}{\rightleftharpoons}} Li_x C$$

Overall reaction

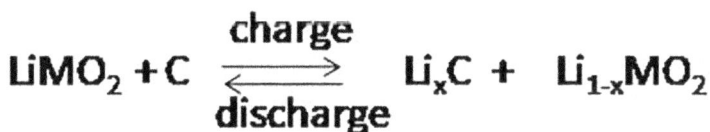

$$LiMO_2 + C \underset{\text{discharge}}{\overset{\text{charge}}{\rightleftharpoons}} Li_x C + Li_{1-x}MO_2$$

FIGURE 6.7 General electrochemical reactions in Li-ion battery.

6.4.2.1 Polymer Nanocomposite for Cathode Material in Li-Ion Batteries

An ideal cathode material should yield high free energy of reaction with Li, should intercalate a large amount of Li, should have high electronic conductivity and should be nontoxic and environmentally friendly (Deng 2015, Gong & Yang 2011). Due to certain concerns about the current existing material, $LiCoO_2$, such as high cost, structural instability, decaying capacity, the presence of toxic Co, a low Li diffusion coefficient and very low electronic conductivity, pushes scientists to explore new nanostructured materials for the cathode (Barpanda et al. 2012, Chung et al. 2012). Different inorganic materials like metal oxides have been hybridized with conducting polymer for developing novel cathode materials. For example, vanadium oxide, which has a layered structure, low cost, high theoretical capacity, ease of synthesis, and abundance and better safety, has been hybridized with conducting polymer for Li-ion battery applications (Ponzio et al. 2007, Zhao et al. 2012, Chao et al. 2014). For example, Chao et al. (2014) fabricated V_2O_5 nanoarray-based positive electrode materials by growing a V_2O_5 nanobelt array on 3D GN foam (also called ultrathin graphene foam [UGF]) followed by a thin mesoporous coating of PEDOT through electro-polymerization. SEM images of the individual materials are shown in Figure 6.8a–f. The nanocomposite electrode showed ultrafast and stable Li-ion storage performance, with higher specific capacitance (265 mAhg^{-1} at 5C) and improved rate and cycling capabilities than the bare V_2O_5 nanobelt arrays on 3D GN foam (Figure 6.8g–h). The role of PEDOT in the composite is to facilitate the charge transfer, and hence, the reaction kinetics, and protect the nanobelt array structure. Yan et al. (2015) fabricated PANI/LiVPO$_4$F composite material via the sol-gel method followed by a self-assembly. The electrochemical results concluded that PANI/LiVPO$_4$F composite exhibits superior performance with an initial discharge capacity of 149.3 mAh g^{-1} at 0.1C in the voltage range of 3.0–4.5 V, which is much higher than that of pure LiVPO$_4$F (127.5m Ahg^{-1}). A recent review by Gong and Yang (2011) and Xu (2014) will be very helpful for further understanding the recent progress in developing new materials for cathode application in Li-ion batteries. However, the development of CPs or their nanocomposite with carbon nanomaterials for cathode materials has attracted considerable research attention due to their light weight, low cost, ease of synthesis and easily tunable properties and flexibility. By using polymer nanocomposites with carbon nanomaterials as an electrode, the dream of a flexible plastic battery can only be realized. But these materials face severe difficulties in terms of chemical stability in the presence of a nonaqueous electrolyte, poor Li-ion conductivity, low energy density and low working voltage window (Novak et al. 1997). Wu et al. prepared 3,4,9,10-perylenetetracarboxylic dianhydride (PTCDA)/carbon nanotube composite (PDTA/CNT) or its corresponding polymer poly(3,4,9,10- perylenetetracarboxylic dianhydride ethylene diamine)/ carbon nanotube composite (PI/CNT) for using as cathode materials in Li-ion batteries. The composite materials showed better electrochemical performance both in terms of rate capability and capacity than pristine PTCDA or its corresponding polymer. Polymerization enhanced the cycling stability of organic cathode materials. The retention of capacity of the PI/CNT was 93% after 300 cycles under a current of 100 mA g^{-1}, while for PTCDA/CNT, the retention was only 74% after 300 cycles.

The superior electrochemical performance was related to the increased electronic conductivity of PTCDA after making a composite with CNTs and lower solubility of the polymer in a nonaqueous electrolyte compared to monomer PTCDA (Wu et al. 2013). Several polymer nanocomposites have been prepared to explore their suitability as cathode material for Li-ion batteries (Wakayama et al. 2016, Posudievsky et al. 2013a). The recent results are also summed up in some recent reviews (Pitchai et al. 2011, Xu et al. 2013, Myung et al. 2015).

6.4.2.2   Polymer Nanocomposites for Anode Material in Li-Ion Batteries

The current standardized, popular material for anode in Li-ion batteries is graphite. Despite several advantages, graphite-based electrode materials have some limitations, such as low theoretical reversible energy density, the introduction of irreversibility in intercalation/de-intercalation process due to the formation of a protective layer at solid-electrolyte interphase (SEI; An et al. 2016, Cheng et al. 2016), low diffusion rate of lithium (Roy & Srivastava 2015) and safety issues related to lithiation potential (0.2 vs Li/Li$^+$), which is close to the lithium stripping voltage (Roy & Srivastava 2015). Transitional metal oxides have been extensively explored as anode material in Li-ion batteries based on a conversion reaction mechanism that generally provides a high gravimetric specific capacity. As mentioned

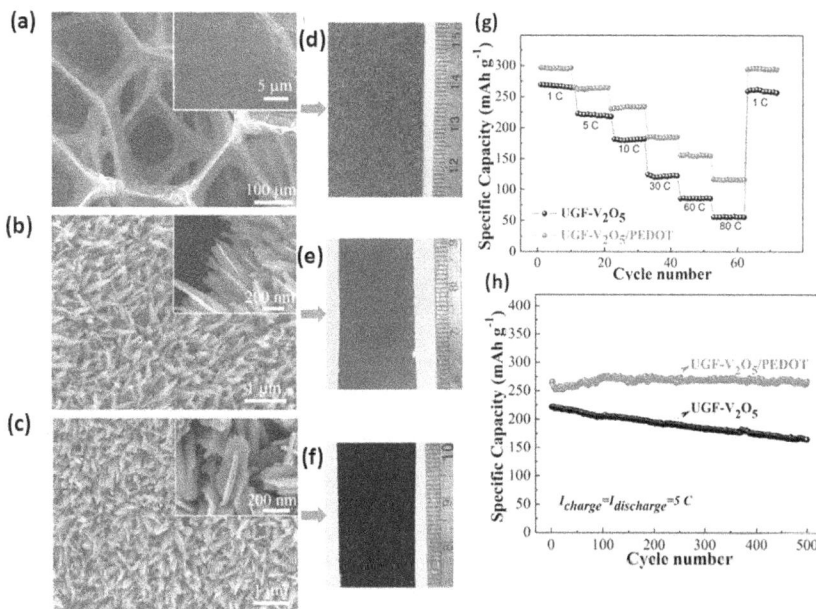

FIGURE 6.8   (a–f) SEM images and photos of UGF (a, d); UGF with V_2O_5 nanobelts (b, e); and UGF–V_2O_5/PEDOT core/shell nanobelts (c, f) (fine structures in insets). (g) Rate capability and (h) cycling performance of UGF with V_2O_5 nanobelts and UGF–V_2O_5/PEDOT core/shell nanobelts

Source: Collected from Chao et al. Adv. Mater. *2014*, 26, 5794–800 with permission

earlier, the major limitation of these inorganic materials is a low ionic conductivity that results in a limited capacity value. An elegant approach to enhance the electronic or ionic conductivity is to make a composite through the formation of a very thin film of conducting polymers (few nanometers) on the surface of the anode material (Li et al. 2013, 2017). For example, PPy-coated coaxial MWNTs@MnO$_2$ was synthesized by in situ polymerization of PPy monomers in the presence of preprepared MWNTs@MnO$_2$ (Figure 6.9A & B). As anode material, the composite exhibited a high reversible capacity of 530 mAh g^{-1} tested at 1000 mA g^{-1} even after 300 cycles and an excellent rate performance (Figure 6.9C). The better electrochemical performance of MWNTs@MnO$_2$@PPy is attributed to the synergistic effect of the MWNT matrix and the highly conductive PPy coating layer. The impedance study (Figure 6.9D) revealed that the charge transfer resistance (R$_{ct}$) value of MWCNT@MnO$_2$@PPy is lower than that of MWCNT@MnO$_2$, indicating that the PPy coating significantly improved the charge transfer resistance of the electrode. MoS$_2$, which has a layered structure, has also been explored for anode materials in Li-ion batteries. For example, a MoS$_2$/PANI composite as anode material showed a high initial capacity of 910 mAh/g and retained an impressive high capacity of 915 mAh/g even after 200 cycles at a current density of 1 A/g (Liu et al.

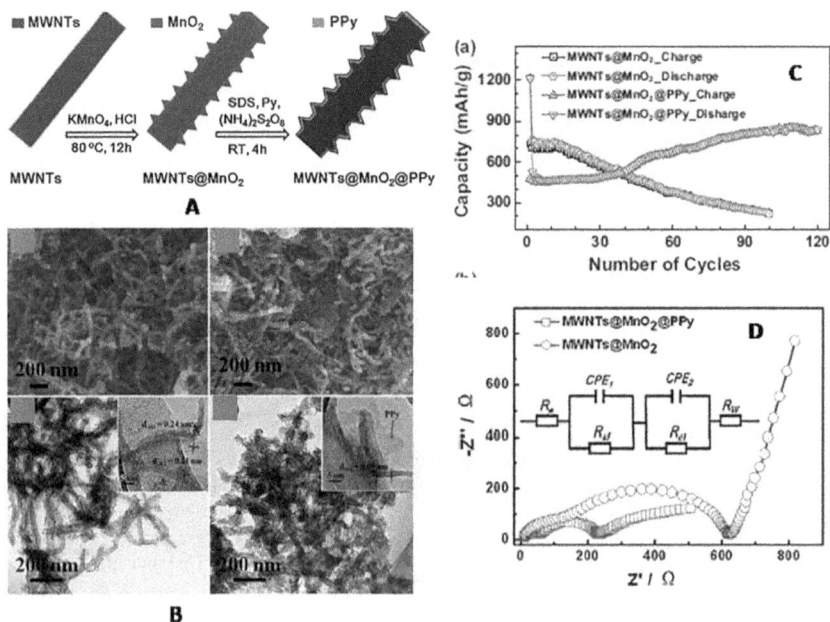

FIGURE 6.9 (A) Schematic illustration of the fabrication of MWNTs@MnO$_2$@PPy, (B) SEM image of MWNTs@MnO$_2$(upper left), MWNTs@MnO$_2$@PPy (upper right) and HR-TEM images of MWNTs@MnO$_2$ (lower left), MWNTs@MnO$_2$@PPy (lower right). (C) Cycling performance and (D) EIS profile.

Source: Collected from Li et al. Electrochim. Acta *2013*, 111, 165–71 with permission

2015). In a different study, an MoS_2/PANI composite with optimum composition (MoS_2:66.7%, polyaniline: 33.1%) exhibited a high charge capacity of 1064 mAh/g at a current density of 100 mA/g where the capacitance was retained up to 90.2% of initial capacitance value after 50 cycles (Yang et al. 2013a). These results clearly indicate the potential of conducting polymer composite materials for developing anode materials for Li-ion batteries.

6.4.3 POLYMER NANOCOMPOSITE MATERIALS AS A SOLID-STATE ELECTROLYTE

Generally, conventional energy storage devices like supercapacitors and batteries operated by liquid electrolyte, but the safety issues related to leakage of the electrolyte limits their importance (Lu et al. 2014). So the replacement of a conventional electrolyte by a solid electrolyte is necessary for development of flexible, light weight and high safety energy storage device (Lu et al. 2014). An ideal electrolyte for an energy storage device should have (1) high ionic conductivity ($>10^{-4}$ S cm^{-1} at room temperature) and very low electronic conductivity, (2) a large Li^+ ion transference number (close to unity; for Li-ion batteries), (3) electrochemical stability in a large voltage window, (4) thermal and chemical stability in the battery environment, (5) mechanical stability, and (vi) nontoxic and environmentally friendly properties (Li et al. 2016). Polymer electrolytes (macromolecules with constitutional units having ionic or ionizable groups) are considered as potential alternatives to liquid electrolyte. Poly(ethylene oxide) (PEO) and its derivatives, the most explored polymer electrolytes, can show ionic conductivity when complexed with alkali metal ions like lithium ions (Cheng et al. 2007). Despite several advantages, their room temperature ionic conductivity is still low, which limits their applications, and different measures have been taken to solve this issue. The development of a polymer electrolyte has undergone three stages (Stephan & Nahm 2006), namely, (1) a solid polymer electrolyte (SPE), (2) a gel polymer electrolyte (GPE) and (3) a polymer composite electrolyte (PCE). SPEs comprise a polymer matrix and a lithium salt and generally show very low ionic conductivity. Several attempts were made by adding different type of lithium salts (like lithium trifluoromethanesulfonate [LiTf], lithium bis(trifluoromethanesulfonimidate) [LiTFSI], lithium bis(trifluoromethanesulfonimide) [LiBETI], lithium perchlorate [$LiClO_4$] and lithium bis(oxalato)borate [LiBOB]) to improve the ionic conductivity of SPE, but it still remains low ($<10^{-4}$ Scm^{-1} at room temperature; Xu 2014). A polymer nanocomposite approach of adding different nanofillers like TiO_2, SiO_2, Fe_3O_4, Al_2O_3 and $SeZrO_2$ to polymer electrolytes results in an increase in conductivity and mechanical and electrochemical properties (Schaefer et al. 2011). For example, poly(methacrylate) (PMA)/poly(ethylene glycol) (PEG)-$LiClO_4$_3 wt% SiO_2 composite polymer electrolyte showed the best ionic conductivity of 0.26 mS cm^{-1} at room temperature (Zhu et al. 2014). The addition of inorganic filler can enhance the conductivity of the polymer composite electrolyte by two ways: (1) decrease the overall crystallinity of the polymer matrix and (2) create additional Li^+ ion-conducting pathways through the filler surface regions (Croce et al. 2001). GPEs are more promising and practical compared to polymer electrolytes as they show higher ionic conductivity (as high as $10^{-3} S cm^{-1}$)

compared to SPEs (Alipoori et al. 2020). In GPEs, in addition to a polymer electrolyte and an ionic compound (metal salt or acid), an external solvent or plasticizer is added. The solvent or the plasticizer can interact to polymer chains either physically or chemically and improves the mobility of ions and hence improve the conductivity without much sacrifice of mechanical strength. Since then, a number of polymers like poly(vinylidene fluoride) (PVdF) polyacrylonitrile (PAN), poly(ethylene oxide) (PEO), poly(methyl methacrylate) (PMMA), thermoplastic polyurethane (TPU), poly(propylene carbonate) (PPC) and polyvinyl alcohol have been used for polymer gel electrolyte preparation (Suthanthiraraj & Johnsi 2017). Recently, different inorganic materials like TiO_2 (Lim et al. 2014), SiO_2 (Gao et al. 2013), Sb_2O_3 (Chen & Dai 2014) and carbon nanomaterial like graphene oxide (GO) (Huang et al. 2014) were introduced into GPEs for further enhancement of the ionic conductivity, mechanical property and environmental stability. It has been observed that the effect of these additional nanofillers on ionic conductivity of the composite gel electrolyte is highly dependent on the surface functionality and specific surface area of these fillers (Yang et al. 2013b). For example, Gao and coworkers investigated the effect of nano-SiO_2 and nano-TiO_2 fillers on different properties like ionic conductivity, dielectric properties, proton density, ionic mobility and environmental stability of silicotungstic acid—H_3PO_4—poly (vinyl alcohol) polymer gel electrolyte (Gao & Lian 2013). The SiO_2 fillers based on gel electrolyte showed higher moisture content, most stable proton transport network and higher proton conductivity. Between the two fillers, SiO_2 addition results in a reduction in the electrode polarization and a lower double-layer capacitance in the supercapacitor. However, TiO_2 exhibited the highest intrinsic dielectric constant and the highest capacitance. Lim et al. fabricated a composite gel electrolyte based on PVA-$LiClO_4$-TiO_2 and investigated the effect of TiO_2 addition on the electrochemical performance of the supercapacitors with activated carbon electrodes (Lim et al. 2014). The composite electrolyte showed a maximum ionic conductivity value of 1.30×10^{-4} S.cm^{-1} with 8 wt% of TiO_2 content. TiO_2 nanoparticles by their acid-base properties and surface oxygen atoms, which act as vacant coordination sites, increase the amorphous nature of PVA and increase the number of mobile Li$^+$ ions. Recently, Zhai et al. developed a series of novel nanocomposite polymer gel electrolytes consisting of Poly(vinylidene fluoride-co-hexafluoropropylene) (PVDF-HFP), ionic liquid of N-methyl-N-methoxyethyl piperidiniumbis(trifluoromethanesulfonyl)imide ($PP_{1201}TFSI$) as a plasticizer and nano-sized TiO_2 as additive (Zhai et al. 2016). The nano-composite polymer electrolyte NCPE-2, with a composition of 10% TiO_2 to PVDF-HFP, showed the best ionic conductivity of 2×10^{-3} S·cm^{-1} with the lowest crystallinity. The LiFePO$_4$/NCPE-2/Li coin-typed cell exhibited a first discharge capacity of 127.9 mAh g^{-1} at 0.5C and retained 97.6% of its initial value after 50th cycle.

6.4.4 POLYMER NANOCOMPOSITE MEMBRANE FOR FUEL CELLS

Fuel cells are electrochemical cells that continuously convert chemical energy of fuels (hydrogen, methanol, ethanol, etc.) and oxidizing agents (often oxygen) into electrical energy through pair of redox reaction (Kim et al. 2015a). A fuel

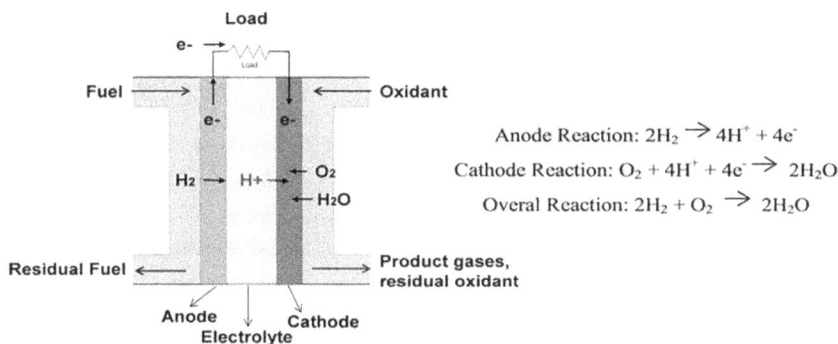

Anode Reaction: $2H_2 \rightarrow 4H^+ + 4e^-$
Cathode Reaction: $O_2 + 4H^+ + 4e^- \rightarrow 2H_2O$
Overal Reaction: $2H_2 + O_2 \rightarrow 2H_2O$

FIGURE 6.10 Basic principle of PEMFC.

cell comprises a cathode and an anode chamber separated by an electrolyte (Figure 6.10). There are six kinds of fuel cells classified on the basis of their mode of operation, operating temperature, and nature of the electrolyte: (1) alkaline fuel cells (AFCs), (2) proton exchange membrane fuel cells (PEMFCs), (3) direct methanol fuel cells (DMFCs), (4) solid oxide fuel cells (SOFCs), (5) molten carbonate fuel cells (MCFCs) and (6) phosphoric acid fuel cells (PACFs). AFCs, PEMFCs and DMFCs use polymer membranes as electrolytes, which is called a polymer electrolyte membrane. Here we mainly discuss the recent development of polymer nanocomposite aimed at their use as polymer electrolyte membranes in fuel cells. The basic principle of PEMFCs is depicted in Figure 6.10 along with electrode reactions. In PEMFCs, hydrogen is used as fuel, and the charge carrier is the hydrogen ion (proton). In the presence of a catalyst at the anode, the supplied hydrogen gas split into electrons and protons through a chemical reaction. The protons then are transported to the cathode through the polymer electrolyte, while the electrons proceed along the external circuit and generate electrical power. Oxygen is used as a common oxidant that combines with the proton and electron to produce water (Figure 6.10; Kim et al. 2015a).

6.4.4.1 Polymer Nanocomposite as Polymer Electrolyte Membranes

An ideal polymer electrolyte membrane should exhibit high ionic conductivity (i.e., proton conductivity more than 10^{-2} S/cm) between the electrodes, superior mechanical stabilities in terms of strength and flexibility, zero electronic conductivity, low gas permeability, low production cost and chemical inertness toward oxidation and reduction (Sebastian & Baglio 2017). Polymer electrolyte membranes also require humidification for continuous ion conduction, which sometimes results in membrane degradation and a loss of cell efficiency (Ji & Wei 2009). The most common and commercially available polymer membrane material for PEMFC is polymer of fluorosulfonic acid ionomers developed by DuPont with the trade name Nafion. The major drawbacks with Nafion is that it has a limited dehydration temperature below 100°C, which can cause slow kinetic reactions, have poor cost-efficiency and cause CO poisoning of the platinum electrode catalysts. The performance of a Nafion membrane in fuel cells is optimized at a relative humidity approaching 100% and an operating temperature of approximately 80°C (Sahu et al. 2009). Different types of polymers like sulfonated poly(ether ether ketone), poly(ether sulfone) and poly(benzimidazole) have been explored in the literature as possible replacements for the Nafion membranes but not much success (Tohidian et al. 2015). To enhance the performance of these proton-conducting polymer electrolytes, nanocomposite of these polymers were prepared by hybridizing with different nanofillers of controllable structure and superior properties (Chen et al. 2012, Koyilapu et al. 2020, 2021, Mukhopadhyay et al. 2020, Kim et al. 2015a). The properties of polymer composite membranes are improved by the synergy between the organic and inorganic material properties. The inorganic component in the polymer matrix can suppress the fuel diffusion and membrane swelling with a simultaneous improvement of mechanical strength and thermal stability of the electrolyte membrane. Already, it has been reported that the addition of nanofillers like inorganic particles, such as metal oxide (Adjemian et al. 2006), silica (Jiang et al. 2006), clay (Koyilapu et al. 2021, Singha et al. 2019), heteropolyacids (Ramani et al. 2006), graphene and graphene oxide (Sahu et al. 2016, Mahmood et al. 2014) and CNTs (Cele et al. 2010), have improved the mechanical, gas barrier and thermal stability of a polymer electrolyte membrane.

6.5   CONCLUSION AND FUTURE PROSPECTIVE

In this chapter, we have tried to summarize the latest development on the area of polymer nanocomposite-based materials for energy storage applications. The dream of light weight, flexible energy storage device for modern and wearable electronics can only be achieved by using polymer nanocomposite as electrode and electrolyte materials. Significant research efforts were made to develop novel polymer nanocomposite-based electrode and electrolyte materials for energy storage applications. Despite many successes, a few challenges are still ahead and need to be tackled in the future. At present, energy storage devices like supercapacitors suffer from lower energy densities, which can be addressed by developing novel electrolytes (especially solid electrolytes for flexible devices and organic electrolytes) with high ionic conductivity and larger electrochemical potential windows. For example, hybrid energy storage devices like lithium/sodium-ion hybrid capacitors comprising both battery-type and capacitor-type electrodes can output high energy and power density. Another major issue with polymer-based materials is the low cycle life of the device, which may be addressed by using a redox-active electrolyte that will promote fast and reversible charge-transfer process between the electrode and the electrolyte and simultaneously lower the extent of unfavorable and irreversible redox processes. The synthetic procedure adopted plays a key role in determining the final structure/morphology of the composite and the dispersion of the filler in the polymer matrix. Despite several established procedures, novel preparation strategies that ensure large-scale production of composite materials, new and reproducible structure/morphology, better interface and degree of dispersion of nanofiller are still needed. A possible synthetic strategy will be to build 3D hybrids with tunable/controlled structures having large specific surface area and good permeability. More theoretical and experimental studies that bring a deeper understanding of the mechanism of energy storage, the interfacial interactions between the electrolyte and the electrode and the structure property relationship are highly required. At the end of the day, a closer collaboration between laboratory research and industry is highly anticipated to solve the challenges and fulfill the ultimate goal of large-scale fabrication of polymer composite–based, efficient, environmentally friendly and renewable storage devices for solving the problems of our daily life.

ACKNOWLEDGEMENTS

Dr. BKK also acknowledges DST SERB project CRG/2020/003132 and BHU IOE project R/Dev/D/IOE/incentive/ 2021–2022/32409 for financial support.

REFERENCES

Abdelhamid, M. E., O'Mullane, A. P., Snook, G. A. 2015. Storing energy in plastics: A review on conducting polymers & their role in electrochemical energy storage. RSC Advances **5**(15):11611–26.

Adjemian, K. T., Dominey, R., Krishnan, L., Ota, H., Majsztrik, P., Zhang, T., et al. 2006. Function and characterization of metal oxide-Nafion composite membranes for elevated-temperature H_2/O_2 PEM fuel cells. Chemistry of Materials **18**:2238–48.

Alipoori, S., Mazinani, S., Hamed, S., Aboutalebi, Sharif, F. 2020. Review of PVA-based gel polymer electrolytes in flexible solid-state supercapacitors: Opportunities and challenges. Journal of Energy Storage **27**:101072.

An, S. J., Li, J., Daniel, C., Mohanty, D., Nagpure, S., Wood, D. L. 2016. The state of understanding of the lithium-ion-battery graphite solid electrolyte interphase (SEI) and its relationship to formation cycling. Carbon **105**:52–76.

Armand, M. 1990. Polymers with ionic conductivity. Advance Materials **2**:278–86.

Arslan, M., Tasdelen, M. A. 2017. Polymer nanocomposites via click chemistry reactions. Polymers **9**:499.

Arya, A., Sharma, A. L. 2017. Polymer electrolytes for lithium ion batteries: A critical study. Ionics **23**(3):497–540.

Aziz, S. B., Woo, T. J., Kadir, M. F. Z., Ahmed, H. M. 2018. A conceptual review on polymer electrolytes and ion transport models. Journal of Science: Advanced Materials and Devices **3**:1–17.

Barpanda, P., Nishimura, S.-I., Yamada, A. 2012. High-voltage pyrophosphate cathodes. Advanced Energy Materials **2**(7):841–59.

Beaudin, M., Zareipour, H., Schellenberglabe, A., Rosehart, W. 2010. Energy storage for mitigating the variability of renewable electricity sources: An updated review. Energy for Sustainable Development **14**(4):302–14.

Belanger, D., Brousse, T., Long, J., Long, W. 2008. Manganese oxides: Battery materials make the leap to electrochemical capacitors. Electrochemical Society Interfaces **17**:49–52.

Bendrea, A. D., Cianga, L., Cianga, I. 2011. Review paper: Progress in the field of conducting polymers for tissue engineering applications. Journal of Biomaterials Application **26**:3–84.

Cele, N. P., Sinha Ray S., Pillai, S. K., Ndwandwe, M., Nonjola S, Sikhwivhilu L, et al. 2010. Carbon nanotubes based Nafion composite membranes for fuel cell applications. Fuel Cells **10**(1):64–71.

Chao, D., Xia, X., Liu, J., Fan, Z., Ng, C. F., Lin, J., et al. 2014. A V_2O_5/conductive-polymer core/shell nanobelt array on three-dimensional graphite foam: A high-rate, ultrastable, and freestanding cathode for lithium-ion batteries. Advanced Materials **26**(33):5794–800.

Chen, J., Guo, Q., Li, D., Tong, J., Li, X. 2012. Properties improvement of SPEEK based proton exchange membranes by doping of ionic liquids and Y_2O_3. Progress in Natural Science: Materials International 22(1):26–30.

Chen, X., Paul, R., Dai, L. 2017. Carbon-based supercapacitors for efficient energy storage. National Science Review 4:453–89.

Chen, Y. P., Liu, B. R., Liu, Q., Wang, J., Liu, J. Y., Zhang, H. S., Hu, S. X., Jing, X. Y. 2015. Flexible all-solid-state asymmetric supercapacitor assembled using coaxial $NiMoO_4$ nanowire arrays with chemically integrated conductive coating. Electrochimica Acta 178:429–38.

Chen, T., Dai, L. 2014. Flexible supercapacitors based on carbon nanomaterials. Journal of Material Chemistry A 2:10756–75.

Cheng, H., Zhu, C., Huang, B., Lu, M., Yang, Y. 2007. Synthesis and electrochemical characterization of PEO-based polymer electrolytes with room temperature ionic liquids. Electrochimica Acta 52(19):5789–94.

Cheng, X.-B., Zhang, R., Zhao, C.-Z., Wei, F., Zhang, J.-G., Zhang, Q. 2016. A review of solid electrolyte interphases on lithium metal anode. Advanced Science 3(3):1500213.

Chung, S.-Y., Bloking, J. T., Chiang, Y.-M. 2012. Electronically conductive phospho-olivines as lithium storage electrodes. Nature Materials 1(2):123–8.

Cong, H.-P., Ren, X.-C., Wang, P., Yu, S.-H. 2013. Flexible graphene—polyaniline composite paper for high-performance supercapacitor. Energy & Environmental Science 6:1185–91.

Conway, B. E. 1999. Introduction and historical perspective. In: Conway BE, editor. Electrochemical supercapacitors: Scientific fundamentals and technological applications. Boston, MA: Springer. pp. 1–9.

Croce, F., Persi, L., Scrosati, B., Serraino-Fiory, F., Plichta, E., Hendrickson, M. 2001. Role of the ceramic fillers in enhancing the transport properties of composite polymer electrolytes. Electrochimical Acta 46(16):2457–61.

Das T. K., Prusty, S. 2012. Review on conducting polymers and their applications. Polymer Plastics Technology and Engineering 51(14):1487–500.

Deng, D. 2015. Li-ion batteries: Basics, progress, and challenges. Energy Science & Engineering 3(5):385–418.

Donne, S. W. 2013. General principles of electrochemistry. In: Béguin F, Frąckowiak E, editors. Supercapacitors. Wiley-VCH Verlag GmbH & Co. Germany pp. 1–68.

Fu, C. P., Zhou, H. H., Liu, R., Huang, Z. Y., Chen, J. H., Kuang, Y. F. 2012. Supercapacitor based on electropolymerized polythiophene and multi-walled carbon nanotubes composites. Material Chemistry & Physics 132:596–600.

Gao, H., Lian, K. 2013. Effect of SiO_2 on silicotungstic acid-H_3PO_4-poly (vinyl alcohol) electrolyte for electrochemical supercapacitors. Journal of Electrochemical Society 160:A505–A510.

Geim, A. K., Novoselov, K. S. 2007. The rise of graphene. Nature Materials 6:183–91.

Gomez, H., Ram, M. K., Alvi, F., Villalba, P., Stefanakos, E., Kumar, A. 2011. Graphene-conducting polymer nanocomposite as novel electrode for supercapacitors. Journal of Power Sources 196:4102–28.

Gong, Z., Yang, Y. 2011. Recent advances in the research of polyanion-type cathode materials for Li-ion batteries. Energy & Environmental Sciences 4(9):3223–42.

Gupta, S., Price, C. 2016. Investigating graphene/conducting polymer hybrid layered composites as pseudocapacitors: Interplay of heterogeneous electron transfer, electric double layers and mechanical stability. Composite Part B: Engineering 105:46–59.

Gupta, V., Miura, N. 2006. Polyaniline/single-wall carbon nanotube(PANI/SWCNT) composites for high performance supercapacitors. Electrochim Acta 52:1721–6.

Han, J., Li, L., Fang, P., Guo, R. 2012. Ultrathin MnO_2 nanorods on conducting polymer nanofibers as a new class of hierarchical nanostructures for high-performance supercapacitors. Journal of Physical Chemistry C 116:5900–15907.

Han, X., Chang, C., Yuan, L., Sun, T., Sun, J. 2007. Aromatic carbonyl derivative polymers as high performance Li-ion storage materials. Advance Materials **19**:1616–21.

Han, Y., Dai, L. 2019. Conducting polymers for flexible supercapacitors. Macromolecular Chemistry & Physics **220**:800355.

Hong, X., Fu, J., Liu, Y., Li, S., Wang, X., Dong, W., Yang, S. 2019. Recent progress on graphene/polyaniline composites for high-performance supercapacitors. Materials **12**:1451.

Huang, K. J., Wang, L., Liu, Y. J., Wang, H. B., Liu, Y. M., Wang, L. L. 2013. Synthesis of polyaniline/2-dimensional graphene analog MoS_2 composites for high-performance supercapacitor. Electrochimica Acta **109**:587–94.

Huang, M., Li, F., Dong, F., Zhang, Y. X., Zhang, L. L. 2015a. MnO_2-based nanostructures for high-performance supercapacitors. Journal of Material Chemistry A **3**:21380–423.

Huang, Y., Huang, Y., Meng, W., Zhu, M., Xue, H., Lee, C.-S., Zhi, C. 2015b. Enhanced tolerance to stretch-induced performance degradation of stretchable MnO_2-based supercapacitors. ACS Applied Materials & Interfaces **7**:2569–74.

Huang, Y.-F., Wu, P.-F., Zhang, M.-Q., Ruan, W.-H., Giannelis, E. P. 2014. Boron crosslinked graphene oxide/polyvinyl alcohol nanocomposite gel electrolyte for flexible solid-state electric double layer capacitor with high performance. Electrochimica Acta **132**:103–11.

Huang, Z., Li, L., Wang, Y., Zhang, C., Liu, T. 2018. Polyaniline/graphene nanocomposites towards high-performance supercapacitors: A review. Composites Communications **8**:83–91.

Ji, M., Wei, Z. 2009. A review of water management in polymer electrolyte membrane fuel cells. Energies **2**:1057–106.

Jiang, R., Kunz, H. R., Fenton, J. M. 2006. Composite silica/Nafion® membranes prepared by tetraethylorthosilicate sol-gel reaction and solution casting for direct methanol fuel cells. Journal of Membrane Science **272**(1):116–24.

Jurewicz, K., Delpeux, S., Bertagna, V., Beguin, F., Frackowiak, E. 2001. Supercapacitors from nanotubes/ polypyrrole composites. Chemical Physics Letter **347**(1–3):36–40.

Kandasamy, S. K., Kandasamy, K. 2018. Recent advances in electrochemical performances of graphene composite (graphene-polyaniline/polypyrrole/activated carbon/ carbon nanotube) electrode materials for supercapacitor: A review. Journal of Inorganic and Organometallic Polymers and Materials **28**:559–584.

Khawas, K., Kumari, P., Daripa, S., Oraon, R., Kuila, B. K. 2017. Hierarchical polyaniline-MnO_2-reduced graphene oxide ternary nanostructures with whiskers-like polyaniline for supercapacitor application. ChemistrySelect **2**:11783–9.

Khosrozadeh, A., Xing, M., Wang, Q. 2015. A high-capacitance solid-state supercapacitor based on free-standing film of polyaniline and carbon particles. Applied Energy **153**:87–93.

Kim, D. J., Jo, M. J., Nam, S. Y. 2015a. A review of polymer—nanocomposite electrolyte membranes for fuel cell application. Journal of Industrial and Engineering Chemistry **21**:36–52.

Kim, M., Lee, C., Seo, Y. D., Cho, S., Kim, J., Lee, G., Kim, Y. K., Jang, J. 2015b. Fabrication of various conducting polymers using graphene oxide as a chemical oxidant. Chemistry of Materials **27**:6238–48.

Koyilapu, R., Singha, S., Kutcherlapati, S. N. R., Jana, T. 2020. Grafting of vinylimidazolium-type poly(ionic liquid) on silica nanoparticle through RAFT polymerization for constructing nanocomposite based PEM. Polymer **195**:122458.

Koyilapu, R., Subhadarshini, S., Singha, S., Jana, T. 2021. An in-situ RAFT polymerization technique for the preparation of poly (N-vinyl imidazole) modified Cloisite nanoclay to develop nanocomposite PEM. Polymer **212**:123175.

Kuila, B. K. 2020. Nanoheterostructured materials based on conjugated polymer and two dimensional materials: Synthesis and applications. 2D Nanoscale Heterostructured Materials **2020**:91–124.

Kuila, B. K. Park, K., Dai, L. 2010. Soluble P3HT-grafted carbon nanotubes: Synthesis and photovoltaic application. Macromolecules **43**(16):6699–705.

Kumar, A., Kumar, V., Awasthi, K. 2018. Polyaniline—carbon nanotube composites: Preparation methods, properties, and applications. Polymer-Plastics Technology and Engineering **57**:70–97.

Kumar, N. A., Baek, J.-B. 2014. Electrochemical supercapacitors from conducting polyaniline—graphene platforms. Chemical Communications **50**:6298–308.

Kumar, N. A., Choi, H.-J., Shin, Y. R., Chang, D. W., Dai, L., Baek, J.-B. 2012. Polyaniline-grafted reduced graphene oxide for efficient electrochemical supercapacitors. ACS Nano **6**:1715–23.

Kumari, P., Khawas, K., Nandy, S., Kuila, B. K. 2016. A supramolecular approach to Polyaniline graphene nanohybrid with three dimensional pillar structures for high performing electrochemical supercapacitor applications. Electrochimica Acta **190**:596–604.

Li, J., Zou, M., Zhao, Y., Lin, Y., Lai, H., Guan, L., et al. 2013. Coaxial MWNTs@ MnO_2 confined in conducting PPy for kinetically efficient and long-term lithium ion storage. Electrochimica Acta **111**:165–71.

Li, Q., Chen, J., Fan, L., Kong, X., Lu, Y. 2016. Progress in electrolytes for rechargeable Li-based batteries and beyond. Green Energy & Environment **1**(1):18–42.

Li, Y., Ye, D., Liu, W., Shi, B., Guo, R., Pei, H. et al. 2017. A three-dimensional core-shell nanostructured composite of polypyrrole wrapped MnO_2/reduced graphene oxide/carbon nanotube for high performance lithium ion batteries. Journal of Colloid and Interface Science **493**:241–8.

Li, Y., Zhu, J., Wei, S., Ryu, J., Sun, L., Guo, Z. 2011. Poly(propylene)/graphene nanoplatelet nanocomposites: Melt rheological behavior and thermal, electrical, and electronic properties. Macromolecular Chemistry and Physics **212**:1951–59.

Lim, C.-S., Teoh, K., Liew, C.-W., Ramesh, S. 2014. Capacitive behavior studies on electrical double layer capacitor using poly (vinyl alcohol)—lithium perchlorate based polymer electrolyte incorporated with TiO_2. Material Chemistry and Physics **143**:661–67.

Liu, H., Zhang, F., Li, W. Y., Zhang, X. L., Lee, C. S., Wang, W. L., Tang, Y. B. 2015. Porous tremella-like MoS_2/polyaniline hybrid composite with enhanced performance for lithium-ion battery anodes. Electrochimica Acta **167**:132–8.

Liu, R., Lee, S. B. 2008. MnO_2/poly(3,4-ethylenedioxythiophene) coaxial nanowires by one-step coelectrodeposition for electrochemical energy storage. Journal of American Chemical Society **130**(10):2942–3.

Liu, Y., Kumar, S. 2014. Polymer/carbon nanotube nanocomposite fibers—a review. ACS Applied Material Interfaces **6**:6069–87.

Long, L., Wang, S., Xiao, M., Meng, Y. 2016. Polymer electrolytes for lithium polymer batteries. Journal of Material Chemistry A **4**(26):10038–69.

Lu, P., Xue, D., Yang, H., Liu, Y. 2013. Supercapacitor and nanoscale research towards electrochemical energy storage. International Journal of Smart and Nano Materials **4**(1):2–26.

Lu, X., Yu, M., Wang, G., Tong, Y., Li, Y. 2014. Flexible Solid-State Supercapacitors: Design, Fabrication and Applications. Energy & Environmental Science **7**:2160–81.

Mahmood, N., Zhang, C., Yin, H., Hou, Y. 2014. Graphene-based nanocomposites for energy storage and conversion in lithium batteries, supercapacitors and fuel cells. Journal of Material Chemistry A **2**(1):15–32.

Mallick, K., Witcomb, M. J., Dinsmore, A., and Scurrell, M. S. 2005. Fabrication of a metal nanoparticles and polymer nanofibers composite material by an in situ chemical synthetic route. Langmuir **21**:7964–7.

Melechko, A. V., Merkulov, V. I., McKnight, T. E., Guillorn, M. A., Klein, K. L., Lowndes, D. H., Simpson, M. L. 2005. Journal of Applied Physics **97**:041301–39.

Meng, Q., Cai, K., Chen, Y., Chen, L. 2017. Research progress on conducting polymer based supercapacitor electrode materials. Nano Energy **36**:268–85.

Mi, H. Y., Zhang, X. G., An, S. Y., Ye, X. G., Yang, S. D. 2007. Microwave-assisted synthesis and electrochemical capacitance of polyaniline/multi-wall carbon nanotubes composite. Electrochemistry Communications **9**:2859–62.

Mini, P. A., Balakrishnan, A., Nair, S. V., Subramanian, K. R. V. 2011. Highly super capacitive electrodes made of graphene/poly(pyrrole). Chemical Communications **47**:5753–5.

Mittal, G., Dhand, V., Rhee, K.Y., Park, S.-J., Lee, W.R. 2015. A review on carbon nanotubes and graphene as fillers in reinforced polymer nanocomposites. Journal of Industrial and Engineering Chemistry **21**:11–25.

Mondal, S., Rana, U., Malik, S. 2017. Reduced graphene oxide/Fe$_3$O$_4$/polyaniline nanostructures as electrode materials for an all-solid-state hybrid supercapacitor, Journal of Physical Chemistry C **121**(14):7573–83.

Mukhopadhyay, S., Das, A., Jana, T., Das, S. K. 2020. Fabricating a MOF Material with polybenzimidazole into an efficient proton exchange membrane. ACS Applied Energy Materials **3**:7964.

Myung, S.-T., Amine, K., Sun, Y.-K. 2015. Nanostructured cathode materials for rechargeable lithium batteries. Journal of Power Sources **283**:219–36.

Novak, P., Muller, K., Santhanam, K. S. V., Haas O. 1997. Electrochemically active polymers for rechargeable batteries. Chemical Review **97**(1):207–82.

Otero, T. F., Cantero, I. 1999. Conducting polymers as positive electrodes in rechargeable lithiumion batteries. Journal of Power Sources **81–82**:838–41.

Oueiny, C., Berlioz, S., Perrin, F.-X. 2014. Carbon nanotube—polyaniline composites. Progress in Polymer Science **39**:707–48.

Paul, D. R., Robeson, L. M. 2008. Polymer nanotechnology: Nanocomposites. Polymer **49**(15):3187–204.

Pitchai, R., Thavasi, V., Mhaisalkar, S. G., Ramakrishna, S. 2011. Nanostructured cathode materials: A key for better performance in Li-ion batteries. Journal of Material Chemistry **21**(30):11040–51.

Ponzio, E. A., Benedetti, T. M., Torresi, R. M. 2007. Electrochemical and morphological stabilization of V$_2$O$_5$ nanofibers by the addition of polyaniline. Electrochim Acta **52**(13):4419–27.

Posudievsky, O. Y., Kozarenko, O. A., Dyadyun, V. S., Jorgensen, S. W., Spearot, J. A., Koshechko, V. G., et al. 2013. Mechanochemically prepared ternary hybrid cathode material for lithium batteries. Electrochimica Acta **109**:866–73.

Quintana, M., Vazquez, E., Prato, M. 2013. Organic functionalization of graphene in dispersions. Account of Chemical Research **46**:138–48.

Ramani, V., Kunz, H. R., Fenton, J. M. 2006. Metal dioxide supported heteropolyacid/ Nafion® composite membranes for elevated temperature/low relative humidity PEFC operation. Journal of Membrane Science **279**(1–2):506–12.

Rastogi, P. K., Ganesan, V., Krishnamoorthi, S. 2014. A promising electrochemical sensing platform based on a silver nanoparticles decorated copolymer for sensitive nitrite determination. Journal of Material Chemistry A **2**:933–43.

Rhazi, M. E., Majid, S., Elbasri, M., Salih, F. E., Oularbi, L., Lafdi, K. 2018. Recent progress in nanocomposites based on conducting polymer: Application as electrochemical sensors. International Nano Letters **8**:79–99.

Roy, P., Srivastava, S. K. 2015. Nanostructured anode materials for lithium ion batteries. Journal of Material Chemistry A **3**(6):2454–84.

Sahu, A. K., Ketpang, K., Shanmugam, S., Kwon, O., Lee, S., Kim, H. 2016. Sulfonated graphene—Nafion composite membranes for polymer electrolyte fuel cells operating under reduced relative humidity. Journal of Physical Chemistry C **120**(29):15855–66.

Sahu, A. K., Pitchumani, S., Sridhar, P., and Shukla, A. K. 2009. Nafion and modified-nafion membranes for polymer electrolyte fuel cells: An overview. Bulletin of Material Science **32**(3):285–94.

Schaefer, J. L., Lu, Y., Moganty, S. S. Agarwal, P., Jayaprakash, N., Archer, L. A. 2011. Electrolytes for high-energy lithium batteries. Applied Nanoscience **2**:91–109.

Scrosati, B., Garche, J. 2010. Lithium batteries: Status, prospects and future. Journal of Power Sources **195**(9):2419–30.

Sebastian, D., and Baglio, V. 2017. Advanced materials in polymer electrolyte fuel cells. Materials **10**:1163.

Shah, S. S., Das, H. T., Barai, H. R., Aziz, M. A. 2022, Boosting the electrochemical performance of polyaniline by one-step electrochemical deposition on nickel foam for high-performance asymmetric supercapacitor. Polymers **14**(2):270.

Sharma, R., Rastogi, A., Desu, S. 2008. Manganese oxide embedded polypyrrole nanocomposites for electrochemical supercapacitor. Electrochimica Acta **53**:7690–5.

Shen, K. W., Ran, F., Zhang, X. X., Liu, C., Wang, N. J., Niu, X. Q., Liu, Y., Zhang, D. J., Kong, L. B., Kang, L., Chen, S. W. 2015. Supercapacitor electrodes based on nano-polyaniline deposited on hollow carbon spheres derived from cross-linked co-polymer. Synthetic Metals **209**:369–76.

Shown I., Ganguly A., Chen, L.-C., Chen, K.-H. 2015. Conducting polymer-based flexible supercapacitor. Energy Science & Engineering **3**(1):2–26.

Sih, B. C., Wolf, M. O. 2005. Metal nanoparticle-conjugated polymer nanocomposites. Chemical Communication 3375–84.

Simon, P., Gogotsi, Y. 2008. Materials for electrochemical capacitors. Nature Materials **7**(11):845–54.

Singh G., Alam M., Kumar Y., Husain, S. 2022. Highly capacitive mesoporous polyaniline spheres as scalable and high electrochemical performance supercapacitor electrode. ChemistrySelect **7**(11):e202200386.

Singha, S., Koyilapu, R., Dana, K., Jana, T. 2019. Polybenzimidazole-clay nanocomposite membrane for PEM fuel cell: Effect of organomodifier structure polymer. Polymer **167**:13–20.

Sk, M. M., Yue, C. Y., Jena, R. K. 2015. Non-covalent interactions and supercapacitance of pseudo-capacitive composite electrode materials (MWCNT single bond COOH/MnO$_2$/PANI). Synthetic Metals **208**:2–12.

Snook, G. A., Kao, P., Best, A. S. 2011. Conducting-polymer-based supercapacitor devices and electrodes. Journal of Power Sources **196**(1):1–12.

Stephan A. M., Nahm, K. 2006. Review on composite polymer electrolytes for lithium batteries. Polymer **47**(16):5952–64.

Sun, X., Gan, M., Ma, L., Wang, H., Zhou, T., Wang, S., et al. 2015. Fabrication of PANI-coated honeycomb-like MnO$_2$ nanospheres with enhanced electrochemical performance for energy storage. Electrochimica Acta **180**:977–82.

Suthanthiraraj, S. A., Johnsi, M. 2017. Nanocomposite polymer electrolytes. Ionics **23**:2531–42.

Tang, H., Wang, J., Yin, H., Zhao, H., Wang, D., Tang, Z. 2015. Growth of polypyrrole ultrathin films on MoS$_2$ monolayers as high-Performance supercapacitor electrodes. Advanced Materials **27**:1117–23.

Tang, P., Han, L., Zhang, L., Wang, S., Feng, W., Xu, G., Zhang, L. 2015. Controlled construction of hierarchical nanocomposites consisting of MnO$_2$ and PEDOT for high performance supercapacitor applications. ChemElectroChem **2**:949–57.

Tohidian, M., Ghaffarian, S. R., Nouri, M., Jaafarnia, E., Haghighi, A. H. 2015. Polyelectrolyte nanocomposite membranes using imidazole-functionalized nanosilica for fuel cell applications. Journal of Macromolecular Science Part B Physics **54**(1):17–31.

Utracki, L. A., Sepehr, M., Boccaleri, E. 2007. Synthetic, layered nanoparticles for polymeric nanocomposites (PNCs). Polymers for Advance Technology **18**(1):1–37.

Wakayama, H., Yonekura, H., Kawai, Y. 2016. Three-dimensional bicontinuous nanocomposite from a self-assembled block copolymer for a high-capacity all-solid-state lithium battery cathode. Chemisstry of Material **28**(12):4453–9.

Wang, H., Lin, J., Shen, Z. X. 2016, Polyaniline (PANi) based electrode materials for energy storage and conversion. Journal of Science: Advanced Materials and Devices **1**(3):225–55.

Wang, J.-G., Yang, Y., Huang, Z.-H., Kang, F. 2012. Interfacial synthesis of mesoporous MnO2/polyaniline hollow spheres and their application in electrochemical capacitors. Journal of Power Sources **204**:236–43.

Wang, S., Ma, L., Gan, M., Fu, S., Dai, W., Zhou, T., Sun, X., Wang, H., Wang, H. 2015. Free-standing 3D graphene/polyaniline composite film electrodes for high-performance supercapacitors. Journal of Power Sources **299**:347–355.

Wang, Y.C., Tao, S.Y., An, Y.L., Wu, S., Meng, C.G. 2013. Bio-inspired high performance electrochemical supercapacitors based on conducting polymer modified coral-like mono-lithic carbon. Journal of Material Chemistry A **1**:8876–87.

Wei, H., Ding, D., Wei, S., Guo, Z. 2013. Anticorrosive conductive polyurethane multiwalled carbon nanotube nanocomposites. Journal of Material Chemistry A **1**:10805–13.

Wei, W., Cui, X., Chen, W. Ivey, D. G. 2011. Manganese oxide-based materials as electro-chemical supercapacitor electrodes. Chemical Society Review **40**:1697–721.

Wu, H., Wang, K., Meng, Y., Lu, K., Wei, Z. 2013a. An organic cathode material based on a polyimide/ CNT nanocomposite for lithium ion batteries. Journal of Material Chemistry A **1**(21):6366–72.

Wu, H., Yu, G., Pan, L., Liu, N., McDowell, M. T., Bao, Z., Cui. Y. 2013b. Stable Li-ion bat-tery anodes by in-situ polymerization of conducting hydrogel to conformally coat silicon nanoparticles. Nature Communication **4**:1943.

Wu, J., Zhang, Q., Wang, J., Huang, X., Bai, H. 2018. A self-assembly route to porous poly-aniline/reduced graphene oxide composite materials with molecular-level uniformity for high-performance supercapacitors. Energy & Environmental Science **11**:1280–6.

Xu, K. 2014. Electrolytes and interphases in Li-ion batteries and beyond. Chemical Review **114:**11503–618.

Xu, X., Lee, S., Jeong, S., Kim, Y., Cho, J. 2013. Recent progress on nanostructured 4 V cathode materials for Li-ion batteries for mobile electronics. Mater Today **16**(12):487–95.

Yan, H., Wu, X., Li, Y. 2015. Preparation and characterization of conducting polyaniline-coated LiVPO$_4$F nanocrystals with core-shell structure and its application in lithium-ion batteries. Electrochimica Acta **182**:437–44.

Yan, J., Wang, Q., Wei, T., Fan, Z. 2014. Recent advances in design and fabrication of elec-trochemical supercapacitors with high energy densities. Advanced Energy Materials **4**(4):1300816.

Yang, C., Wei, H., Guan, L., Guo, J., Wang, Y., Yan, X., et al. 2015. Polymer nanocomposites for energy storage, energy saving, and anticorrosion. Journal of Material Chemistry A **3**(29):14929–41.

Yang, J., Liu, Y., Liu, S., Li, L., Zhang, C., Liu, T. 2017. Conducting polymer composites: Material synthesis and applications in electrochemical capacitive energy storage. Material Chemistry Frontiers **1**(2):251–68.

Yang, L. C., Wang, S. N., Mao, J. J., Deng, J. W., Gao, Q. S., Tang, Y., Schmidt, O. G. 2013a. Hierarchical MoS$_2$/polyaniline nanowires with excellent electrochemical performance for lithium-ion batteries. Advanced Materials **25**:1180–4.

Yang, S., Meng, D., Sun, J., Huang, Y., Huang, Y. and Geng, J. 2014. Composite films of poly(3-hexylthiophene) grafted single-walled carbon nanotubes for electrochemical detection of metal ions. ACS Applied Material and Interfaces **6**:7686–94.

Yang, X., Zhang, F., Zhang, L., Zhang, T., Huang, Y., Chen, Y. 2013b. A high-performance graphene oxide-doped ion gel as gel polymer electrolyte for all-solid-state supercapacitor applications. Advanced Functional Materials **23**:3353–60.

Yanilmaz, M., Dirican, M., Asiri, A. M., Zhang, X. 2019. Flexible polyaniline-carbon nano-fiber supercapacitor electrodes. Journal of Energy Storage **24**:100766.

Yin, Z., Ding, Y., Zheng, Q., Guan, L. 2012. CuO/polypyrrole core—shell nanocomposites as anode materials for lithium-ion batteries. Electrochemistry Communications **20**:40–3.

Zarrintaj, P., Khalili, R., Vahabi, H., Saeb, M. R., Ganjali, M. R., Mozafari, M. 2019. Chapter 8 — Polyaniline/metal oxides nanocomposites. In: Fundamentals and emerging applications of polyaniline, 1st edition. Elsevier, 131–41.

Zhai, W., Zhang, Y.-W. Wang, L., Cai, F., Liu, X.-M., Shi, Y.-J., Yang, H. 2016. Study of nano-TiO2 composite polymer electrolyte incorporating ionic liquid $PP_{1201}TFSI$ for lithium battery. Solid State Ionics **286**:111–11.

Zhang, A. Q., Xiao, Y. H., Lu, L. Z., Wang, L. Z., Li, F. 2013. Polypyrrole/MnO_2 composites and their enhanced electrochemical capacitance. Journal of Applied Polymer Science **128**:1327–31.

Zhang, H.Q., Hu, Z.Q., Li, M., Hu, L.W., Jiao, S.Q. 2014. A high-performance supercapacitor based on a polythiophene/multiwalled carbon nanotube composite by electropolymerization in an ionic liquid microemulsion. Journal of Material Chemistry A **2**:17024–30.

Zhang, L. L., Zhao, X. S. 2009. Carbon-based materials as supercapacitor electrodes. Chemical Society Review **38**(9):2520–31.

Zhang, X., Samorì, P. 2017. Graphene/polymer nanocomposites for supercapacitors. ChemNanoMat **3**:362–72.

Zhang, X. Y., Hou, L. L., Cnossen, A., Coleman, A. C., Ivashenko, O., Rudolf, P., van Wees, B. J., Browne, W. R., Feringa, B. L. 2011. One-pot functionalization of graphene with porphyrin through cycloaddition reactions. Chemistry European Journal **17**:8957–64.

Zhao, H., Yuan, A., Liu, B., Xing, S., Wu, X., Xu, J. 2012. High cyclic performance of V_2O_5@PPy composite as cathode of recharged lithium batteries. Journal of Applied Electrochemistry **42**(3):139–44.

Zhu, Z., Hong, M., Guo, D., Shi, J., Tao, Z., Chen, J. 2014. All-solid-state lithium organic battery with composite polymer electrolyte and pillar[5]quinone cathode. Journal of American Chemical Society **136**:16461–4.

7 Carbon Quantum Dots

Karan Surana and Bhaskar Bhattacharya

7.1 INTRODUCTION

Quantum dots (QDs) in general are zero-dimensional (0D) semiconducting nano-sized entities whose optical and electronic behavior is intimately linked to their size. Controlling the size of the QDs allows us to tailor its bandgap and thereby its absorption characteristics. Owing to their quantum confinement effect and the ability to generate multiple excitons from a single photon, they are classified separately from the usual league of nanomaterials (Surana et al., 2014; Chamarro et al., 1996; Shabaev et al., 2006). The well-known and studied QDs belong to the group II–VI and III–V semiconductors, which have been exploited well for their diverse applications. Recently, a new league of carbon (C) and graphene (G) QDs or carbon nanodots (NDs) has caused a major stir in the scientific community (Surana et al., 2022; Ghaffarkhah et al., 2022). These colorful forms of carbon are being actively pursued for their strongly tunable absorbance and fluorescent characteristics. Furthermore, they exhibit excellent biocompatibility, tunable surface functionalities, and simple synthesis routes, which have further increased their popularity (Figure 7.1; Lim et al., 2015). Due to low cytotoxicity and chemical inertness compared to the chalcogenide QDs, these 0D carbons have found application in biochemical sensing, bioimaging, photovoltaics (PVs), catalysis, and drug delivery (Pan et al., 2020; Luo et al., 2013; Wang et al., 2016; Li et al., 2012; Kumar et al., 2013).

Despite the association of 0D carbon and graphene with the term *quantum dots*; their quantum confinement effect is yet to be proven (if it exists at all; Cao et al., 2019). Interestingly enough, the photoluminescence (PL) property of these NDs arises from the surface passivation or the dopant material rather than the C core. The passivation materials in turn become responsible for its dynamic characteristics. The existence of C QDs was first proposed by Xu et al. (2004), who discovered them during the purification of carbon nanotubes (CNTs), while the first true synthesis of C QDs was carried out by Sun et al. (2006) from graphite powder which showed luminescence only on surface passivation with an organic species. However, the first occurrence of fluorescent G QDs from graphene oxide (GO) was reported by Pan et al. (2010).

In this chapter, we present briefly the synthesis strategies utilized to date along with the realized applications.

7.2 SYNTHESIS OF C QDS

Similar to the synthesis of any other nanomaterial, C NDs can be prepared by either the top-down or the bottom-up approach.

7.2.1 Top-Down Approach

As the name suggests, the top-down approach entails breaking down large or bulk carbon structures such as graphene oxide, graphite, activated carbon, carbon soot, CNTs, and others by processes, such as electrochemical oxidation, arc discharge, and laser ablation. In fact, the first reported appearance of C QDs by Xu et al. (2004) involved the purification of CNTs prepared by the arc discharge method. The C QDs prepared by Li et al. (2010) in daylight and ultraviolet (UV) light has been depicted in Figure 7.2.

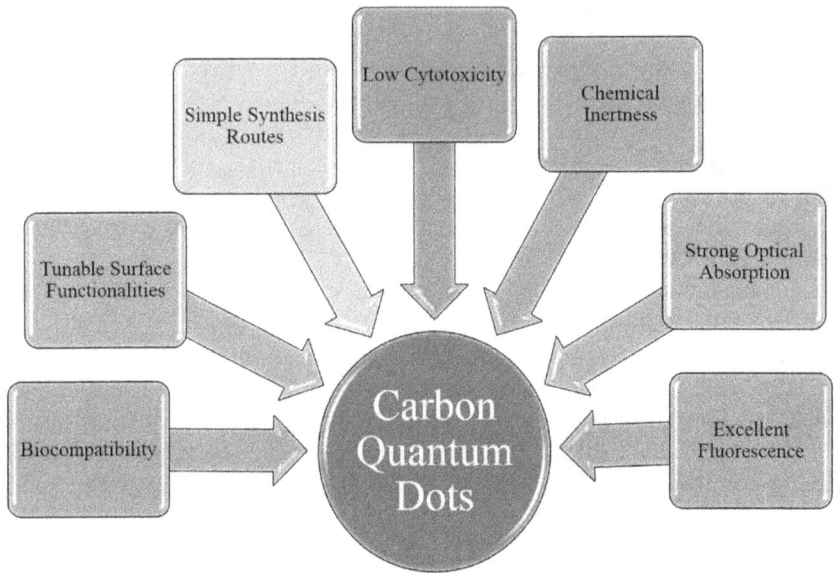

FIGURE 7.1   Properties of carbon QDs.

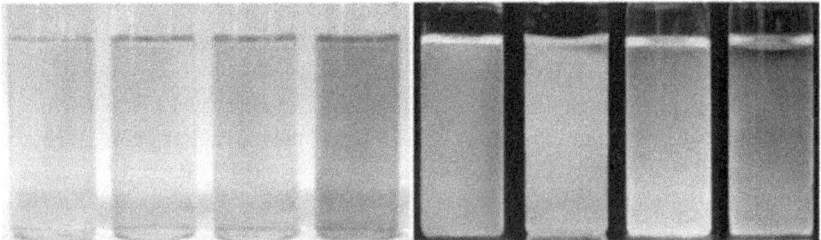

FIGURE 7.2   The optical images of C QDs under daylight (left) and UV light (right).

Source: Li et al. (2010)

7.2.2 Bottom-Up Approach

7.2.2.1 Hydrothermal/Solvothermal Route

One of the most commonly reported routes for the synthesis of colorful C NDs is the hydrothermal or solvothermal method. Since the only difference between the hydrothermal and the solvothermal method is the main solvent, hence for convenience, the name 'hydrothermal' is used in most cases. Figure 7.3 shows C QDs prepared by hydrothermal method in our laboratory. Zhang et al. (2010) reported the formation of C QDs with a single precursor, wherein ascorbic acid was mixed with water and ethanol and heated at 180 °C for 4 h in a stainless steel autoclave. The finally purified yellow-colored solution had an average particle size of approximately 2.5 nm. Liang et al. (2016) used purified chitosan powder and acetic acid for preparing C QDs by mixing and subjecting them to hydrothermal carbonization at 200 °C for 12 h. The identified average particle size was approximately 5 nm.

A number of papers have reported the use of green precursors for synthesis. For instance, Sharma et al. (2019) reported the use of rose petals as a precursor for C QDs. The extract from the petals was mixed properly with ethylenediamine and L-cysteine, followed by hydrothermal treatment at 180 °C for 5 h. Owing to the reaction time, particles of <5 nm were prepared that had a quantum yield (QY) of 9.6%. Lotus root powder as the carbon source was used by Liu et al. (2019). An aqueous solution of the powder was subjected to heating at 170 °C for 7 h. For inducing S and N dopants, sulfuric acid and ammonium hydroxide were added before hydrothermal treatment. Green, yellow, and blue luminescence was observed in undoped, S-doped, and N-doped C QDs, respectively, with an average size of approximately 5 nm. Yang et al. (2017) used an aqueous solution of strawberry powder for the hydrothermal

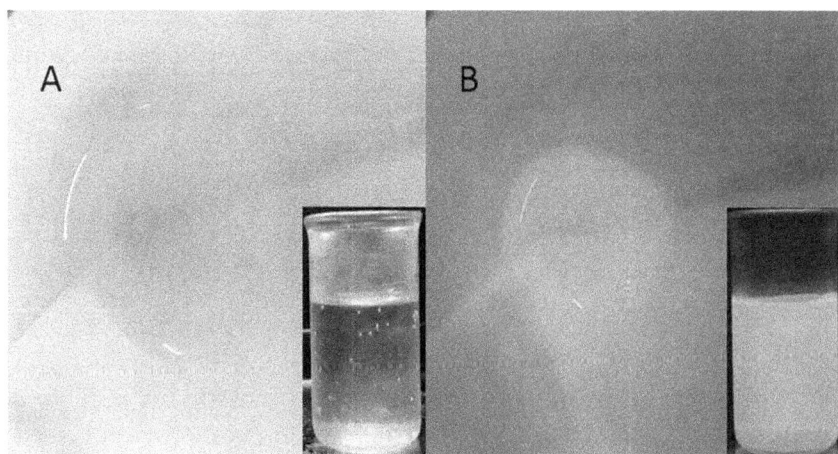

FIGURE 7.3 C QDs on a filter paper (A), daylight (B), and UV light. (Inset) C QDs in a beaker.

treatment at 170 °C for various times ranging from 3–20 h. The luminescence intensity increased gradually, with an increased reaction time, and the average particle size recorded was approximately 2.5 nm. Zhao et al. (2018) also employed strawberry powder as a carbon precursor but added ammonia for nitrogen doping. Briscoe et al. (2015) used a chitin alcoholic solution of chitin, chitosan, and glucose as precursors for the solvothermal treatment at 200 °C for 6 h. The obtained average particle size was approximately 15 nm, 8 nm, and 2.5 nm, respectively, while their QY was 11.6%, 13.4%, and 1.4%, respectively.

Guo et al. (2017) used three different precursors for preparing C dots, namely, bee pollen, citric acid, and glucose, which were mixed in water and treated at 180 °C for 4 h. The average particle size of the dots obtained was 2.4, 2.4 and 5.2 nm, respectively, which were used as sensitizers in dye sensitized solar cell (DSSC). De & Karak (2013) reported the use of banana for C QD synthesis. After separating the pulp, banana juice was mixed with ethanol and transferred to a glass bottle capped

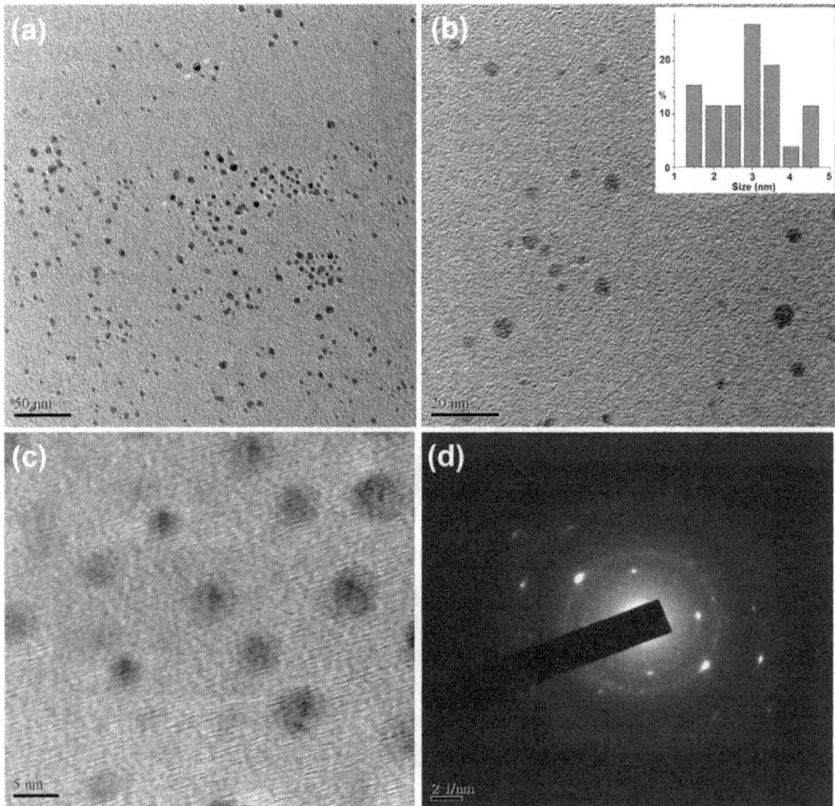

FIGURE 7.4 Transmission electron microscopy (TEM) images at different magnifications (a) 50 nm, (b) 20 nm (with size distribution, inset), and (c) 5 nm and (d) SAED pattern of the carbon dots.

Source: De & Karak (2013)

with a cotton cork for heating at 150 °C for 4 h. The obtained dark brown residue was mixed with water and ethanol for filtration and centrifuged to collect the C QDs. The change in pH did not bring about any shift in the emission wavelength, and the average particle size obtained was approximately 3 nm (Figure 7.4).

7.2.2.2 Microwave Route

The microwave route for synthesis is quite a popular method of the wet-chemical synthesis family as it prepares the end product in a noticeably short time compared to other methods. Furthermore, the ability to regulate the size and shape of the end product by modifying the reaction parameters; the ultra-fast volumetric reduction and high reaction rate aid in its popularity. In a paper by Singh et al. (2018), citric acid and urea were chosen as the carbon and nitrogen sources, respectively, and were mixed in water while maintaining pH 7. The solution was subjected to microwave irradiation at 750 W for 5 min, which resulted in yellow-colored C QDs in powder form. With absorption in the UV region, the average particle size obtained was approximately 6.5 nm. The QDs were used for selective detection of the neurotoxin quinolinic acid in human serum. This same process was initially reported by Qu et al. (2012) where a maximum QY of 40% was obtained at 515 nm. The prepared dots were used as fluorescent ink. Citric acid has been a chosen carbon source for many studies; for instance, Lim et al. (2018) prepared a homogeneous solution of citric acid with ethylenediamine for microwave irradiation at 750 W for 8 min. The dialyzed and freeze-dried C QDs had an average size of about 10 nm, a QY of 9.84%.

In a fascinating report by Zhu et al. (2009), PEG-200 and saccharides were dissolved in water. The solution was microwaved at 500 W for a few minutes, leading to the formation of yellow to dark brown C QDs. The recorded average PL lifetime of the C dots was 8.70 ns with an average particle size below 5 nm. Choi et al. (2017) reported a single-step formation of C QDs in which lysine dissolved in distilled water was pyrolyzed at 750 W for 5–6 min, resulting in the dark brown QD solution. The average PL lifetime was 1.6 ns, and the particle size ranged between 5–10 nm. A green precursor was used by Raji et al. (2019) wherein dried jackfruit seeds were powdered and mixed with o-phosphoric acid and subjected to microwave irradiation at 600 W for 1.5 min. A clear yellow solution of N-doped C QDs was obtained with a PL lifetime of 1.75 ns and an average particle size of 5 nm. Macairan et al. (2019) prepared C QD for sensing application by first sonicating glutathione in formamide for a few minutes. The resultant solution was subjected to microwave treatment at 180 °C for 5 min. Dialysis and centrifugation were carried on for days to collect the C dots. The prepared dots had an average particle size of 5–9 nm and a QY of 6.49%.

7.2.2.3 Pyrolysis Method

Pyrolysis is an effective way to decomposing a material to its constituents under heat since it is carried out in the absence of oxygen. Dager et al. (2019) heated finely powdered fennel seeds for 3 h at 500 °C, thereby carbonizing the green-colored powder to dark gray. The obtained powder was dissolved in deionized water followed by centrifugation and dialysis for obtaining the uniform-sized C QDs. Monodispersed C QDs of an average size of 3.9 nm were found, and no effect of the excitation wavelength was reported, as the QDs gave PL emission peaks at the same position. Wang et al. (2016) treated an aqueous mixture of citric acid and ammonia at 200 °C for 3 h for preparing N-doped C QDs. The average size of the QDs was 10.8 nm, and the PL peak showed red shifting, with an increase in excitation wavelength.

7.2.2.4 Wet-Chemical Method

One of the most versatile methods for synthesizing materials is the wet chemical route as it does not require any fancy equipment. Tan et al. (2016) reported the synthesis of oil-soluble C QDs whereby 1,2-Hexadecanediol was mixed with octadecene and heated at 100 °C for 30 min, followed by heating at 300 °C for 3 h under argon gas. The doping of nitrogen and boron was done by the addition of melamine and sodium borohydride. The obtained spherical QDs had an average size of about 2.4 nm.

7.3   APPLICATIONS OF C QDS

7.3.1   Chemical Sensing

One of the versatile applications of C QD lies in the arena of chemical sensing. Among the known mechanisms, one of the basic ways is to determine the change in the fluorescence (FL) intensity under the influence of any physical or chemical stimuli. C QDs have been successfully used in selectively detecting Hg^{2+} ions in aqueous solutions and live cells (Guo et al., 2013; Huang et al., 2013; Yan et al., 2014). It was reported that the FL emission of a C QD solution and C QDs immobilized in sol-gel are sensitive to the presence of Hg^{2+} (Gonçalves et al., 2010). The FL intensity of the C QD was quenched by micromolars of Hg^{2+} ions with a Stern–Volmer constant of 1.3×10^5 M^{-1}. It was hypothesized that a stable nonfluorescent complex was formed between C QDs and Hg^{2+} ions, which led to the static quenching. Sensitivity in the order of nanomolars was obtained by using N-doped C QDs (Barman & Sadhukhan, 2012).

Apart from Hg^{2+}, the detection of Cu^{2+}, Pb^{2+}, and Ag^+ has also been successfully reported (Salinas-Castillo et al., 2013; Wee et al., 2013; Qian et al., 2014). For instance, Liu et al. (2012) modified the C QDs with lysine, and bovine serum albumin was used for the selective detection of Cu^{2+} in tap water. The coordination reaction of Cu^{2+} with the carboxylic and amine group led to the sensing of Cu^{2+} ions.

7.3.2   Biosensing

C QDs have been successfully explored as fluorescent labels owing to their biocompatibility and low toxicity (Cao et al., 2007; Bhunia et al., 2013). Sun et al. (2006) presented the first report for *in vitro* and *in vivo* bioimaging (Yang et al., 2009a). They recorded the optical images of E. Coli ATCC 25922 labeled with PEGylated C QD at various excitation wavelengths. In another work by Yang et al. (2009a), an aqueous solution of PEGylated C QDs was intravenously injected into mice and the FL images were obtained at different excitation wavelengths. Adequate image contrast was obtained in both red and green emissions as shown in Figure 7.5. Similar results were obtained by Tao et al. (2012) when FL imaging was carried out at different excitation wavelengths from 455 to 704 nm. Li et al. (2011) explored the method

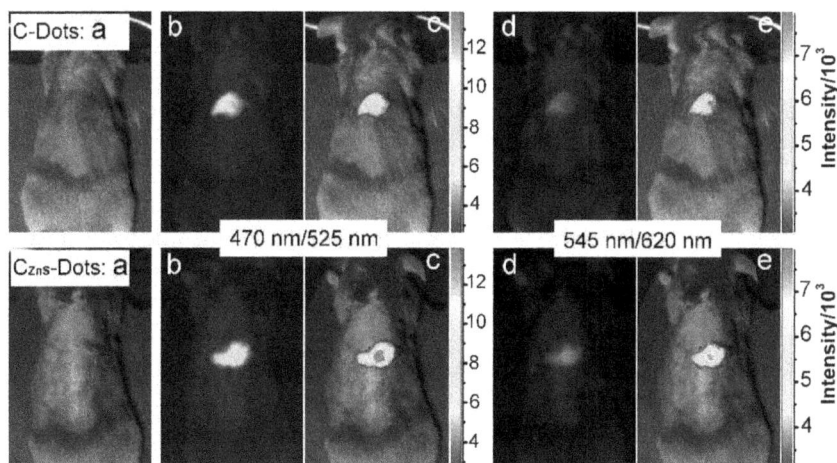

FIGURE 7.5 Intravenous injection: (a) bright field, (b, d) as-detected fluorescence (excitation/emission wavelengths indicated), and (c, e) color-coded images, as reported by Yang et al. (2009a; C QD optical imaging in vivo).

of FL quenching for the detection of nucleic acids. A high level of selectivity was obtained, such that even a single-base mismatch could be identified. Additionally, C QDs have also shown positive results as fluorescent probes for detecting small bioanalytes like antibacterial drugs.

Niu & Gao (2014) prepared N-C QDs via a one-step pyrolysis method using glutamic acid as the precursor. The prepared N-C QDs were utilized for the detection of amoxicillin, a drug commonly used for treating bacterial infections. It was observed that the amoxicillin molecules were able to isolate the N-C QDs from each other, which helped in reducing the frequency of nonradiative transitions, thereby leading to an increase in the FL intensity. The label-free detection of dopamine was demonstrated by Qu et al. (2013), who used dopamine as the carbon precursor. The FL of the quenched Fe^{3+}-CQD complex was effectively recovered by dopamine. The increment in FL was directly proportional to the increment in dopamine concentration up to a certain molar concentration with a detection limit of 68 nM.

7.3.3 BIOIMAGING

The optical properties of C QDs make them extremely desirable for biological systems, both *in vitro* and *in vivo*. Generally, the C core of C QDs is nontoxic, and any cytotoxicity, if present, is due to the surface passivation agents (Lim et al., 2015). To facilitate *in vivo* imaging, surface passivation agents having low cytotoxicity can be utilized at high concentrations. For instance, no toxic effects were traced in mice for up to 28 days when 8–40 mg.kg^{-1} of PEG-passivated C QDs were intravenously injected. The mice subjected to various doses of C QDs and the NaCl control exhibited similar levels of physiological indicators, hence confirming the nontoxic effect of C QDs at various exposure levels and times. Although the harvested organs showed no abnormalities, the amount of C QDs obtained in the liver and spleen was higher than in other organs (Yang et al., 2009b). Additionally, the cells were treated with various amounts of C QD for assessing cell viability. For C QD concentrations up to 18 mg.mL^{-1}, an average cell viability of greater than 95% was obtained, which conclusively indicates that C QD are more biocompatible than other semiconductor QDs (Baker & Baker, 2010; Bourlinos et al., 2008).

The study of Hsu et al. (2013) revealed that C QDs are commonly localized in the cytoplasm and cell membrane. C QDs bounded in an amphiphilic biocompatible polymer were injected in a Chinese hamster and were observed to settle in the cytosol after passing through the ovary cell membrane (Fowley et al., 2012). C QDs coated with b-PEI by Hu et al. (2014) were observed to have distributed uniformly through the cytoplasm. These studies suggest that the localization of C QD is strongly dependent on the utilized surface passivation agent used along with its mode.

7.3.4 PHOTOVOLTAICS

Mirtchev et al. (2012) used the dip-coating method to sensitize the TiO$_2$-coated substrate with C QDs prepared from γ-butyrolactone for 48 h. A sandwich-structured quantum dot sensitized solar cell (QDSSC) was prepared with Pt CE and iodine electrolyte, yielding an efficiency of 0.13% with a V$_{oc}$ of 0.38 V and J$_{sc}$ of 0.53 mA/cm^2. Since iodine could be a potential corrosive agent to QDs, polysulfide electrolyte might serve for better PV characteristics.

Lim et al. (2018) used C QD prepared from citric acid and ethylenediamine for depositing on polyethylenimineethoxylated-coated indium doped tin oxide (ITO) glass followed by spin-coating PTB7:PC$_{71}$BM solution and sequential thermal deposition of MoO$_3$ and Ag. The organic photovoltaic (OPV) device gave a V$_{oc}$ of 0.76 V, a J$_{sc}$ of 16.7 mA/cm^2, and a η of 8.34% (Figure 7.6). The addition of C QDs also helped in increasing the external quantum efficiency (EQE) of the device from 46.8% to

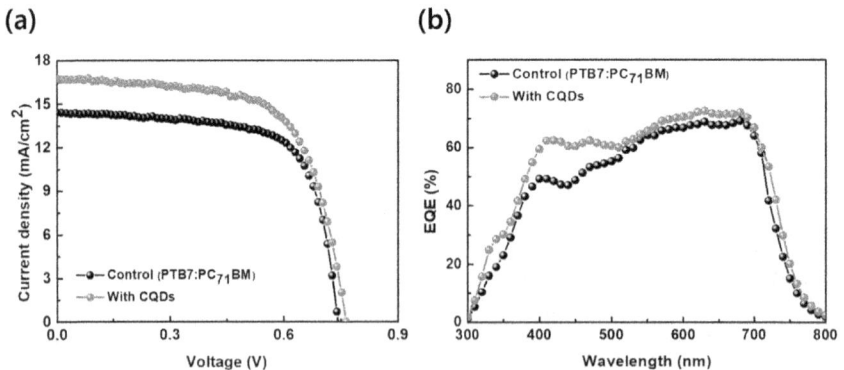

FIGURE 7.6 (a) The current density-voltage (J-V) characteristics of devices; (b) EQE spectra of devices.

Source: Lim et al. (2018)

60.9%. Chava et al. (2017) prepared green luminescent C QDs via a hydrothermal approach and used them for sensitizing ZnO photoelectrodes. A maximum η of 1.18% was achieved with iodine-based electrolyte and a sensitization time of 15 min.

The study of G QDs as part of an ionic liquid (IL) and iodide-based electrolyte was reported by Porfarzollah et al. (2020). The use of only IL or G QDs as electrolyte for conventional DSSC resulted in J_{sc} of 16–17 mA/cm^2; however, the combination of G QDs and IL rose the J_{sc} to 19.57 mA/cm^2, with a η of 4.57%.

Zou et al. (2018) reported the study of perovskite solar cells (PSCs) having C QDs embedded within the perovskite layer. The optimized addition of 0.1% C QDs led to an increment of J_{sc} to 16.40 mA/cm^2. The device without C QDs yielded an efficiency of 4.4% while that with C QDs resulted in 7.62%.

7.3.5 PHOTOCATALYSIS

As the name suggests, photocatalysis involves the acceleration of electrons generated from photons in the presence of a catalyst. It is used in disintegrating many environmental pollutants, such as the conversion of CO_2 into small molecular fuels by utilizing solar energy. Li et al. (2013) reported that the selective oxidation of alcohols to benzaldehydes with 100% selectivity and good conversion efficiency (92%) was achieved by small-sized C QDs (< 5 nm). The QDs play the role of an effective near infra-red (NIR) light-driven photocatalyst, owing to their NIR light-driven electron transfer property and superior catalytic activity for H_2O_2 decomposition. Similarly, Liu et al. (2014) demonstrated that the nanocomposite of AuNP-CQD yielded a conversion efficiency of 63.8% with a selectivity of 99.9% for the oxidation of cyclohexane to cyclohexanone under the influence of H_2O_2 as an oxidant under visible light.

The photocatalytic activities of the C QDs can be controlled by doping the QDs and by tailoring the surface passivation agents (Ma et al., 2012; Hu et al., 2013). Alternatively, the C QDs of a size of 5–10 nm were used as an acid catalyst in order to catalyze a number of organic transformations under visible light in aqueous media, owing to their light-induced photon properties in the solution phase (Li et al., 2014).

7.4 CONCLUSION

The field of C QDs has only started, and there would be many surprises on the way. The excellent absorption and FL properties of C QDs have already ensured their application in many diverse areas of science and technology. Among the various synthesis routes, the chemical route is simpler and yields nearly monodispersed QDs. Since these QDs do not contain any carcinogenic or toxic elements, C QDs are among the best choices for applications in living beings for detection. Apart from the mentioned applications, many more shall appear in the near future. The field being relatively nascent, it is open for significant exploration and needs more focused research for possible applications.

ACKNOWLEDGMENT

The financial assistance as Incentive to Senior Faculties under the Institute of Eminence Scheme (Dev. Scheme No. 6031) from Banaras Hindu University is gratefully acknowledged by the author B.B. The author K.S. is thankful to UGC, India for providing the UGC-DS Kothari Postdoc Fellowship (No. F. 4–2/2006 (BSR)/CH/20–21/0247).

REFERENCES

Baker, S. N., & Baker, G. A. (2010). Luminescent carbon nanodots: emergent nanolights, *Angewandte Chemie International Edition*, 49, 6726–6744.

Barman, S., & Sadhukhan, M. (2012). Facile bulk production of highly blue fluorescent graphitic carbon nitride quantum dots and their application as highly selective and sensitive sensors for the detection of mercuric and iodide ions in aqueous media, *Journal of Materials Chemistry*, 22, 21832–21837.

Bhunia, S. K., Saha, A., Maity, A. R., Ray, S. C., & Jana, N. R. (2013). Carbon nanoparticle-based fluorescent bioimaging probes, *Scientific Reports*, 3, 1–7.

Bourlinos, A. B., Stassinopoulos, A., Anglos, D., Zboril, R., Georgakilas, V., & Giannelis, E. P. (2008). Photoluminescent carbogenic dots, *Chemistry of Materials*, 20(14), 4539–4541.

Briscoe, J., Marinovic, A., Sevilla, M., Dunn, S., & Titirici, M. (2015). Biomass-derived carbon quantum dot sensitizers for solid-state nanostructured solar cells, *Angewandte Chemie International Edition*, 54(15), 4463–4468.

Cao, L., Shiral Fernando, K. A., Liang, W., Seilkop, A., Monica Veca, L., Sun, Y. P., & Bunker, C. E. (2019). Carbon dots for energy conversion applications, *Journal of Applied Physics*, 125(22), 220903.

Cao, L., Wang, X., Meziani, M. J., Lu, F., Wang, H., Luo, P. G., Lin, Y., Harruff, B. A., Veca, L. M., Murray, D., & Xie, S. Y. (2007). Carbon dots for multiphoton bioimaging, *Journal of the American Chemical Society*, 129(37), 11318–11319.

Chamarro, M., Gourdon, C., Lavallard, P., Lublinskaya, O., & Ekimov, A. I. (1996). Enhancement of electron-hole exchange interaction in CdSe nanocrystals: A quantum confinement effect, *Physical Review B*, 53(3), 1336.

Chava, R. K., Im, Y., & Kang, M. (2017). Nitrogen doped carbon quantum dots as a green luminescent sensitizer to functionalize ZnO nanoparticles for enhanced photovoltaic conversion devices, *Materials Research Bulletin*, 94, 399–407.

Choi, Y., Thongsai, N., Chae, A., Jo, S., Kang, E. B., Paoprasert, P., Park, S. Y., & In, I. (2017). Microwave-assisted synthesis of luminescent and biocompatible lysine-based carbon quantum dots, *Journal of Industrial and Engineering Chemistry*, 47, 329–335.

Dager, A., Uchida, T., Maekawa, T., & Tachibana, M. (2019). Synthesis and characterization of mono-disperse carbon quantum dots from fennel seeds: photoluminescence analysis using machine learning, *Scientific Reports*, 9(1), 1–12.

De, B., & Karak, N. (2013). A green and facile approach for the synthesis of water soluble fluorescent carbon dots from banana juice, *RSC Advances*, 3(22), 8286–8290.

Fowley, C., McCaughan, B., Devlin, A., Yildiz, I., Raymo, F. M., & Callan, J. F. (2012). Highly luminescent biocompatible carbon quantum dots by encapsulation with an amphiphilic polymer, *Chemical Communications*, 48(75), 9361–9363.

Ghaffarkhah, A., Hosseini, E., Kamkar, M., Sehat, A. A., Dordanihaghighi, S., Allahbakhsh, A., van der Kuur, C., & Arjmand, M. (2022). Synthesis, applications, and prospects of graphene quantum dots: a comprehensive review, *Small*, 18(2), 2102683.

Gonçalves, H. M., Duarte, A. J., & da Silva, J. C. E. (2010). Optical fiber sensor for Hg (II) based on carbon dots, *Biosensors and Bioelectronics*, 26(4), 1302–1306.

Guo, X., Zhang, H., Sun, H., Tade, M. O., & Wang, S. (2017). Green synthesis of carbon quantum dots for sensitized solar cells, *ChemPhotoChem*, 1(4), 116–119.

Guo, Y., Wang, Z., Shao, H., & Jiang, X. (2013). Hydrothermal synthesis of highly fluorescent carbon nanoparticles from sodium citrate and their use for the detection of mercury ions, *Carbon*, 52, 583.

Hsu, P. C., Chen, P. C., Ou, C. M., Chang, H. Y., & Chang, H. T. (2013). Extremely high inhibition activity of photoluminescent carbon nanodots toward cancer cells, *Journal of Materials Chemistry B*, 1(13), 1774–1781.

Hu, L., Sun, Y., Li, S., Wang, X., Hu, K., Wang, L., Liang, X. J., & Wu, Y. (2014). Multifunctional carbon dots with high quantum yield for imaging and gene delivery, *Carbon*, 67, 508–513.

Hu, S., Tian, R., Dong, Y., Yang, J., Liu, J., & Chang, Q. (2013). Modulation and effects of surface groups on photoluminescence and photocatalytic activity of carbon dots, *Nanoscale*, 5(23), 11665–11671.

Huang, H., Lv, J. J., Zhou, D. L., Bao, N., Xu, Y., Wang, A. J., & Feng, J. J. (2013). One-pot green synthesis of nitrogen-doped carbon nanoparticles as fluorescent probes for mercury ions, *RSC Advances*, 3(44), 21691–21696.

Kumar, V., Toffoli, G., & Rizzolio, F. (2013). Fluorescent carbon nanoparticles in medicine for cancer therapy, *ACS Medicinal Chemistry Letters*, 4(11), 1012–1013.

Li, H., He, X., Kang, Z., Huang, H., Liu, Y., Liu, J., Lian, S., Tsang, C. H. A., Yang, X., & Lee, S. T. (2010). Water-soluble fluorescent carbon quantum dots and photocatalyst design, *Angewandte Chemie International Edition*, 49(26), 4430–4434.

Li, H., Kang, Z., Liu, Y., & Lee, S. T. (2012). Carbon nanodots: synthesis, properties and applications, *Journal of Materials Chemistry*, 22(46), 24230–24253.

Li, H., Liu, R., Kong, W., Liu, J., Liu, Y., Zhou, L., Zhang, X., Lee, S. T., & Kang, Z. (2014). Carbon quantum dots with photo-generated proton property as efficient visible light controlled acid catalyst, *Nanoscale*, 6(2), 867–873.

Li, H., Liu, R., Lian, S., Liu, Y., Huang, H., & Kang, Z. (2013). Near-infrared light controlled photocatalytic activity of carbon quantum dots for highly selective oxidation reaction, *Nanoscale*, 5(8), 3289–3297.

Li, H., Zhang, Y., Wang, L., Tian, J., & Sun, X. (2011). Nucleic acid detection using carbon nanoparticles as a fluorescent sensing platform, *Chemical Communications*, 47(3), 961–963.

Liang, Z., Kang, M., Payne, G. F., Wang, X., & Sun, R. (2016). Probing energy and electron transfer mechanisms in fluorescence quenching of biomass carbon quantum dots, *ACS Applied Materials & Interfaces*, 8(27), 17478–17488.

Lim, H., Liu, Y., Kim, H. Y., & Son, D. I. (2018). Facile synthesis and characterization of carbon quantum dots and photovoltaic applications, *Thin Solid Films*, 660, 672–677.

Lim, S. Y., Shen, W., & Gao, Z. (2015). Carbon quantum dots and their applications, *Chemical Society Reviews*, 44(1), 362–381.

Liu, J. M., Lin, L. P., Wang, X. X., Lin, S. Q., Cai, W. L., Zhang, L. H., & Zheng, Z. Y. (2012). Highly selective and sensitive detection of Cu^{2+} with lysine enhancing bovine serum albumin modified-carbon dots fluorescent probe, *Analyst*, 137(11), 2637–2642.

Liu, L., Yu, X., Yi, Z., Chi, F., Wang, H., Yuan, Y., Li, D., Xu, K., & Zhang, X. (2019). High efficiency solar cells tailored using biomass-converted graded carbon quantum dots, *Nanoscale*, 11(32), 15083–15090.

Liu, R., Huang, H., Li, H., Liu, Y., Zhong, J., Li, Y., Zhang, S., & Kang, Z. (2014). Metal nanoparticle/carbon quantum dot composite as a photocatalyst for high-efficiency cyclohexane oxidation, *ACS Catalysis*, 4(1), 328–336.

Luo, P. G., Sahu, S., Yang, S. T., Sonkar, S. K., Wang, J., Wang, H., LeCroy, G. E., Cao, L., & Sun, Y. P. (2013). Carbon "quantum" dots for optical bioimaging, *Journal of Materials Chemistry B*, 1(16), 2116–2127.

Ma, Z., Ming, H., Huang, H., Liu, Y., & Kang, Z. (2012). One-step ultrasonic synthesis of fluorescent N-doped carbon dots from glucose and their visible-light sensitive photocatalytic ability, *New Journal of Chemistry*, 36(4), 861–864.

Macairan, J. R., Jaunky, D. B., Piekny, A., & Naccache, R. (2019). Intracellular ratiometric temperature sensing using fluorescent carbon dots, *Nanoscale Advances*, 1(1), 105–113.

Mirtchev, P., Henderson, E. J., Soheilnia, N., Yip, C. M., & Ozin, G. A. (2012). Solution phase synthesis of carbon quantum dots as sensitizers for nanocrystalline TiO_2 solar cells, *Journal of Materials Chemistry*, 22(4), 1265–1269.

Niu, J., & Gao, H. (2014). Synthesis and drug detection performance of nitrogen-doped carbon dots, *Journal of Luminescence*, 149, 159–162.

Pan, D., Zhang, J., Li, Z., & Wu, M. (2010). Hydrothermal route for cutting graphene sheets into blue-luminescent graphene quantum dots, *Advanced Materials*, 22(6), 734–738.

Pan, M., Xie, X., Liu, K., Yang, J., Hong, L., & Wang, S. (2020). Fluorescent carbon quantum dots—synthesis, functionalization and sensing application in food analysis, *Nanomaterials*, 10(5), 930.

Porfarzollah, A., Mohammad-Rezaei, R., & Bagheri, M. (2020). Ionic liquid-functionalized graphene quantum dots as an efficient quasi-solid-state electrolyte for dye-sensitized solar cells, *Journal of Materials Science: Materials in Electronics*, 31(3), 2288–2297.

Qian, Z., Ma, J., Shan, X., Feng, H., Shao, L., & Chen, J. (2014). Highly luminescent N-doped carbon quantum dots as an effective multifunctional fluorescence sensing platform, *Chemistry—A European Journal*, 20(8), 2254–2263.

Qu, K., Wang, J., Ren, J., & Qu, X. (2013). Carbon dots prepared by hydrothermal treatment of dopamine as an effective fluorescent sensing platform for the label-free detection of iron (III) ions and dopamine, *Chemistry—A European Journal*, 19(22), 7243–7249.

Qu, S., Wang, X., Lu, Q., Liu, X., & Wang, L. (2012). A biocompatible fluorescent ink based on water-soluble luminescent carbon nanodots, *Angewandte Chemie International Edition*, 51(49), 12215–12218.

Raji, K., Ramanan, V., & Ramamurthy, P. (2019). Facile and green synthesis of highly fluorescent nitrogen-doped carbon dots from jackfruit seeds and its applications towards the fluorimetric detection of Au^{3+} ions in aqueous medium and in in vitro multicolor cell imaging, *New Journal of Chemistry*, 43(29), 11710–11719.

Salinas-Castillo, A., Ariza-Avidad, M., Pritz, C., Camprubí-Robles, M., Fernández, B., Ruedas-Rama, M. J., Megia-Fernández, A., Lapresta-Fernández, A., Santoyo-Gonzalez, F., Schrott-Fischer, A., & Capitan-Vallvey, L. F. (2013). Carbon dots for copper detection with down and upconversion fluorescent properties as excitation sources, *Chemical Communications*, 49(11), 1103–1105.

Shabaev, A. L., Efros, A. L., & Nozik, A. J. (2006). Multiexciton generation by a single photon in nanocrystals, *Nano letters*, 6(12), 2856–2863.

Sharma, V., Singh, S. K., & Mobin, S. M. (2019). Bioinspired carbon dots: from rose petals to tunable emissive nanodots, *Nanoscale Advances*, 1(4), 1290–1296.

Singh, R., Kashayap, S., Singh, V., Kayastha, A. M., Mishra, H., Saxena, P. S., Srivastava, A., & Singh, R. K. (2018). QPRTase modified N-doped carbon quantum dots: a fluorescent bioprobe for selective detection of neurotoxin quinolinic acid in human serum, *Biosensors and Bioelectronics*, 101, 103–109.

Sun, Y. P., Zhou, B., Lin, Y., Wang, W., Fernando, K. S., Pathak, P., Meziani, M. J., Harruff, B. A., Wang, X., Wang, H., & Luo, P. G. (2006). Quantum-sized carbon dots for bright and colorful photoluminescence, *Journal of the American Chemical Society*, 128(24), 7756–7757.

Surana, K., Mehra, R. M., Soni, S. S., & Bhattacharya, B. (2022). Real-time photovoltaic parameters assessment of carbon quantum dots showing strong blue emission, *RSC Advances*, 12(3), 1352–1360.

Surana, K., Singh, P. K., Rhee, H. W., & Bhattacharya, B. (2014). Synthesis, characterization and application of CdSe quantum dots, *Journal of Industrial and Engineering Chemistry*, 20(6), 4188–4193.

Tan, L., Huang, G., Liu, T., Fu, C., Zhou, Y., Zhu, Z., & Meng, X. (2016). Synthesis of highly bright oil-soluble carbon quantum dots by hot-injection method with N and B Co-Doping, *Journal of Nanoscience and Nanotechnology*, 16(3), 2652–2657.

Tao, H., Yang, K., Ma, Z., Wan, J., Zhang, Y., Kang, Z., & Liu, Z. (2012). In vivo NIR fluorescence imaging, biodistribution, and toxicology of photoluminescent carbon dots produced from carbon nanotubes and graphite, *Small*, 8(2), 281–290.

Wang, H., Sun, P., Cong, S., Wu, J., Gao, L., Wang, Y., Dai, X., Yi, Q., & Zou, G. (2016). Nitrogen-doped carbon dots for "green" quantum dot solar cells, *Nanoscale Research Letters*, 11(1), 1–6.

Wee, S. S., Ng, Y. H., & Ng, S. M. (2013). Synthesis of fluorescent carbon dots via simple acid hydrolysis of bovine serum albumin and its potential as sensitive sensing probe for lead (II) ions, *Talanta*, 116, 71–76.

Xu, X., Ray, R., Gu, Y., Ploehn, H. J., Gearheart, L., Raker, K., & Scrivens, W. A. (2004). Electrophoretic analysis and purification of fluorescent single-walled carbon nanotube fragments, *Journal of the American Chemical Society*, 126(40), 12736–12737.

Yan, F., Zou, Y., Wang, M., Mu, X., Yang, N., & Chen, L. (2014). Highly photoluminescent carbon dots-based fluorescent chemosensors for sensitive and selective detection of mercury ions and application of imaging in living cells, *Sensors and Actuators B: Chemical*, 192, 488–495.

Yang, J., Tang, Q., Meng, Q., Zhang, Z., Li, J., He, B., & Yang, P. (2017). Photoelectric conversion beyond sunny days: all-weather carbon quantum dot solar cells, *Journal of Materials Chemistry A*, 5(5), 2143–2150.

Yang, S. T., Cao, L., Luo, P. G., Lu, F., Wang, X., Wang, H., Meziani, M. J., Liu, Y., Qi, G., & Sun, Y. P. (2009a). Carbon dots for optical imaging in vivo, *Journal of the American Chemical Society*, 131(32), 11308–11309.

Yang, S. T., Wang, X., Wang, H., Lu, F., Luo, P. G., Cao, L., Meziani, M. J., Liu, J. H., Liu, Y., Chen, M., & Huang, Y. (2009b). Carbon dots as nontoxic and high-performance fluorescence imaging agents, *The Journal of Physical Chemistry C*, 113(42), 18110–18114.

Zhang, B., Liu, C. Y., & Liu, Y. (2010). A novel one-step approach to synthesize fluorescent carbon nanoparticles, *European Journal of Inorganic Chemistry*, 2010(28), 4411–4414.

Zhao, Y., Duan, J., He, B., Jiao, Z., & Tang, Q. (2018). Improved charge extraction with N-doped carbon quantum dots in dye-sensitized solar cells, *Electrochimica Acta*, 282, 255–262.

Zhu, H., Wang, X., Li, Y., Wang, Z., Yang, F., & Yang, X. (2009). Microwave synthesis of fluorescent carbon nanoparticles with electrochemiluminescence properties, *Chemical Communications*, 34, 5118–5120.

Zou, H., Guo, D., He, B., Yu, J., & Fan, K. (2018). Enhanced photocurrent density of HTM-free perovskite solar cells by carbon quantum dots, *Applied Surface Science*, 430, 625–631.

8 Mesoporous Silica-Based Materials for Energy Storage Applications

Jianren Wang, Neus Vilà and Alain Walcarius

8.1 INTRODUCTION

Nanotechnology has been booming since the middle of the last century. The resulting nanomaterials/nanostructures (1–100 nm in one or more dimensions) with inherent functionalities make them competitive in various fields owing to their unique physicochemical features (Kankala et al., 2020; Li et al., 2012). Among nanomaterials, the ordered mesoporous silica-based ones have gained strong research interest due to their easily controlled topology, large surface area, and organized porous architectures, as well as their rather low cost (Yan & Zhao, 2007). The first mesoporous silica sieve, labeled Mobil Composition of Matter No. 41 (MCM-41), was reported in 1992 by Kresge and his co-workers in Mobil Corporation (Kresge et al., 1992). This pioneering work opens the door for the controlled synthesis of mesoporous silica materials by the micelle-templating method based on sol-gel chemistry. At first, cationic surfactants are used as structure-directing agents, leading to the M41-series made of MCM-41 (two-dimensional [2D] hexagonal, *p6mm*), MCM-48 (3D cubic, *Ia3d*) and MCM-50 (lamellar, *p2*; Beck et al., 1992). Then, meso-structured silica materials have sprung up, and its family embraces many new members, such as the anionic surfactant–templated mesoporous silica series, the Santa Barbara Amorphous series, the Institute of Bioengineering and Nanotechnology series, and Korean Institute of Science and Technology series, among others (Han & Che, 2013; Han & Ying, 2004; Hussain et al., 2014; Lee et al., 2009).

Several classical procedures have been developed for the preparation of nano-structured silica, including Stöber (Teng et al., 2013), biphase stratification (Shen et al., 2014), or evaporation-induced aggregating assembly (Wei et al., 2011) methods for getting mesoporous silica particles and evaporation-induced self-assembly (Brinker et al., 2004), electrochemically assisted self-assembly (Walcarius et al., 2007; Walcarius, 2021), or Stöber growth (Teng et al., 2012) methods for mesoporous silica thin films. A key point in getting good-quality ordered mesoporous silica materials is the interplay between the spatial arrangement of the soft template and its interactions with the silica source. During their preparation (illustrated in Figure 8.1), the micelles formed by surfactants at their critical micelle concentration (CMC) play an important role in obtaining the long-range ordered structure (Wan & Zhao, 2007). Many of them have been verified to cooperatively self-assemble with

DOI: 10.1201/9781003208709-8

the silica network precursors, such as cationic-type (cetyltrimethylammonium chloride or cetyltrimethylammonium bromide, CTAB), anionic-type (phosphoric acid, N-myristoyl-l-alanine, sodium dodecyl sulfate, sulfonic acid, and alkyl carboxylic acids), and nonionic-type (block copolymers based on polyethylene oxide [PEO] and polypropylene oxide [PPO], and Pluronic F123, as well as F127 and Brij 30) (Kankala et al., 2020). Another important parameter affecting the long-range mesoscopic periodic structure is the molar ratio between the silica precursors (usually tetraalkoxysilanes or sodium metasilicate) and the surfactant in the starting sol so that the silica frameworks can form around liquid crystal mesophases (Wan & Zhao, 2007). They have found applications in various fields (Ciriminna et al., 2013), including electrochemistry (Walcarius, 2013). If focusing on the particular example of drug delivery, for instance, a variety of mesostructured silica architectures with different pore sizes (2–30 nm; Butler et al., 2016; Trewyn et al., 2007; Vallet-Regi et al., 2001), structure (hexagonal, cubic, concentric, radial; Chen et al., 2013; Knežević & Lin, 2013; Zhu et al., 2014; Zahiri et al., 2020), morphology (core-shell, yolk-shell, sphere, hollow; Dai et al., 2015; Fang et al., 2014; Gao et al., 2011; Wu et al., 2011), or surface features (roughness, charge, hydrophilic; Angiolini et al., 2018; Luo et al., 2013; Zhang et al., 2015), have been successfully prepared in recent years. From the electrochemical point of view, the most attractive configuration relies on mesoporous silica thin films that can be successfully prepared on solid electrode supports thanks to the versatility of the sol-gel process (Feng et al., 2013; Walcarius, 2013, 2015). Of particular interest are the ordered mesoporous silica films enabling fast transport rates and easy access to a high number of active sites (Etienne et al., 2007). Among them, nanostructured coatings exhibiting vertically aligned nanochannels are expected to

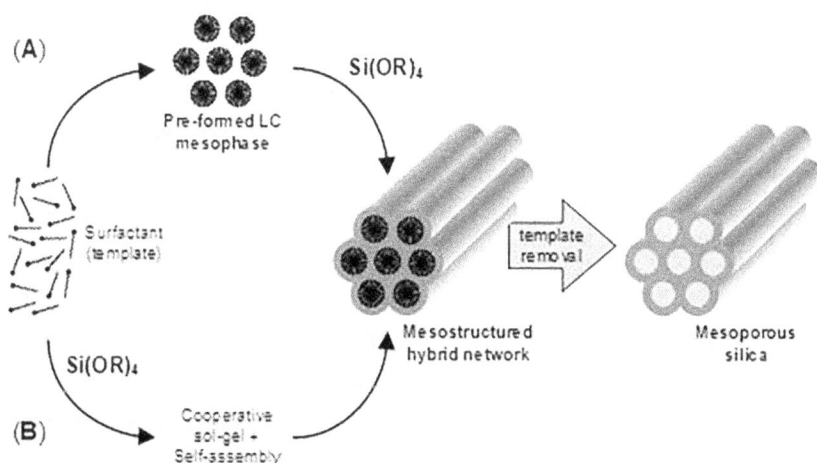

FIGURE 8.1 Schematic pathway for preparing surfactant-templated mesoporous silicas, illustrating a formation mechanism based on a preformed liquid crystal (LC) mesophase (route A) or a cooperative process (route B).

Source: Reproduced from Walcarius (2013).

be very promising (Urbanova & Walcarius, 2014). Even if the vertical growth of sol-gel-derived mesoporous silica thin films is not favored thermodynamically (Brinker et al., 1999), few methods are nowadays available to generate highly ordered silica films with meso-channels perpendicular to a solid support (e.g., an electrode surface), that is, the electrochemically assisted self-assembly (Goux et al., 2009), the Stöber solution growth process (Teng et al., 2012), or other approaches (Kao et al., 2015). Mesoporous silica films with worm-like channels are also of interest for fast electrocatalytic transformations (Ahoulou et al., 2020).

The functionalization/hybridization of such mesoporous frameworks to tailor their inherent properties have also gained great research interest (Lin et al., 2014; Hoffmann et al., 2006; Walcarius et al., 2005), notably in order to fulfill the requirements from diverse applications involving energy harvesting/storage, separation, drug delivery, catalysis, and the like. Their tailorable surficial functional groups are one of the main reasons that make them easy to introduce the target species into the framework via either physical or chemical interactions like hydrophilic/hydrophobic interaction (Butler et al., 2016; Trewyn et al., 2007), covalent bonding (Vilà et al., 2014; Vilà et al., 2016), van der Waals interactions (Yoo & Woo, 2018), and electrostatic interaction (Rakibuddin & Kim, 2020). Up to now, different types of advanced silica-based hybrid materials have been synthesized, which can be classified into the following categories:

1. *Functional organic molecules attached to silica frameworks.* Two general methods, namely, covalent conjugation and electrostatic/hydrogen interactions, can be used to achieve such functionalization. For the covalent conjugation method, the organic molecules can be introduced onto a silica surface by either the condensation of organosilanes with silica source in the starting sol or post-synthesis grafting of organosilanes of other derivatization procedures like click chemistry (Vilà et al., 2014; (Brühwiler, 2010; Schlossbauer et al., 2008; Etienne et al., 2009). The abundant silanol groups on the silica surface also make it possible to immobilize target molecules through electrostatic/hydrogen interactions (Zhou et al., 2015) or via entrapment of supramolecular species (Ahoulou et al., 2019). In recent years, different types of electroactive molecules have been successfully anchored on the surface of mesoporous silica films, such as metallocene derivatives (Wang et al., 2020), iron-triazole complexes (Ahoulou et al., 2019) or tetrazine (Vilà et al., 2015), for instance, to implement the applications in different fields.

2. *Polymer–silica composites.* In addition to small molecules, the long-chain polymers can also couple with mesoporous silica to form organic–inorganic hybrids, either by means of *in situ* polymerization of preloaded monomers inside silica channels (Peng et al., 2006), or the space-confined growth of polymer via electropolymerization inside silica channels (Ullah et al., 2021), or even by coating polymer outside the silica architecture to form a core-shell composite structure (Paris et al., 2015). As a result, a wide variety of advanced composites have been synthesized by properly combining mesoporous silica and various polymers with different physicochemical properties, including polyaniline (Ullah et al., 2021), poly(acrylic acid) (Chen et al., 2013), polydopamine (Zheng et al., 2016), or poly(2-vinylpyridine) (Niedermayer et al., 2015), among others.

3. *Inorganic material–silica composites.* Inorganic materials, either in a pure phase or in composite forms, always show many enhanced activities toward specific applications when their size goes down to the nanoscale (Gleiter, 2000). However, the non-silica periodically ordered inorganic materials, such as mesoporous metal (oxides), carbon, C_3N_4, and others, are difficult to prepare via similar protocols as that for the mesoporous silica (i.e., using soft templates; Walcarius, 2013). Utilizing mesoporous silica as hard templates to guide the growth of inorganic materials could be a versatile strategy to extend the scope of nanostructured porous materials. Some successful examples of ordered mesoporous inorganic materials (or the composites with silica), including chromium oxide, cobalt oxide, manganese oxide, carbon, platinum, and others, have been prepared by the so-called nanocasting method with silica template (Walcarius, 2013; Kong et al., 2016). Besides, the rich surface chemistry of silica also permits its durable attachment to different types of inorganic materials, giving rise to other hybrid composites (Gushikem & Rosatto, 2001).

Energy issues are becoming one of the themes in this era due to the rapid progress of society and the explosive consumption of non-renewable energy sources, such as oil and coal. To avoid the potential upcoming energy crisis, people started to turn their attention to utilizing clean energy resources, like wind energy, solar energy, and hydropower, among others, and develop energy storage devices to store such renewable but intermittent/geographically uneven energy sources (Lin et al., 2018; Simon et al., 2014). The intriguing properties of mesoporous silica materials mentioned earlier enable them to become one of the key components for the energy harvesting/storage devices for various reasons, such as (1) the nanoscale ordered structure could enhance their activities in terms of the specific energy-related applications since the existence of so-called nano-effect (Wan et al., 2015); (2) their tailorable surface properties and large surface area are likely to immobilize various guest species, which could largely extend the versatility of the final devices; (3) their homogeneous and perfectly ordered structure could also build an ideal platform to further reveal the fundamentals of the energy harvesting/storage process and let people better understand the structure–function relationship. As a result, silica-based materials for energy-related applications are gradually becoming a hot research topic. The percentage of papers in energy-related fields grew from 9% to 15% over the past decades in the category of silica-related research, according to the Web of Science. The tremendous advancements in mesoporous silica materials have largely accelerated the development of energy-related devices and will have a significant impact on them from a long-term perspective. This chapter gives a brief introduction of silica-based materials in the field of electrochemical energy storage.

8.2 LITHIUM-ION BATTERIES

Lithium-ion batteries (LIBs) have already dominated the energy storage market in a wide range of portable electronics, electric vehicles, and renewable energy storage systems (Entwistle et al., 2020). Achieving higher energy density is one of the most important goals for the further development of LIBs. However, the most commonly used graphite anodes in the present market show relatively low theoretical gravimetric and volumetric capacities of about 372 mA hg^{-1} and approximately 840 mA hcm^{-3}, respectively (Wang et al., 2018). Therefore, the discovery and optimization of advanced anode materials are critical aspects to achieve higher energy densities of LIBs. Silica has been considered to be a promising anode material because of its high theoretical capacity of 1965 mA hg^{-1}, low discharge potential (<0.5 V *vs.* Li$^+$/ Li), and abundance in the earth's crust (major constituent of sand; Yan et al., 2013). But some severe problems finally restrict the practical application of silica anode in the market. To be specific, the real capacity of silica is always far away from the theoretical value and fades very fast during the long-term cycling test due to its poor conductivity (insulating), slow lithium diffusion rate inside the silica bulk phase, and pulverization caused by the huge volume expansion (~200%) during the lithiation process. Properly designing nanostructures of silica anodes to migrate the active sites from bulk to surface/near-surface region and buffer the strain during lithiation–delithiation has proven to be effective in alleviating the existent problems of silica, which will improve their energy storage behaviors.

The construction of silica hollow structures with nano-level skeletons could be an effective approach to improve the capacity and to relieve the volume change during the charge-discharge processes. For example, Yan et al. (2013) prepared a kind of hollow porous SiO$_2$ nanocubes by utilizing Co$_3$[Co(CN)$_6$]$_2$ nanoparticles as the hard template, and the schematic illustration of the preparation process is shown in Figure 8.2A. After etching the templates using acid, the uniform silica nanocubes with a shell thickness of 20 nm and a cavity diameter of 100 nm can be prepared, as clearly shown in the corresponding transmission electron microscopy (TEM) and scanning electron microscopy (SEM) images (Figure 8.2B).The intercalation of lithium ions into the silica nanocubes occurs at a relatively low potential (below 0.5V *vs.* Li$^+$/Li), and their initial charge capacity of the nano-level silica shell can reach 1457 mA h g^{-1} (74% of theoretical capacity), which is much higher than that of graphite anode (372 mA h g^{-1}; Figure 8.2C&D). Despite the thickness of the silica shell expanding to 40 nm after the discharge process, the hollow structure can still tolerate the volume change and maintain its original shape (Figure 8.2E). As a result, the capacity can maintain 63% after 30 consecutive cycles. Similarly, silica nanotubes can be prepared by using anodic aluminum oxide as the hard template (Favors et al., 2014), as illustrated in Figure 8.2F. After etching the template, silica nanotubes with a diameter of 200 nm and a shell thickness of 20 nm can be obtained (Figure 8.2G). The silica nanotubes can also deliver a capacity of 1040 mA h g^{-1} at the current density of 50 mA g^{-1} and even a little bit higher capacity of 1247 mAhg^{-1} after 100 cycles (Figure 8.2H).

Properly coupling silica together with some high-conductive materials, like carbon or MXenes, for instance, is also one of the promising approaches to improve energy storage performance. In this case, the conductive components will not only circumvent the inherent poor electrical conductivity of silica but also act as the spacer to restrain the large volume variation during their charge-discharge processes. For instance, Mu et al. (2020) successfully embedded silica nanoparticles into the interlamination of MXene layers and firmly anchored them on the surface of MXene through chemical bonds (Si-C bond, or Si-O bond with OH terminals of MXene).

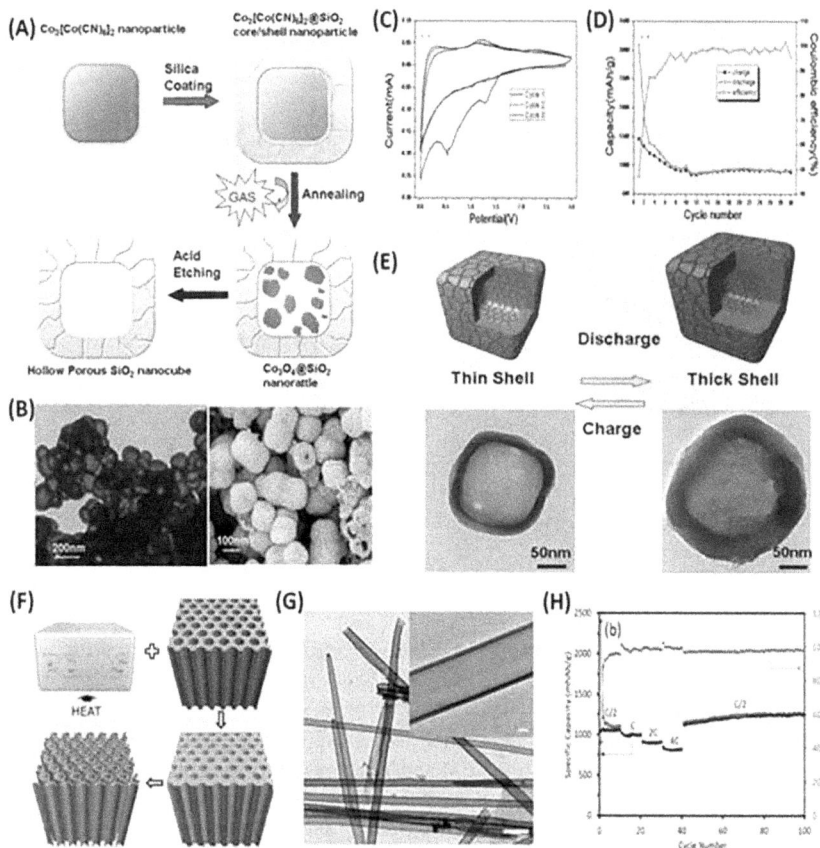

FIGURE 8.2 (A–E) The preparation of hollow porous SiO_2 nanocubes and their lithium storage properties: (A) The schematic illustration of the hollow porous SiO_2 nanocubes formation process; (B) TEM and SEM images of the hollow porous SiO_2 nanocubes; (C) their cyclic voltammetry (CV) curves recorded between 3 and 0 V at a scan rate of 0.1 mV s^{-1} and (D) cycling performance at a current density of 100 mAg^{-1}; the morphology comparison between the charge state and discharge state of the hollow porous SiO_2 nanocubes. Reproduced from Yan et al. (2013). (F–H) The preparation of SiO_2 nanotubes and their lithium storage properties: (F) The schematic illustration of the SiO_2 nanotubes formation process; (G) TEM images of the SiO_2 nanotubes with a scale bar of 400 nm (50 nm for the inset); and (H) the rate performance and the cycling performance at a current density of 50 mA h g^{-1} of silica nanotubes. Reproduced from Favors et al. (2014).

The illustration of the preparation process is given in Figure 8.3A, and the corresponding high-resolution SEM and TEM images confirm the conceived MXene–SiO_2–MXene laminated structure (Figure 8.3B&C). MXene is a new type of highly conductive 2D nanosheets with the highest recorded conductivity of approximately 10^4 S cm^{-1}. Therefore, the MXene nanosheets could improve the poor conductivity of the electrode and restrain the volume expansion of the embedded silica nanoparticles during the charge–discharge processes. From the CV curves of the composites at a

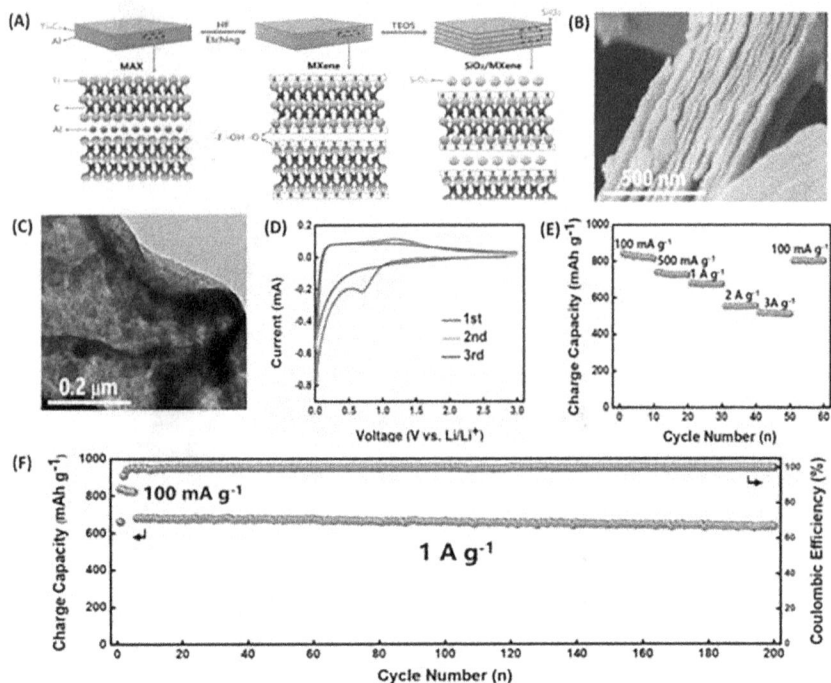

FIGURE 8.3 (A) The schematic illustration of the preparation process of silica/MXene composites; (B&C) the high-resolution SEM and TEM images of the composite; (D) the corresponding CV curves in a voltage window of 0 to about 3.0 V at a scan rate of 0.1 mV s^{-1}; (E) the rate performance of the composites at current densities ranging from 100 mA h g^{-1} to 3000 mA h g^{-1}; (F) the cycling performance with initial 10 cycles at the current density of 100 mA h g^{-1} and the following 180 cycles at the current density of 1000 mA h g^{-1}.

Source: Reproduced from Mu et al. (2020)

scan rate of 0.1 mV s^{-1}, a sharp reduction peak at about 0 V represents the lithiation reaction of SiO$_2$ (Mu et al., 2020), and the oxidation peaks at about 0.15 and 1.3 V are the dealloying of Li$_x$Si to Si (Yan et al., 2013) and the delithiation of Li$_2$Si$_2$O$_5$ to SiO$_2$ (Chang et al., 2012), respectively (see Figure 8.3D). As shown in Figure 8.3E, the silica/MXene composite can achieve a good rate of performance with a capacity of 840 mA h g^{-1} at a current density of 0.1 A g^{-1} and 517 mA h g^{-1} at a high current density of 3 A g^{-1}. Moreover, this composite can remain at its 97% original capacity after 100 consecutive cycles at a current density of 1 A g^{-1} (Mu et al., 2020).

8.3 LITHIUM–SULFUR BATTERIES

In contrast to the capacity range of about 150–200 mA h g^{-1}for the LIB cathodes (Whittingham, 2004), the high theoretical capacity of the sulfur cathode (1675 mA h g^{-1}) and its low cost make lithium–sulfur (Li-S) batteries a promising energy storage system for the scenarios sensitive to volumetric and gravimetric capacity (Peng et al., 2017). However, the intermediate lithium polysulfides (Li_2S_n, $4 \leq n \leq 8$) between the starting sulfur and the final Li_2S product could dissolve into the electrolyte or even across the separator to reach the anode, rendering a fast capacity fading and low charge efficiency due to the, namely, polysulfides' "shuttle" effect (Mikhaylik & Akridge, 2004). In addition, some other problems of the sulfur cathode, including its low electronic conductivity (5×10^{-30} S cm^{-1} at 25 °C), serious volumetric expansion during lithiation ($\sim$80%), and the morphology change caused by the redeposition of polysulfides from solution phase, seriously impede the practical application of Li-S batteries (Peng et al., 2017).

8.3.1 SILICA AS THE SULFUR HOST IN LI-S BATTERIES

Instead of using sulfur alone as the cathode, constructing nanostructured hosts to stabilize sulfur inside is an effective way to extend the cycling performance and even enhance the reaction kinetics during the sulfur lithiation/delithiation processes (Choudhury et al., 2017; Jayaprakash et al., 2011). Carbon-based materials are the most studied porous hosts that can facilitate electron transport to improve sulfur conversion (Wang et al., 2011), but the nonpolar carbon surface is not an ideal substrate for absorbing the dissolved polysulfides with polar properties due to weak interactions (Xie et al., 2019). Silica has proven to be a good adsorbent to trap polysulfides via physical adsorption and (or) chemical binding with the functional groups on the surface of silica, that is, hydroxyl, and the rigid and highly open structure of mesoporous silica can alleviate the volume expansion of sulfur and facilitate the lithium ions' transfer during the charge–discharge processes. Following this idea, dual composites prepared by engineering silica together with some conductive materials, or even pure mesoporous silica, have been claimed to be effective to achieve improved energy storage behaviors as host materials for the sulfur cathode (Wu et al., 2017).

As illustrated in Figure 8.4A, Choi et al. (2018) chose tetraethyl orthosilicate (TEOS) as a silica source and triblock copolymer surfactant as a template to prepare pomegranate-like silica with a large surface area and high pore volume, which is further used as the sulfur host to obtain the silica/sulfur composite for the cathode of Li-S batteries. The mesoporous nature of the as-prepared silica architecture can be confirmed by the corresponding TEM image (Figure 8.4B) and the hysteresis loop of the nitrogen adsorption/desorption isotherms (Figure 8.4C). The abundant mesopores not only permit a large surface area of 710.1 m^2 g^{-1} but also provide enough space to accommodate sulfur. Through a melt infiltration method, the sulfur can be homogeneously loaded inside the mesoporous silica, which is proven by the corresponding energy dispersive spectroscopy (EDS) mapping results (Si and S) of the silica/sulfur composite, shown in Figure 8.4D. The thermogravimetric curve of the composite (Figure 8.4E) also reveals that the weight ratio of sulfur is as high as about 66.3%, thanks to the ubiquitous mesopores. As shown in Figure 8.4F, the mesoporous silica host also gives a strong adsorption ability toward polysulfide (Li_2S_6), and the Li_2S_6 solution becomes almost transparent after 24 h adsorption, which makes the mesoporous silica an ideal host to stabilize the sulfur cathode by relieving the loss of soluble polysulfides during the consecutive charge–discharge processes. As a result, the silica/sulfur composite can deliver a high coulombic efficiency of 93.7% at the

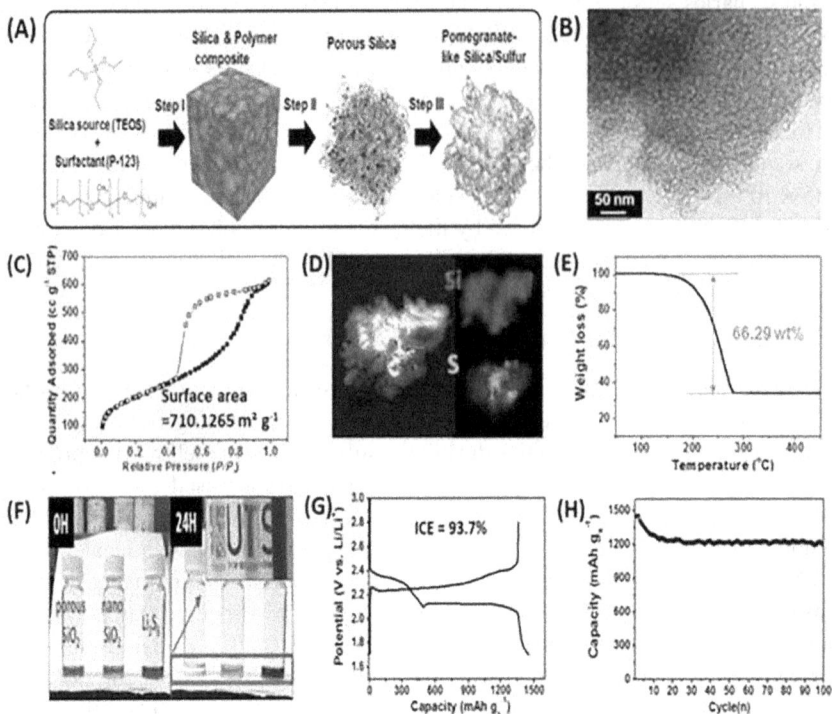

FIGURE 8.4 Synthesis of pomegranate-structured silica and sulfur composite materials as the cathode for an Li-S battery. (A) The illustration of the composite preparation process; (B&C) the TEM image and the nitrogen adsorption/desorption isotherms of the obtained porous silica; (D) EDS mapping image of a porous silica/S composite particle, showing the distribution of Si and S; (E) thermogravimetric analysis curve of the porous silica/S composite; (F) polysulfide (Li_2S_6) adsorption ability of silica materials: digital photos of both porous silica particles and silica nanoparticles immersed in Li_2S_6 solution after 0 h (left) and after 24 h (right); (G&H) the first charge–discharge voltage profile at the current density of 1.6 Ag^{-1} and the cycling performance of the porous silica/S composite at the current density of 0.8 A g^{-1}.

Source: Reproduced from Choi et al. (2018)

current density of 1.6 A g^{-1} for the initial charge–discharge cycle, and a robust cycling performance, as its capacity can still maintain 82.9% (from 1358.6 to 1126.3 mA h g^{-1}) after 100 consecutive cycles at the current density of 0.8 A g^{-1} (Choi et al., 2018).

Similarly, Wu et al. argued the crucial role of silica as the sulfur host on the basis of a composite material made of hierarchical-porous graphene/C/SiO_2 hollow microspheres (G/C/SiO_2) (Wu et al., 2017). In detail, the spray drying technique has been adopted to prill the aqueous ethanol solution containing a certain amount of sucrose, graphene oxide and SiO_2, giving the G/C/SiO_2 microspheres after pyrolysis at 900 °C , as illustrated in Figure 8.5A. The SEM/scanning transmission electron microscopy (STEM) images and the corresponding EDS mapping results reveal the highly porous structure of the sphere and the homogeneous distribution of C, O, and

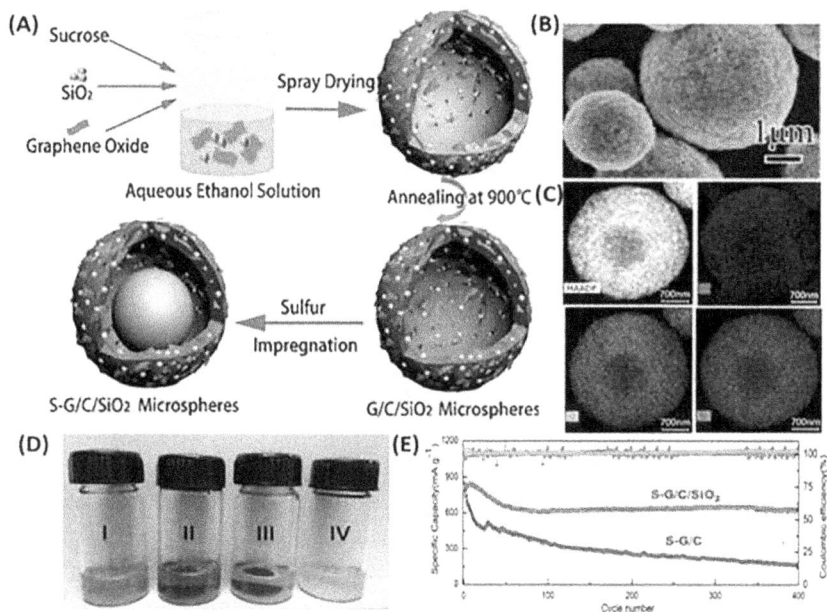

FIGURE 8.5 (A) The illustration of the preparation process of hierarchical-porous graphene/C/SiO$_2$ hollow microspheres (G/C/SiO$_2$) as the sulfur host for Li-S batteries; (B&C) the SEM and STEM images of the G/C/SiO$_2$ and the corresponding EDS elemental distribution of C, O, and Si; (D) the polysulfide (Li$_2$S$_6$) adsorption measurement of different samples: I: pure Li$_2$S$_6$, II: Li$_2$S$_6$+G/C-S, III: Li$_2$S$_6$+S-G/C/SiO$_2$, IV: Li$_2$S$_6$+SiO$_2$; (E) the cyclic performance of the G/C-S and S-G/C/SiO$_2$ at the current density of 0.8 A g^{-1}over 400 cycles.

Source: Reproduced from Wu et al. (2017)

Si (Figure 8.5B&C). The silica component inside the sample plays a vital role to stabilize sulfur during the long-term charge–discharge test, owing to its strong attraction toward Li$_2$S$_6$; this can be seen in Figure 8.5D via the very transparent color of vial III (Li$_2$S$_6$+S-G/C/SiO$_2$) and IV (Li$_2$S$_6$+SiO$_2$) in comparison to vial I (pure Li$_2$S$_6$) and II (Li$_2$S$_6$+S-G/C, prepared by removing the silica in the starting solution) after a 6-h adsorption test. As a result, the S-G/C/SiO$_2$ cathode can deliver a high capacity retention of 79.6% (from 786.7 to 626.3 mA h g^{-1}) after 400 cycles at the current density of 0.8 A g^{-1} (Figure 8.5E). This result is much higher than that of S-G/C cathode that gives a 19.3% capacity retention under the same condition (Figure 8.5E), which further confirms that silica could be an ideal host for the Li-S batteries.

8.3.2 Silica as the Component of the Separator in Li-S Batteries

A separator is also a key component in energy storage devices, especially for Li-S batteries (Peng et al., 2017). In addition to its role in preventing an internal short-circuit while maintaining the ions diffusion pathway, the separator in Li-S batteries should also have the function to avoid the crossing of the polysulfides from the sulfur cathode to the lithium anode, as illustrated in Figure 8.6A (Huang et al., 2015). Because polysulfides in contact with the anode surface would corrode the lithium anode by parasitic reactions and even irreversibly deposit on the anode surface

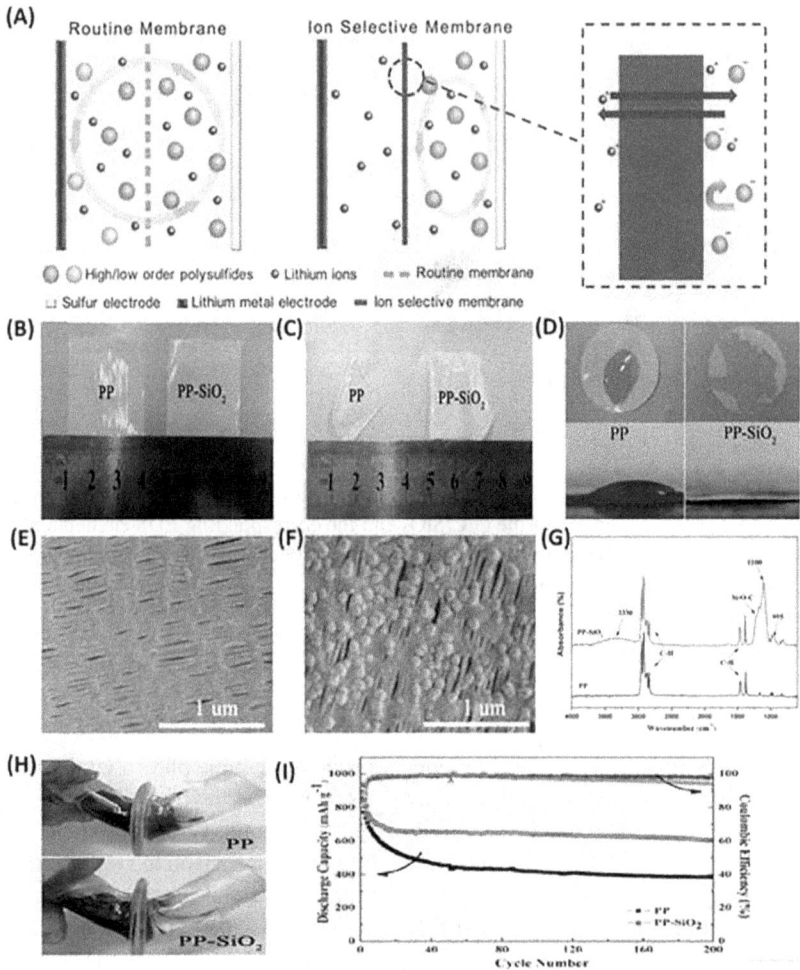

FIGURE 8.6   (A) Scheme to show the difference between the normal separator and functionalized separator in Li-S battery. Reproduced from Huang et al. (2015). Digital photos of the PP and the PP-SiO$_2$ separators before (B) and after thermal treatment (C) at 150 °C for 2 h; (D) spreading of the electrolyte droplet and the corresponding contact angles on the PP and the PP-SiO$_2$ separators; the SEM images of PP separator (E) and PP-SiO$_2$ separator (F); (G) the FTIR spectra of PP separator and PP-SiO$_2$ separator; (H) diffusion test of polysulfides (Li$_2$S$_8$) in V-shaped permeating matrix with the PP and the PP-SiO$_2$ separator after 1 h; (I) the cyclic performance of the Li-S cell with the PP and the PP-SiO$_2$ separator. Reproduced from Li et al. (2017).

during the charge–discharge processes, the formation of "dead" sulfur-containing species would contribute to passivating the lithium anode. This is one of the most important reasons resulting in the fast capacity fading of the final devices. The deliberately designed separators/interlayers are able to trap/block the polysulfides to reach the anode, which could improve energy storage performance regarding coulombic efficiency, self-discharge rate, and cyclic stability of Li-S cells (Li et al., 2015).

Recent years have witnessed the fast development of the separator in Li-S batteries, and different types of advanced membranes including polymer-modified, carbon-modified, or metal oxide–modified separators and their composites, have been prepared to achieve better energy storage performance (Huang et al., 2015). On the basis of experimental evidence and theoretical simulation, SiO_2 can serve as the active component of the separator to suppress the polysulfides transfer via an adsorption mechanism, thanks to the high binding energy between polysulfides and an oxygen atom (Zhu et al., 2016). Besides, doping the silica component of the separators can also adjust their physiochemical properties such as thermal stability, hydrophilic/hydrophobic balance of the interface, and so on (Li et al., 2017). For instance, a mesoporous SiO_2-decorated polypropylene (PP) separator (PP-SiO_2) can be simply prepared by immersing the PP separator in a hydrolyzed TEOS solution with the assistance of Tween-80 as a template, and the digital photos are shown in Figure 8.6B. After decorating PP with SiO_2, the composite shows much better thermal stability, proved by the attenuated shrinkage of the film after a thermal treatment at 150 °C for 2 h (close to the softening point of a PP matrix, ca.165 °C), as illustrated on Figure 8.6C. The largely improved thermal stability of the PP-SiO_2 could enhance the endurability of Li-S batteries, especially at high operating temperatures. Besides, as shown in Figure 8.6D, the silica component can also make the composite more hydrophilic, as the PP-SiO_2 can be fully wetted while there is still an electrolyte droplet on the PP substrate. The improved wettability and high electrolyte uptake ability of the separator could benefit interfacial compatibility, reduce the electrolyte filling time, and facilitate Li-ion migration, leading to an enhanced rate of performance. The microstructure of the PP and the PP-SiO_2 are analyzed with SEM, and the corresponding images are shown in Figure 8.6E and Figure 8.6F, respectively. In contrast to the smooth morphology of the PP separator, the silica particles, with a diameter ranging from tens to hundreds of nanometers, can be observed on the surface of PP-SiO_2 sample. The cross-linked silica networks on the side of the PP substrate lead to new peaks located at 1100 cm^{-1} and 955 cm^{-1} in Fourier transform infra-red (FTIR) spectrum of the PP-SiO_2 separator (Figure 8.6G), originated from the asymmetrical stretching of the Si-O-Si and the stretching of Si-OH groups. All these results demonstrate that silica has been successfully introduced on the PP membrane surface for the PP-SiO_2 sample. The diffusion rate of polysulfides (Li_2S_8) through the PP-SiO_2 separator is much slower than that through the pristine PP separator, as shown by the shallow color on the right part of the V-shaped permeating setup when choosing PP-SiO_2 as the separator (Figure 8.6H). As a result, the discharge capacity of Li-S battery increases from 852.3 mAh g^{-1} to 937 mAh g^{-1} when the battery is assembled with the PP-SiO_2 separator instead of the PP separator (Figure 8.6I). The PP-SiO_2 separator also extends the life span of the Li-S battery, and 603.5 mAh g^{-1} can be retained, which is higher than that of the cell configured with the PP separator (383.6 mAh g^{-1}; Li et al., 2017).

8.4   ALKALI METAL BATTERIES

The alkali metals are regarded as ideal metallic anode materials in battery systems due to the step-change theoretical-specific capacity (Li: 3,860 mA h g^{-1}, Na: 1,166 mA h g^{-1}, Zn: 820mAh g^{-1}) and low redox potential (Li: −3.040 V, Na: −2.714 V, Zn: −0.76 V *vs.* the standard hydrogen electrode). However, the alkali metal anodes encounter great challenges in practical application scenarios due to a short life span and short-circuiting easily (Kang et al., 2018). One of the main reasons causing these problems is the formation of hundreds of microns of metal dendrites/protrusions due to the uneven metal electro-stripping/plating process. The surface-formed dendrites could possibly be peeled off from the anode into the electrolyte during the striping process, resulting in the loss of active materials, or continuously grow across the separator membrane to touch the cathode, inducing a short-circuit (Li et al., 2019).

Recently, developing an artificial interfacial layer between the electrolyte and Li-metal electrode has attracted tremendous attention in lithium–metal batteries (LMBs). The interfacial layer can prevent side reactions, enable fast Li-ion diffusion, and suppress the Li dendrite growth for the efficient operation of the Li-metal anode. Recently, developing an artificial interfacial layer between the electrolyte and the Li-metal electrode has attracted tremendous attention in LMBs. The interfacial layer can prevent side reactions, enable fast Li-ion diffusion, and suppress the Li dendrite growth for the efficient operation of Li metal anode. Recently, developing an artificial interfacial layer between the electrolyte and Li metal electrode has attracted tremendous attention in LMBs. The interfacial layer can prevent side reactions, enable fast Li-ion diffusion, and suppress the Li dendrite growth for the efficient operation of Li-metal anode.

Constructing an artificial interfacial layer between the electrolyte and the alkali metal electrode could enable breaking this challenge, notably by preventing side reactions, ensuring fast Li-ion diffusion, and suppressing the growth of Li dendrite, contributing to more efficient operation of the metal anode (Chen et al., 2020). The mesoporous silica layer has also been identified as an active coating layer to suppress the growth of metal dendrites for different types of metal anodes, including lithium (Pathak et al., 2019), zinc (Kang et al., 2018), and sodium (Jiang et al., 2019), via regulating the flux of metal ions and/or even reacting with the dendrites during the charge–discharge processes. For example, Jiang et al. (2019) constructed a porous nano-SiO_2 auxiliary layer on the surface of the sodium–metal disk (Na@Silica), by a simple brushing and rolling method, for inhibiting the dendrites growth during charge–discharge processes in sodium-metal batteries. Instead of the free diffusion of sodium ions during the metal deposition process, always leading to the formation of dendrites, the insulating and highly porous silica layer on the sodium disk surface is likely to play the role of water-permeable bricks to disperse and guide the Na$^+$ flux, finally restraining the growth of sodium dendrites. The illustration of the difference between the free growth and silica-assisted growth of sodium is shown in Figure 8.7A. The optical photo and SEM image of the bare Na disk (Figure 8.7B) manifest its smooth, flat, and shiny metallic surface. After coating the porous nano-SiO_2 onside, the electrode becomes dim, and the SEM image confirms that there is a highly porous

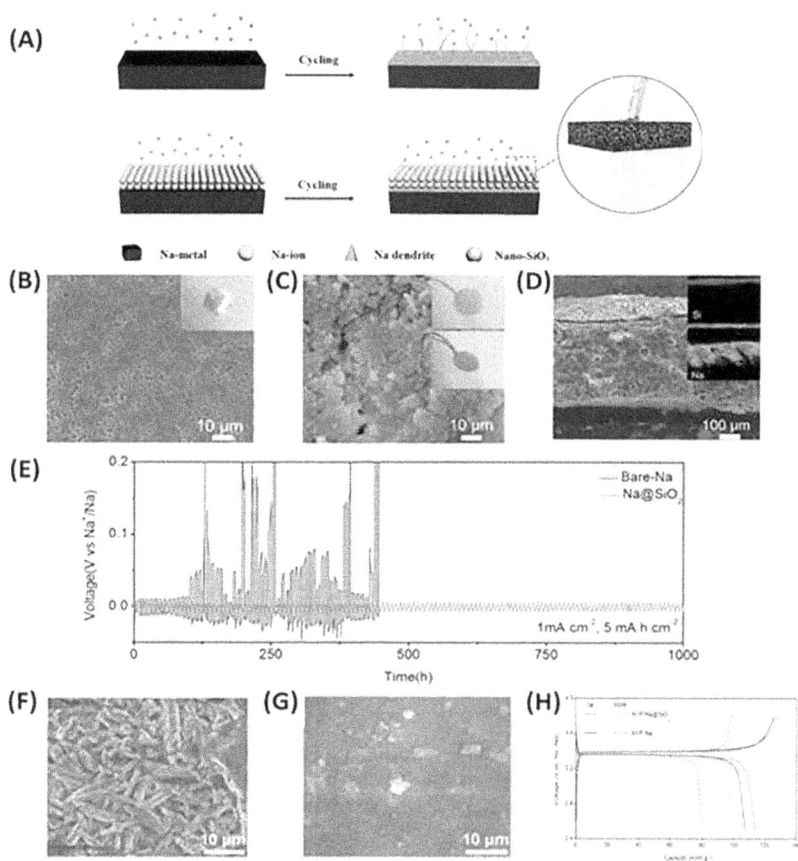

FIGURE 8.7 (A) Schematic illustrations of the Na stripping/plating behavior on (a) bare and (b) nano SiO_2-coated Na electrodes; (B) top-view SEM image of bare Na electrode and the corresponding digital image (inset); (C) top-view SEM image of Na@SiO_2 electrode and the corresponding digital images (insets); (D) cross-sectional SEM image of the Na@SiO_2 electrode and the corresponding EDX elemental mapping images of Si and Na (insets); (E) a comparison of cycling stability of Na@SiO_2 composite electrodes and bare Na symmetrical cells at 1 mA cm^{-2} with as tripping/plating capacity of 5 mA h cm^{-2} in 1MNaPF$_6$/diglyme electrolyte; (F, G) morphologies of Na electrode (F) and Na@SiO_2 electrode (G) after 100 stripping/plating cycles at a current density of 1 mA cm^{-2}; (H) voltage profiles of the full batteries with Na or Na@SiO_2 as anodes and Na$_3$V$_2$(PO$_4$)$_3$ as cathode at 118 mA g^{-1}.

Source: Reproduced from Jiang et al. (2019).

silica layer on the surface of Na@Silica electrode (Figure 8.7C). The corresponding cross-sectional view (Figure 8.7D) also reveals the thickness of the SiO_2 layer at about 100 μm, and the EDS elemental mapping results show the homogeneous distribution of the silicon and sodium in adjoining layers. Figure 8.7E depicts the electrochemical behaviors of the symmetric cells with two pieces of Na@SiO_2

electrodes or two pieces of Na electrodes in 1 M $NaPF_6$/diglyme electrolyte at the current density of 1 mA cm^{-2} for a stripping/plating capacity of 5 mA h cm^{-2}. The results clearly show that the $Na@SiO_2$ symmetric cell is able to deliver a stable and low voltage hysteresis even after 1000 consecutive cycles, indicative of the even and highly reversible plating/striping of sodium. On the contrary, the asymmetric cell gives a large voltage hysteresis, and the voltage even exceeds 0.2 V (the maximum voltage window) after about 400 consecutive cycles. The poor stability of the silica-free Na cell originates from the generation of dendrites due to unstable plating/striping processes, which can be proved by SEM images of the electrodes after long-term cycling, where abundant sodium dendrites can be observed for the bare Na electrode (Figure 8.7F), but this is not the case for the $Na@SiO_2$ electrode (Figure 8.7G). As a result, the battery constructed with $Na@SiO_2$//$Na_3V_2(PO_4)_3$ also gives a better cycling ability than its counterpart of Na//$Na_3V_2(PO_4)_3$, as shown in Figure 8.7H.

Similar to sodium metal electrochemical cells, the artificial silica layer on the surface of a lithium metal plate scan also inhibits the growth of lithium dendrites during charge–discharge processes. Pathak et al. (2019) reported the coating of a lithium metal plate with a thin layer of graphite-silica by radio frequency (RF) sputtering method. Such a layer possesses attractive features playing multiple functions, such as a conductive skeleton to uniform the electric field, lithium storage sites to reduce the lithium deposition current, and a rigid clapboard to spatially restrain the dendrite growth. The different lithium plating behaviors on bare lithium plate and graphite-silica coated lithium plate are illustrated in Figure 8.8A. According to the atomic force microscope (AFM) results (Figure 8.8B&C), the surface of a lithium plate is relatively rough with an average surface root mean square (RMS) of 520.1 nm, whereas this value decreases to 324.5 nm after coating the graphite-silica layer. The higher RMS value is unfavorable to the homogeneous growth of the lithium and always results in large protuberances on the electrode surface (Guan et al., 2018). Besides, such artificial layer turns the surface more hydrophilic (the contact angle decreases from 43° on the lithium plate to 0° on the graphite-silica-coated lithium plate due to the presence of polar groups on the silica surface), which may facilitate the electrolyte transport. More than the mechanism to prevent the growth of sodium dendrites, such graphite-silica layer is also active for the lithium ions storage (see Section 8.2), which can be confirmed by the peaks (~0.15 V) of graphite-silica-coated electrode during CV tests in the symmetrical cell, and the two semicircles observed by impedance in the high-to-middle-frequencies region (Figure 8.8D&E). Therefore, the even lithium plating can be observed for the graphite-silica-coated lithium plate, whereas numerous lithium dendrites can be observed for the bare lithium–metal plate (Figure 8.8F). As a result, the symmetrical cell with graphite-silica-coated lithium as the electrode shows much stability and small voltage hysteresis than that of a pure lithium electrode (Figure 8.8G).

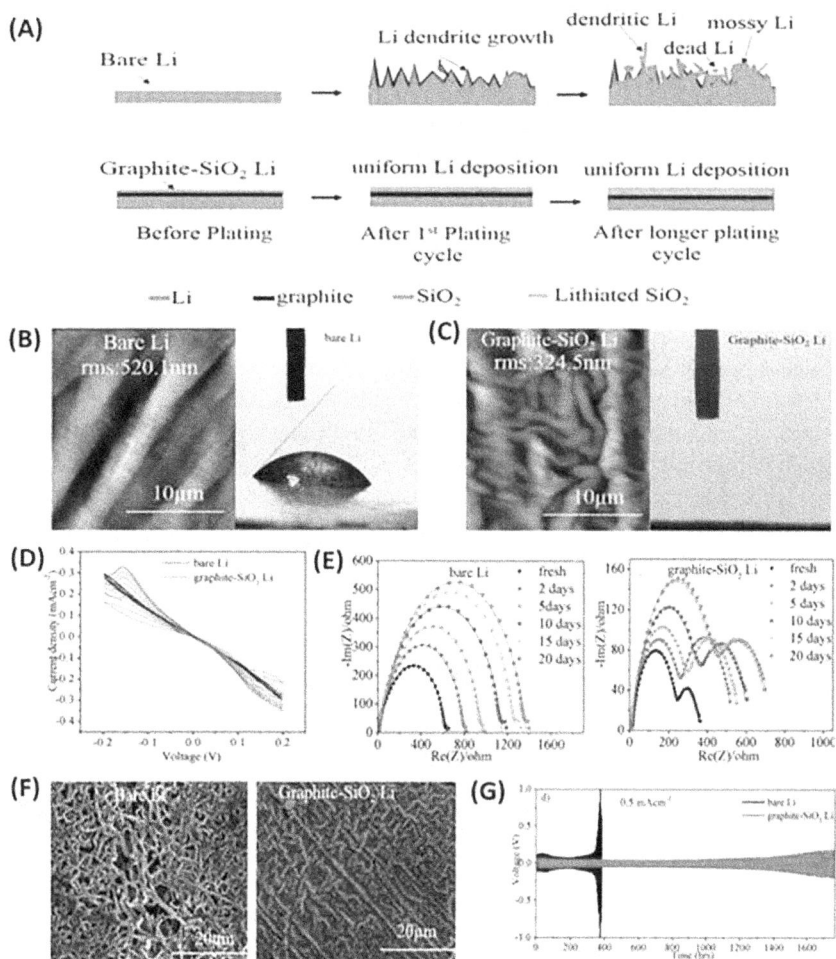

FIGURE 8.8 (A) Schematic illustration of Li deposition on bare Li and on graphite–SiO$_2$ bilayer-modified Li; (B&C) AFM topography and contact angle measurement of bare Li plate and graphite–SiO$_2$-coated Li plate; (D&E) CV measurements and the impedance test (with a 10-mV amplitude AC signal and frequencies ranging from 100 kHz to 0.1 Hz) of bare Li and graphite SiO$_2$-coated Li in a symmetric cell; (F) SEM images of bare Li and graphite–SiO$_2$-coated Li after one cycle of plating at a current density of 0.5 mA cm^{-2} for a capacity of 1 mAh cm^{-2}; (G) long-term cycling performance and voltage hysteresis of symmetrical cells of bare Li electrode or graphite–SiO$_2$-coated Li electrode at the current density of 0.5 mA cm^{-2} for the capacity of 1 mAh cm^{-2}.

Source: Reproduced from Pathak et al. (2019).

8.5 PSEUDOCAPACITORS BASED ON SURFACE FUNCTIONALIZATION OF MESOPOROUS SILICA WITH REDOX MOLECULES

Electrical conductivity has always been a fundamental element for energy storage electrodes, and the insulating character of silica somehow confines its utilization to some extent. However, in contrast to the carbon-based materials (widely used for energy storage), mesoporous silica-based solids still hold some irreplaceable advantages, including well-organized structure, large surface area, and high flexibility for their surface functionalization (Wan & Zhao, 2007; Wan et al., 2007). If one can find an approach to overcome the lack of conductivity, it could widely broaden the applied ranges of silica-based materials in the field of energy storage.

Electron-hopping, a particular electron conduction mechanism, allows electrons to transfer "by jumping" among adjacent and close enough redox molecules, directly without requiring any conductive properties of the supporting materials (Gibbons & Spear, 1966). The electron-hopping mechanism is also able to deliver fast redox kinetics. For example, Amatore et al. (2007) prepared two monolayer systems to study the electron-hopping process, and both molecules contain one terminal ferrocene molecule but with different lengths of hydrocarbon. The redox kinetics could reach about 10^6 s^{-1} for both systems, which is much faster than that of traditional faradic materials. This unique charge transfer feature should make possible the use of insulating materials for energy storage after functionalizing their surface with redox-active molecules. Recently, our group reported a ferrocene functionalized mesoporous silica film (Fc-MS) on an indium-tin oxide (ITO) electrode substrate (Wang et al., 2020), which can be manufactured by combining the Electro-Assisted Self-Assembly method and Copper-catalyzed Azide-Alkyne Cycloaddition click reaction, as illustrated in Figure 8.9A. The TEM image (Figure 8.9B) of Fc-MS reveals that such film is highly porous and composed of fairly ordered hexagonal mesopores, as also identified by the hexagonal diffraction spots of the corresponding Fast Fourier transformation (FFT) diffractogram (inset image). The cross-sectional view of the sample (Figure 8.9C) further reveals the vertical channels run directly from the top to the bottom of the film, and the thickness of the film is approximately 105 nm. Before the click reaction, there are numerous azide groups on the internal surface of silica channels due to the presence of an azide-containing silane in the starting sol, and the corresponding sample is labeled as Az-MS. After the click reaction, most azide groups are converted into ferrocene moieties by the reaction between the azide groups and the ethynyl groups of ethynyl ferrocene molecules, which can be evidenced by the evolution of azide signature (2095 cm^{-1}) in the FTIR spectra (Figure 8.9D). These ferrocene molecules on the surface of the vertically aligned silica meso-channels are capable of conducting the redox reaction through an electron-hopping process even without the help of the insulating silica substrate, as illustrated in Figure 8.9E, and a pair of symmetrical ferrocene redox peaks can be observed during the CV tests performed at various scan rates (Figure 8.9F). The Fc-MS electrode can also give an almost linear potential response during the galvanostatic charge-discharge processes at various current densities from 0.4 to 20 A cm^{-3}, indicative of the surface-controlled behavior of the electron-hopping governed redox reaction (Figure 8.9G). Calculated from these curves, the capacity that can be delivered by such film is 105 C cm^{-3} (1.10 mC cm^{-2}) at a current density of 0.4 A cm^{-3} and the capacity retention can achieve 48% (50.4 C cm^{-3}) at the current density of 20 A cm^{-3} (Figure 8.9H). The covalently bonded ferrocene molecules on the silica wall also show a robust cycling performance, with a capacity retention of 75% after 1000 consecutive cycles (Figure 8.9I). Not only limited to the ferrocene molecules, but

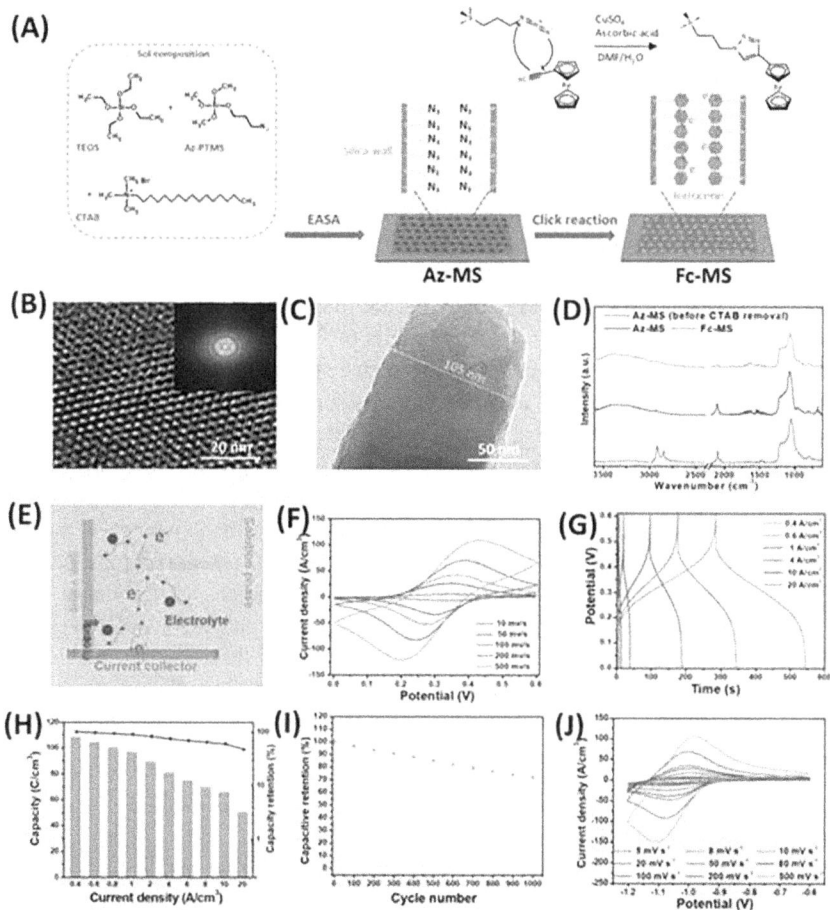

FIGURE 8.9 (A) Schematic illustration of the preparation process of the ferrocene-functionalized vertically-aligned mesoporous silica thin films (Fc-MS) on the ITO electrode; (B&C) the top view and corresponding FFT pattern (inset) and the cross-sectional view of the Fc-MS; (D) IR spectra of Az-MS (before and after the removal of CTAB surfactant) and Fc-MS; (E) the illustration of the electron-hopping process among the ferrocene molecules on the surface of silica; (F–I) the CV curves at different scan rates, charge–discharge curves and corresponding capacities at different current densities, and the cycling performance at the scan rate of 200 mV s^{-1} of Fc-MS in 1 M LiCl; (J) the CV curves of cobaltocenium at different scan rates in 1 M LiCl.

Source: Reproduced from Wang et al. (2020)

the energy storage centers can also be extended to other redox molecules, like cobalto-cenium (Figure 8.9J), indicative of the universal property of such silica film.

The electrochemically assisted generation of silica films is a method that can be applied to uniform deposition on various types of conductive supports, even with very complex morphology, such as microporous metal electrodes (Qu et al., 2011), metal microfibers (Mazurenko et al., 2011), or graphene nanosheets (Fang et al., 2019), for instance. This feature makes it possible the principle of combining the use of ferrocene functionalized silica films (Fc-MS) with advanced substrates of more complex morphology and larger surface than flat ITO electrodes, in order to increase the loading amount of Fc-MS for the sake of increasing capacity. This is what we did (Figure 8.10A) by generating Fc-MS on a freestanding graphene foam (GO) prepared by an electro-exfoliation method (Wang et al., 2021), with the idea to exploit the extremely large theoretical surface area of this type of 2D carbon material (2630 m^2 g^{-1} for graphene). According to the SEM image of GO (Figure 8.10B), the most distributed thickness is 9 ± 6 nm, whereas the sheets' thickness increases to 75 ± 15 nm after coating the ferrocene functionalized silica film on the surface of graphene oxide (GO@Fc-MS), implying the thickness of silica layer is about 60 nm (Figure 8.10C). The EDX mapping results of GO@Fc-MS (Figure 8.10D) also prove the uniform coating of Fc-MS on the surface of graphene, based on the homogeneous distribution of C, Si, and Fe elements. In comparison to that of GO, the CV curve of GO@Fc-MS shows a pair of well-defined and intense ferrocene redox peaks (Figure 8.10E), confirming the redox-active property of the silica film onside. Besides, the current responses of GO and GO@Fc-MS are almost overlapped in the non-faradic region (0.1 V to 0.2V), demonstrating that the existence of mesoporous silica thin film on its surface does not cause obvious resistance for electrolyte transport through the silica-on-graphene oxide layer. The charge-discharge curves of EG@Fc-MS (Figure 8.10F) are distorted but without an obvious plateau, which means the redox reaction is surface-controlled. Based on these curves, the capacity of EG@Fc-MS can reach 196 mC cm^{-2} (or capacitance of 326 mF cm^{-2}) at a current density of 2 mA cm^{-2}, which is higher by about two orders of magnitude in comparison to that of Fc-MS on flat ITO substrate (Wang et al., 2021, 2020).

8.6 CONCLUSION

Although not electrically conductive, the ordered porous silica materials are promising for applications in the field of electrochemical energy storage (batteries and supercapacitors) owing to their attractive properties, such as an easily accessible regular mesostructure, very high specific surface area, and good mechanical stability. They can be used as lithium-insertion materials in Li-ion batteries if properly designed as silica nanostructures likely to accelerate lithium diffusion and support volume expansion during lithiation. Restraining the large volume variation during charge–discharge processes can be improved by coupling silica with other nanomaterials, and if conducting, such nanomaterials can also provide good conductivity to the resulting hybrids. In the field of Li-S batteries, the nanostructured silica materials are interesting hosts to accommodate sulfur/polysulfide compounds and even accelerate the sulfur lithation/delithiation processes, thus contributing

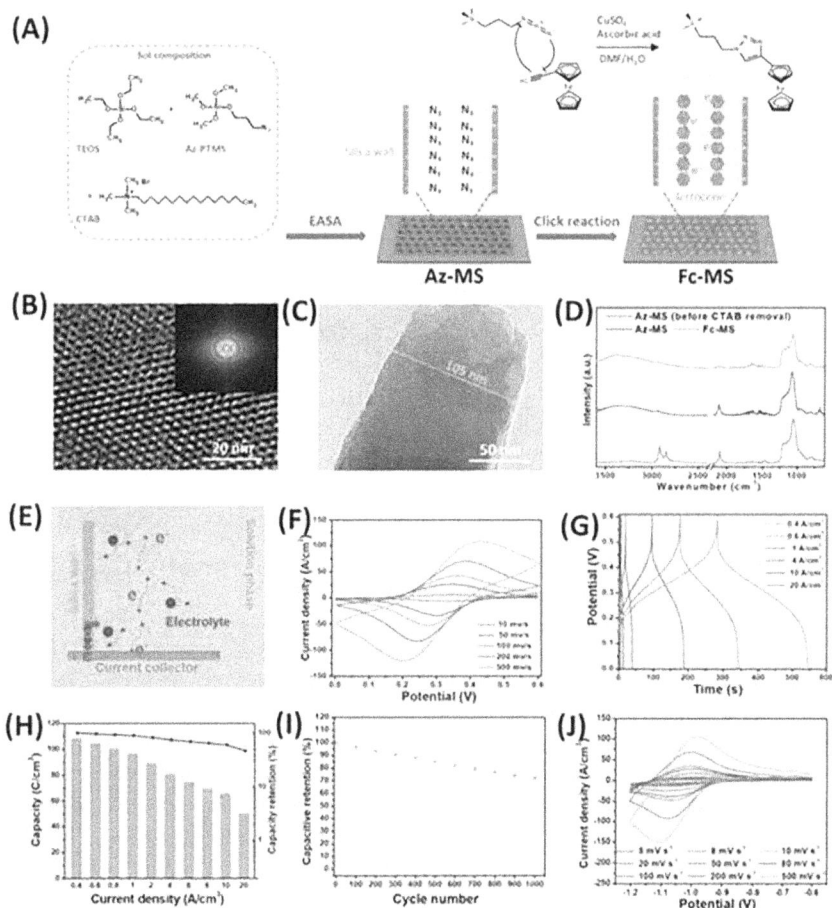

FIGURE 8.10 (A) Schematic illustration of the formation of graphene oxide coated with ferrocene functionalized mesoporous silica film (GO@Fc-MS); (B) high-resolution SEM image of the graphene oxide electrode (GO) and the corresponding sheets thickness distribution; (C) high-resolution SEM image of GO@Fc-MS and the corresponding sheets thickness distribution; (D) the EDX mapping results (C, Si, Fe) of GO@Fc-MS; (E) CV curves of graphite, GO and GO@Fc-MS at a scan rate of 50 mV s^{-1} in 1 M H$_2$SO$_4$; (F) charge/discharge curves of GO@Fc-MS obtained at various current densities, from 2 to 100 mA cm^{-2}; (G) the capacities of GO and GO@Fc-MS at different current densities from 2 to 100 mA cm^{-2}.

Source: Reproduced from Wang et al. (2021)

to robust cycling performance. They can also be used as an active component of the separator to prevent the ingress of polysulfides, resulting in better durability of Li-S batteries. Related stabilization/protection role of a mesoporous silica layer coated onto the surface of metallic anodes can be observed in alkali metal batteries, by means of suppressing the growth of metal dendrites usually occurring during the charge/discharge processes, thus leading to improved cycling ability. Finally, recent investigations on mesoporous silica films decorated with redox-active moieties covalently attached in the interior of the nanochannels reveal effective charge propagation by electron-hopping, which can be exploited for supercapacitive energy storage applications. Such novel pseudocapacitive materials can be easily coated onto the surface of various electrode supports, even with complex topography, by electrochemically assisted deposition methods, opening the door to further developments in the near future.

ACKNOWLEDGMENTS

JW acknowledges a financial support from the China Scholarship Council for his doctoral studies.

REFERENCES

S. Ahoulou, N. Vilà, S. Pillet, D. Schaniel and A. Walcarius, 2019, Coordination polymers as template for mesoporous silica films: a novel composite material Fe(Htrz)$_3$@SiO$_2$ with remarkable electrochemical properties, Chemistry of Materials, 31, 5796–5807.

S. Ahoulou, N. Vilà, S. Pillet, D. Schaniel and A. Walcarius, 2020, Non-covalent immobilization of iron-triazole (Fe(Htrz)$_3$) molecular mediator in mesoporous silica films for the electrochemical detection of hydrogen peroxide, Electroanalysis, 32, 690–697.

C. Amatore, E. Maisonhaute, B. Schöllhorn and J. Wadhawan, 2007, Ultrafast voltammetry for probing interfacial electron transfer in molecular wires, ChemPhysChem, 8, 1321–1329.

L. Angiolini, S. Valetti, B. Cohen, A. Feiler and A. Douhal, 2018, Fluorescence imaging of antibiotic clofazimine encapsulated within mesoporous silica particle carriers: relevance to drug delivery and the effect on its release kinetics, Physical Chemistry Chemical Physics, 20, 11899–11911.

J.S. Beck, J.C. Vartuli, W.J. Roth, M.E. Leonowicz, C.T. Kresge, K.D. Schmitt, C.T.W. Chu, D.H. Olson, E.W. Sheppard, S.B. McCullen, J.B. Higgins and J.L. Schlenker, 1992, A new family of mesoporous molecular sieves prepared with liquid crystal templates, Journal of the American Chemical Society, 114, 10834–10843.

C.J. Brinker, Y. Lu, A. Sellinger and H. Fan, 1999, Evaporation-induced self-assembly: functional nanostructures made easy, Advanced Materials, 11, 579–585, doi: 10.1557/mrs2004.183.

D. Brühwiler, 2010, Postsynthetic functionalization of mesoporous silica, Nanoscale, 2, 887–892, doi: 10.1039/C0NR00039F.

K.S. Butler, P.N. Durfee, C. Theron, C.E. Ashley, E.C. Carnes and C.J. Brinker, 2016, Protocells: modular mesoporous silica nanoparticle-supported lipid bilayers for drug delivery, Small, 12, 2173–2185.

W.S. Chang, C.M. Park, J.H. Kim, Y.U. Kim, G. Jeong and H.J. Sohn, 2012, Quartz (SiO$_2$): a new energy storage anode material for Li-ion batteries, Energy an Environmental Sciences, 5, 6895–6899.

L. Chen, L. Li, L. Zhang, S. Xing, T. Wang, Y.A. Wang, C. Wang and Z. Su, 2013, Designed fabrication of unique eccentric mesoporous silica nanocluster-based core-shell nanostructures for pH-responsive drug delivery, ACS Applied Materials and Interfaces, 5, 7282–7290.

X. Chen, R. Zhang, R. Zhao, X. Qi, K. Li, Q. Sun, M. Ma, L. Qie and Y. Huang, 2020, A "dendrite-eating" separator for high-areal-capacity lithium-metal batteries, Energy Storage Materials, 31, 181–186.

S. Choi, D. Su, M. Shin, S. Park and G. Wang, 2018, Pomegranate-structured silica/sulfur composite cathodes for high-performance lithium—sulfur batteries, Chemistry An Asian Journal, 13, 568–576.

S. Choudhury, M. Zeiger, P. Massuti-Ballester, S. Fleischmann, P. Formanek, L. Borchardt and V. Presser, 2017, Carbon onion-sulfur hybrid cathodes for lithium-sulfur batteries, Sustaining Energy and Fuels, 1, 84–94.

R. Ciriminna, A. Fidalgo, V. Pandarus, F. Beland, L.M. Ilharco and M. Pagliaro, 2013, The sol-gel route to advanced silica-based materials and recent applications, Chemical Reviews, 113, 1303–1340.

L. Dai, Q. Zhang, H. Gu and K. Cai, 2015, Facile synthesis of yolk-shell silica nanoparticles for targeted tumor therapy, Journal of Materials Chemistry B, 3, 8303–8313, doi: 10.1039/c5tb01620g.

J.E. Entwistle, S.G. Booth, D.S. Keeble, F. Ayub, M. Yan, S.A. Corr, D.J. Cumming and S.V. Patwardhan, 2020, Insights into the electrochemical reduction products and processes in silica anodes for next-generation lithium-ion batteries, Advanced Energy Materials, 10, 2001826.

M. Etienne, A. Goux, E. Sibottier and A. Walcarius, 2009, Oriented mesoporous organosilica films on electrode: a new class of nanomaterials for sensing, Journal of Nanoscience and Nanotechnology, 9, 2398–2406.

M. Etienne, A. Quach, D. Grosso, L. Nicole, C. Sanchez and A. Walcarius, 2007, Molecular transport into mesostructured silica thin films: electrochemical monitoring and comparison between p6m, P63/mmc, Pm3n structures, Chemistry of Materials, 19, 844–856, doi: 10.1021/cm0625068.

L. Fang, Q.Q. He, M.J. Zhou, J.P. Zhao and J.M. Hu, 2019, Electrochemically assisted deposition of sol-gel films on graphene nanosheets, Electrochemistry Communications, 109, 106609.

Y. Fang, G. Zheng, J. Yang, H. Tang, Y. Zhang, B. Kong, Y. Lv, C. Xu, A.M. Asiri, J. Zi, F. Zhang and D. Zhao, 2014, Dual-pore mesoporous carbon@silica composite core-shell nanospheres for multidrug delivery, Angewandte Chemie International Edition, 53, 5366–5370.

Z. Favors, W. Wang, H.H. Bay, A. George, M. Ozkan and C.S. Ozkan, 2014, Stable cycling of SiO$_2$ nanotubes as high-performance anodes for lithium-ion batteries, Scientific Reports, 4, 1–7.

D. Feng, J. Wei, M. Wang, Q. Yue, Y. Deng, A.M. Asiri and D. Zhao, 2013, Advances in mesoporous thin films via self-assembly process. Advanced Porous Materials, 1, 164–186.

Y. Gao, Y. Chen, X. Ji, X. He, Q. Yin, Z. Zhang, J. Shi and Y. Li, 2011, Controlled intracellular release of doxorubicin in multidrug-resistant cancer cells by tuning the shell-pore sizes of mesoporous silica nanoparticles, ACS Nano, 5, 9788–9798.

D.J. Gibbons and W.E. Spear, 1966, Electron hopping transport and trapping phenomena in orthorhombic sulphur crystals, Journal of Physics and Chemistry of Solids, 27, 1917–1925, doi: 10.1016/0022-3697(66)90122-3.

H. Gleiter, 2000, Nanostructured materials: basic concepts and microstructure, Acta Materialia, 48, 1–29, doi: 10.1016/S1359-6454(99)00285-2.

A. Goux, M. Etienne, E. Aubert, C. Lecomte, J. Ghanbaja and A. Walcarius, 2009, Oriented mesoporous silica films obtained by electro-assisted self-assembly (EASA), Chemistry of Materials, 21, 731–741.

X. Guan, A. Wang, S. Liu, G. Li, F. Liang, Y.W. Yang, X. Liu and J. Luo, 2018, Controlling nucleation in lithium metal anodes, Small, 14, 1–21.

Y. Gushikem and S.S. Rosatto, 2001, Metal oxide thin films grafted on silica gel surfaces: recent advances on the analytical application of these materials, Journal of the Brazilian Chemical Society, 12, 695–705, doi: 10.1590/S0103-50532001000600002.

L. Han and S. Che, 2013, Anionic surfactant templated mesoporous silicas (AMSs), Chemical Society Reviews, 42, 3740–3752, doi: 10.1039/c2cs35297d.

Y. Han and J.Y. Ying, 2004, Generalized fluorocarbon-surfactant-mediated synthesis of nanoparticles with various mesoporous structures, Angewandte Chemie International Edition, 44, 288–292, doi: 10.1002/anie.200460892.

F. Hoffmann, M. Cornelius, J. Morell and M. Froeba, 2006, Silica-based mesoporous organic-inorganic hybrid materials, Angewandte Chemie International Edition, 45, 3216–3251.

J.Q. Huang, Q. Zhang and F. Wei, 2015, Multi-functional separator/interlayer system for high-stable lithium-sulfur batteries: progress and prospects, Energy Storage Materials, 1, 127–145.

M. Hussain, D. Fino and N. Russo, 2014, Development of modified KIT-6 and SBA-15-spherical supported Rh catalysts for N_2O abatement: from powder to monolith supported catalysts, Chemical Engineering Journal, 238, 198–205.

N. Jayaprakash, J. Shen, S.S. Moganty, A. Corona and L.A. Archer, 2011, Porous hollow carbon@sulfur composites for high-power lithium-sulfur batteries, Angewandte Chemie International Edition, 50, 5904–5908.

F. Jiang, T. Li, P. Ju, J. Sun, C. Liu, Y. Li, X. Sun and C. Chen, 2019, Nano-SiO_2 coating enabled uniform Na stripping/plating for dendrite-free and long-life sodium metal batteries, Nanoscale Advances, 1, 4989–4994.

L. Kang, M. Cui, F. Jiang, Y. Gao, H. Luo, J. Liu, W. Liang and C. Zhi, 2018, Nanoporous $CaCO_3$ coatings enabled uniform Zn stripping/plating for long-life zinc rechargeable aqueous batteries, Advanced Energy Materials, 8, 1–8.

R.K. Kankala, Y.H. Han, J. Na, C.H. Lee, Z. Sun, S. Bin Wang, T. Kimura, Y.S. Ok, Y. Yamauchi, A.Z. Chen and K.C.W. Wu, 2020, Nanoarchitectured structure and surface biofunctionality of mesoporous silica nanoparticles, Advanced Materials, 32, 1–27, doi: 10.1002/adma.201907035.

K.-C. Kao, C.-H. Lin, T.-Y. Chen, Y.-H. Liu and C.-Y. Mou, 2015, A general method for growing large area mesoporous silica thin films on flat substrates with perpendicular nano-channels, Journal of the American Chemical Society, 137, 3779–3782.

N.Ž. Knežević and V.S.Y. Lin, 2013, A magnetic mesoporous silica nanoparticle-based drug delivery system for photosensitive cooperative treatment of cancer with a mesopore-capping agent and mesopore-loaded drug, Nanoscale, 5, 1544–1551, doi: 10.1039/c2nr33417h.

B. Kong, J. Tang, Y. Zhang, T. Jiang, X. Gong, C. Peng, J. Wei, J. Yang, Y. Wang, X. Wang, G. Zheng, C. Selomulya and D. Zhao, 2016, Incorporation of well-dispersed sub-5-nm graphitic pencil nanodots into ordered mesoporous frameworks, Nature Chemistry, 8, 171–177.

C.T. Kresge, M.E. Leonowicz, W.J. Roth, J.C. Vartuli and J.S. Beck, 1992, Ordered mesoporous molecular sieves synthesized by a liquid-crystal template mechanism, Nature, 359, 710–712.

C.H. Lee, T.S. Lin, C.Y. Mou, 2009, Mesoporous materials for encapsulating enzymes, Nano Today, 4, 165–179.

G.C. Li, H.K. Jing, Z. Su, C. Lai, L. Chen, C.C. Yuan, H.H. Li and L. Liu, 2015, A hydrophilic separator for high performance lithium sulfur batteries, Journal of Materials Chemistry A, 3, 11014–11020.

J. Li, Y. Huang, S. Zhang, W. Jia, X. Wang, Y. Guo, D. Jia and L. Wang, 2017, Decoration of silica nanoparticles on polypropylene separator for lithium-sulfur batteries, ACS Applied Materials and Interfaces, 9, 7499–7504.

S. Li, Q. Liu, J. Zhou, T. Pan, L. Gao, W. Zhang, L. Fan and Y. Lu, 2019, Hierarchical Co_3O_4 nanofiber—carbon sheet skeleton with superior Na/Li-philic property enabling highly stable alkali metal batteries, Advanced Functional Materials, 29, 1–10.

Z. Li, J.C. Barnes, A. Bosoy, J.F. Stoddart and J.I. Zink, 2012, Mesoporous silica nanoparticles in biomedical applications, Chemical Society Reviews, 41, 2590–2605, doi: 10.1039/c1cs15246g.

Y. Lin, W. Yantasee and G.E. Fryxell, 2014, Electrochemical sensors: functionalized silica, In S.E. Lyshevski (Ed.): Dekker Encyclopedia of Nanoscience and Nanotechnology (3rd Edition, Vol. 2), CRC Press, pp. 1283–1293.

Z. Lin, E. Goikolea, A. Balducci, K. Naoi, P.L. Taberna, M. Salanne, G. Yushin and P. Simon, 2018, Materials for supercapacitors: when Li-ion battery power is not enough, Materials Today, 21, 419–436.

G.F. Luo, W.H. Chen, Y. Liu, J. Zhang, S.X. Cheng, R.X. Zhuo and X.Z. Zhang, 2013, Charge-reversal plug gate nanovalves on peptide-functionalized mesoporous silica nanoparticles for targeted drug delivery, Journal of Materials Chemistry B, 1, 5723–5732.

I. Mazurenko, M. Etienne, R. Ostermann, B.M. Smarsly, O. Tananaiko, V. Zaitsev and A. Walcarius, 2011, Controlled electrochemically-assisted deposition of sol-gel biocomposite on electrospun platinum nanofibers, Langmuir, 27, 7140–7147.

Y.V. Mikhaylik and J.R. Akridge, 2004, Polysulfide shuttle study in the Li/S battery system, Journal of the Electrochemical Society, 151, A1969, doi: 10.1149/1.1806394.

G. Mu, D. Mu, B. Wu, C. Ma, J. Bi, L. Zhang, H. Yang and F. Wu, 2020, Microsphere-like SiO_2/MXene hybrid material enabling high performance anode for lithium ion batteries, Small, 16, 1–9, doi: 10.1002/smll.201905430.

S. Niedermayer, V. Weiss, A. Herrmann, A. Schmidt, S. Datz, K. Müller, E. Wagner, T. Bein and C. Bräuchle, 2015, Multifunctional polymer-capped mesoporous silica nanoparticles for pH-responsive targeted drug delivery, Nanoscale, 7, 7953–7964.

J.L. Paris, M.V. Cabanas, M. Manzano and M. Vallet-Regí, 2015, Polymer-grafted mesoporous silica nanoparticles as ultrasound-responsive drug carriers, ACS Nano, 9, 11023–11033.

R. Pathak, K. Chen, A. Gurung, K.M. Reza, B. Bahrami, F. Wu, A. Chaudhary, N. Ghimire, B. Zhou, W.H. Zhang, Y. Zhou and Q. Qiao, 2019, Ultrathin bilayer of graphite/SiO_2 as solid interface for reviving Li metal anode, Advanced Energy Materials, 9, 1–10, doi: 10.1002/aenm.201901486.

H. Peng, J. Tang, L. Yang, J. Pang, H.S. Ashbaugh, C.J. Brinker, Z. Yang and Y. Lu, 2006, Responsive periodic mesoporous polydiacetylene/silica nanocomposites, Journal of the American Chemical Society, 128, 5304–5305.

H.J. Peng, J.Q. Huang, X.B. Cheng and Q. Zhang, 2017, Review on high-loading and high-energy lithium—sulfur batteries, Advanced Energy Materials, 7, 1–54.

F. Qu, R. Nasraoui, M. Etienne, Y.B. Saint Côme, A. Kuhn, J. Lenz, J. Gajdzik, R. Hempelmann and A. Walcarius, 2011, Electrogeneration of ultra-thin silica films for the functionalization of macroporous electrodes, Electrochemistry Communications, 13, 138–142.

M. Rakibuddin and H. Kim, 2020, Sol-gel derived Fe_3O_4 quantum dot decorated silica composites for effective removal of arsenic (III) from water, Materials Chemistry and Physics, 240, 122245, doi: 10.1016/j.matchemphys.2019.122245.

A. Schlossbauer, D. Schaffert, J. Kecht, E. Wagner and T. Bein, 2008, Click chemistry for high-density biofunctionalization of mesoporous silica, Journal of the American Chemical Society, 130, 12558–12559.

D. Shen, J. Yang, X. Li, L. Zhou, R. Zhang, W. Li, L. Chen, R. Wang, F. Zhang and D. Zhao, 2014, Biphase stratification approach to three-dimensional dendritic biodegradable mesoporous silica nanospheres, Nano Letters, 14, 923–932.

P. Simon, Y. Gogotsi and D. Bruce, 2014, Where do batteries end and supercapacitors begin? Science, 343, 1210–1211.

Z. Teng, X. Su, Y. Zheng, J. Sun, G. Chen, C. Tian, J. Wang, H. Li, Y. Zhao and G. Lu, 2013, Mesoporous silica hollow spheres with ordered radial mesochannels by a spontaneous self-transformation approach, Chemistry of Materials, 25, 98–105.

Z. Teng, G. Zheng, Y. Dou, W. Li, C.Y. Mou, X. Zhang, A.M. Asiri and D. Zhao, 2012, Highly ordered mesoporous silica films with perpendicular mesochannels by a simple Stöber-solution growth approach, Angewandte Chemie International Edition, 51, 2173–2177, doi: 10.1002/anie.201108748.

B.G. Trewyn, S. Giri, I.I. Slowing and V.S.Y. Lin, 2007, Mesoporous silica nanoparticle based controlled release, drug delivery, and biosensor systems, ChemCommun, 3236–3245.

W. Ullah, G. Herzog, N. Vilà and A. Walcarius, 2021, Electrografting and electropolymerization of nanoarrays of PANI filaments through silica mesochannels, Electrochemistry Communications, 122, 106896.

V. Urbanova and A. Walcarius, 2014, Vertically-aligned mesoporous silica films, Zeitschrift für anorganische und allgemeinem, Chemie, 640, 537–546, doi: 10.1002/zaac.201300442.

M. Vallet-Regi, A. Rámila, R.P. Del Real and J. Pérez-Pariente, 2001, A new property of MCM-41: drug delivery system, Chemistry of Materials, 13, 308–311, doi: 10.1021/cm0011559.

N. Vilà, C. Allain, P. Audebert and A. Walcarius, 2015, Tetrazine-functionalized and vertically-aligned mesoporous silica films with electrochemical activity and fluorescence properties, Electrochemistry Communications, 59, 9–12.

N. Vilà, J. Ghanbaja, E. Aubert and A. Walcarius, 2014, Electrochemically assisted generation of highly ordered azide-functionalized mesoporous silica for oriented hybrid films, Angewandte Chemie International Edition, 53, 2945–2950.

N. Vilà, J. Ghanbaja and A. Walcarius, 2016, Clickable bifunctional and vertically aligned mesoporous silica films, Advanced Materials Interfaces, 3, 1500440.

A. Walcarius, 2013, Mesoporous materials and electrochemistry, Chemical Society Reviews, 42, 4098–4140, doi: 10.1039/c2cs35322a.

A. Walcarius, 2015, Mesoporous materials-based electrochemical sensors, Electroanalysis, 27, 1303–1340, doi: 10.1002/elan.201400628.

A. Walcarius, 2021, Electroinduced surfactant self-assembly driven to vertical growth of oriented mesoporous films, Accounts of Chemical Research, 54, 3563–3575.

A. Walcarius, D. Mandler, J.A. Cox, M. Collinson and O. Lev, 2005, Exciting new directions in the intersection of functionalized sol-gel materials with electrochemistry, Journal of Materials Chemistry, 15, 3663–3689.

A. Walcarius, E. Sibottier, M. Etienne and J. Ghanbaja, 2007, Electrochemically assisted self-assembly of mesoporous silica thin films, Nature Materials, 6, 602–608.

F. Wan, X.L. Wu, J.Z. Guo, J.Y. Li, J.P. Zhang, L. Niu and R.S. Wang, 2015, Nanoeffects promote the electrochemical properties of organic $Na_2C_8H_4O_4$ as anode material for sodium-ion batteries, Nano Energy, 13, 450–457.

Y. Wan, Y. Shi and D. Zhao, 2007, Designed synthesis of mesoporous solids via nonionic-surfactant-templating approach, Chemical Communications, 897–926.

Y. Wan and D. Zhao, 2007, On the controllable soft-templating approach to mesoporous silicates, Chemical Reviews, 107, 2821–2860, doi: 10.1021/cr068020s.

H. Wang, X. Yang, Q. Wu, Q. Zhang, H. Chen, H. Jing, J. Wang, S.B. Mi, A.L. Rogach and C. Niu, 2018, Encapsulating silica/antimony into porous electrospun carbon nanofibers with robust structure stability for high-efficiency lithium storage, ACS Nano, 12, 3406–3416.

H. Wang, Y. Yang, Y. Liang, J.T. Robinson, Y. Li, A. Jackson, Y. Cui and H. Dai, 2011, Graphene-wrapped sulfur particles as a rechargeable lithium-sulfur battery cathode material with high capacity and cycling stability, Nano Letters, 11, 2644–2647.

J. Wang, N. Vila and A. Walcarius, 2020, Redox-active vertically aligned mesoporous silica thin films as transparent surfaces for energy storage applications, ACS Applied Materials and Interfaces, 12, 24262–24270.

J. Wang, N. Vilà and A. Walcarius, 2021, Electroactive organically modified mesoporous silicates on graphene oxide-graphite 3D architectures operating with electron-hopping for high rate energy storage, Electrochimica Acta, 366, 137407, doi: 10.1016/j.electacta.2020.137407.

J. Wei, H. Wang, Y. Deng, Z. Sun, L. Shi, B. Tu, M. Luqman and D. Zhao, 2011, Solvent evaporation induced aggregating assembly approach to three-dimensional ordered mesoporous silica with ultralarge accessible mesopores, Journal of the American Chemical Society, 133, 20369–20377.

M.S. Whittingham, 2004, Lithium batteries and cathode materials, Chemical Reviews, 104, 4271–4301, doi: 10.1021/cr020731c.

H. Wu, G. Liu, S. Zhang, J. Shi, L. Zhang, Y. Chen, F. Chen and H. Chen, 2011, Biocompatibility, MR imaging and targeted drug delivery of a rattle-type magnetic mesoporous silica nanosphere system conjugated with PEG and cancer-cell-specific ligands, Journal of Materials Chemistry, 21, 3037–3045.

H. Wu, Q. Tang, H. Fan, Z. Liu, A. Hu, S. Zhang, W. Deng and X. Chen, 2017, Dual-confined and hierarchical-porous graphene/C/SiO_2 hollow microspheres through spray drying approach for lithium-sulfur batteries, Electrochimica Acta, 255, 179–186.

J. Xie, B.Q. Li, H.J. Peng, Y.W. Song, M. Zhao, X. Chen, Q. Zhang and J.Q. Huang, 2019, Implanting atomic cobalt within mesoporous carbon toward highly stable lithium—sulfur batteries, Advanced Materials, 31, 1–7.

N. Yan, F. Wang, H. Zhong, Y. Li, Y. Wang, L. Hu and Q. Chen, 2013, Hollow porous SiO_2 nanocubes towards high-performance anodes for lithium-ion batteries, Scientific Reports, 3, 1–6.

H. Yoo and K. Woo, 2018, Direct hybridization of hydrophobic nanocrystals with colloidal silica via van der Waals force, Journal of Physical Chemistry Letters, 9, 2106–2112. doi: 10.1021/acs.jpclett.8b00695.

M. Zahiri, M. Babaei, K. Abnous, S.M. Taghdisi, M. Ramezani and M. Alibolandi, 2020, Hybrid nanoreservoirs based on dextran-capped dendritic mesoporous silica nanoparticles for CD133-targeted drug delivery, Journal of Cellular Physiology, 235, 1036–1050, doi: 10.1002/jcp.29019.v.

H. Zhang, H. Xu, M. Wu, Y. Zhong, D. Wang and Z. Jiao, 2015, A soft-hard template approach towards hollow mesoporous silica nanoparticles with rough surfaces for controlled drug delivery and protein adsorption, Journal of Materials Chemistry B, 3, 6480–6489.

X. Zheng, F. Chen, J. Zhang and K. Cai, 2016, Silica-assisted incorporation of polydopamine into the framework of porous nanocarriers by a facile one-pot synthesis, Journal of Materials Chemistry B, 4, 2435–2443.

Z. Zhou, W. Guo, L. Xu, Q. Yang and B. Su, 2015, Two orders-of-magnitude enhancement in the electrochemiluminescence of $Ru(bpy)_3^{2+}$ by vertically ordered silica mesochannels, Analytica Chimica Acta, 886, 48–55.

J. Zhu, E. Yildirim, K. Aly, J. Shen, C. Chen, Y. Lu, M. Jiang, D. Kim, A.E. Tonelli, M.A. Pasquinelli, P.D. Bradford and X. Zhang, 2016, Hierarchical multi-component nanofiber separators for lithium polysulfide capture in lithium-sulfur batteries: an experimental and molecular modeling study, Journal of Materials Chemistry A, 4, 13572–13581.

W. Zhu, L. Wan, C. Zhang, Y. Gao, X. Zheng, T. Jiang and S. Wang, 2014, Exploitation of 3D face-centered cubic mesoporous silica as a carrier for a poorly water soluble drug: Influence of pore size on release rate, Materials Science and Engineering C, 34, 78–85.

9 Metal–Organic Frameworks

Dr. Murugavel Kathiresan and
Dr. M. Anbu Kulandainathan

9.1 INTRODUCTION

Metal–organic frameworks (MOFs) are a subclass of coordination polymers (hybrid organic–inorganic materials) made by the coordination of metal ions with appropriate organic linkers (Figure 9.1; Yaghi, O'Keeffe et al. 2003). Typically they form one-, two- or three-dimensional structures and are often porous. The organic linkers are characteristically mono-, di-, tri-, or tetravalent ligands that tend to form network structures upon coordination. Metal ions form nodes, which connect the linkers' arms to form a cage-like structure. Because of their hollow structure, MOFs have a very large internal surface area. Thus, the proper selection of metal ions and organic linkers becomes essential as it directs the structure and properties of the MOF. In comparison to other porous materials, MOFs have a high degree of structural complexity, homogeneity at the atomic level, structural uniformity, pore structures, tunable porosity, and adaptability in network topology, dimension, and geometry, as well as chemical functionalities (Kitagawa, Kitaura et al. 2004). These features allow researchers to precisely tune the topology, porosity, and functionality of frameworks in real time for a variety of applications. In addition, these structural frameworks can also be modified via post-synthetic modification strategies to improve their performance toward specific applications (Dang, Zhu et al. 2017). Consequently, MOFs have a great potential for effective integration and exploration in a wide range of physical, chemical, and biological applications. MOFs, like Lego bricks, can be assembled randomly, and in terms of flexibility, they may outclass previously known classes of materials.

MOFs are interesting examples of how the peculiar existence of hollow-structured materials could provide a slew of advantages, including improved surface-to-volume ratio, low density, microenvironment, and greater loading capacities, due to the synergistic effects of structures and compositions. MOFs are a rapidly developing research field that was pioneered by Prof. Omar Yaghi at the University of California, Berkeley in the late 1990s (Li, Eddaoudi et al. 1999). More than 90,000 different MOF structures have been published so far, with the number steadily increasing (Moosavi, Nandy et al. 2020). Further, the structural, as well as chemical diversity of MOFs, opens up distinctive prospects for catalysis, electrocatalysis, gas storage and separation, drug delivery, sensing, optoelectronics, energy storage applications, and

 DOI: 10.1201/9781003208709-9

more (Wang and Astruc 2020). Important electrochemical applications of MOFs are in sensors, energy storage and conversion, corrosion inhibition, and electrocatalysis (electrocatalytic reduction of toxic metal salts, degradation of toxic compounds, etc.; Figure 9.2). By definition, electrochemistry involves the transmission or storing of electrons at the electrode–electrolyte interface. As far as electrochemical applications are concerned, the development of new materials exhibiting high performance and low cost is often essential. Most MOFs display poor electronic conductivity and therefore cannot be used as electrode materials. To overcome their insulating properties, pristine MOFs are often modified for electrochemical applications; that is, they are either composited with conducting materials such as graphene, carbon

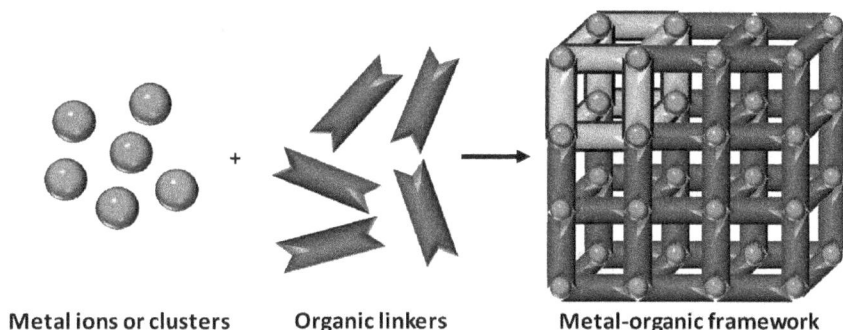

Metal ions or clusters **Organic linkers** **Metal-organic framework**

FIGURE 9.1 Scheme for the preparation of a MOF. Different metal ions or clusters are mixed with organic linkers using a convenient solvent. Coordination polymerization takes place between the precursors, resulting in a cross-linked network showing potential voids.

Source: Adapted with permission from MDPI 2018, Ref. Carrasco (2018)

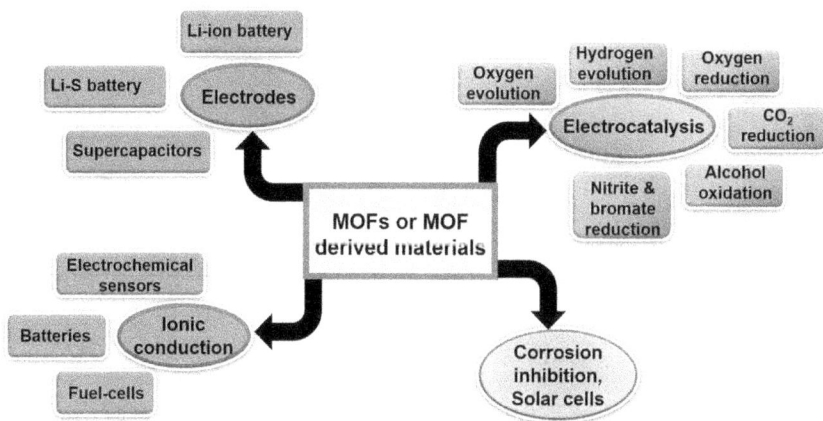

FIGURE 9.2 Electrochemical applications of MOFs or MOF-derived materials.

nanotubes, conducting polymers, and the like or carbonized under inert conditions to yield heteroatom-metal-doped carbon matrices subject to the choice of application (Scheme 9.1). The redox behavior of metal ions also plays an important role in providing a path for electrons.

The top-left panel shows the application as electrodes in Li-ion batteries, Li-S batteries, and supercapacitors. The top-right panel shows the application as electrocatalysts in oxygen evolution, hydrogen evolution, oxygen reduction, CO_2 reduction, alcohol oxidation, and nitrite/bromate reduction reactions. The bottom-left panel shows the use as ionic conductor in electrochemical sensors, batteries, and fuel cells, and the bottom-right panel shows corrosion inhibition and solar cell applications.

Some of the important electrochemical applications of MOFs are discussed in detail in the following sections.

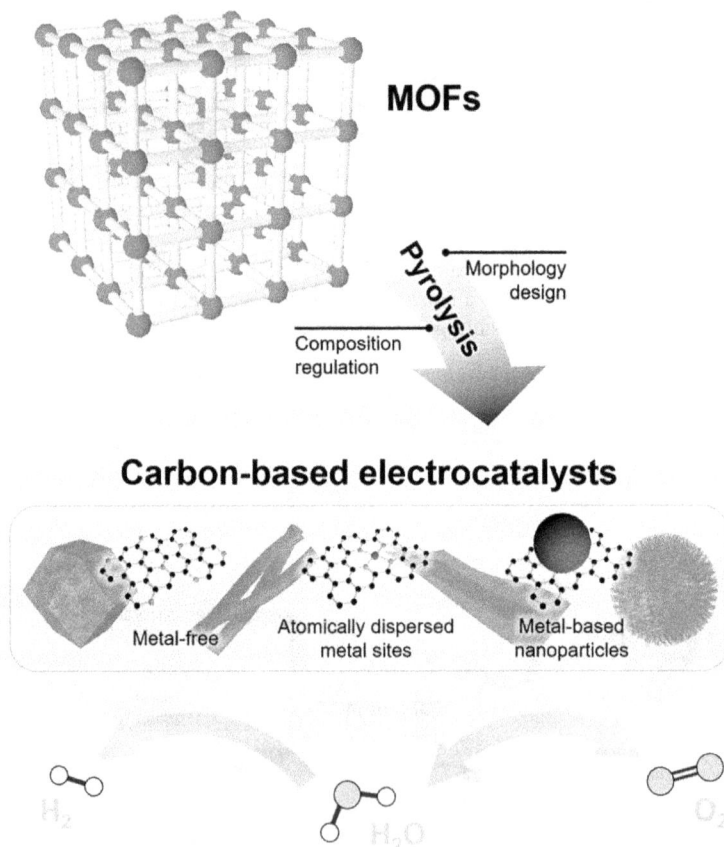

SCHEME 9.1 Design of MOF-derived carbon-based materials with various morphology and composition for the electrocatalysis of oxygen reduction, oxygen evolution, and hydrogen evolution reactions. Reproduced with permission from Wang, Chen et al. (2020).

9.2 OXYGEN EVOLUTION REACTION

Oxygen evolution reaction (OER) is an anodic reaction in recharging aqueous metal–air batteries and water electrolyzers. OER kinetics indicates that it is a very sluggish reaction owing to the participation of 4 electrons for each water molecule. Hence, this reaction typically involves high electric potential in the course of electrolysis (1.6–2 V *vs.* reversible hydrogen electrode [RHE]). IrO_2, a rare metal oxide, and very expensive catalyst is the benchmark catalyst. Effectual OER electrocatalysis is critical for electrochemical water splitting, rechargeable metal–air batteries, and so on (Wang, Chen et al. 2020). The crucial assessment parameters for OER comprise onset potential, potential to attain 10 mA cm^{-2} current density, Tafel slope, and so on. MOFs with customizable pore sizes, compositions, and morphologies have abundant catalytic sites for OER. Despite their substandard conductivity instigated by the insulating behavior of organic ligands and the meager overlap among their p-orbitals and d-orbitals, which impedes their electrochemical response (Du, Li et al. 2021). Alteration of metal centers with secondary metal can alter their d-band center, enhance their electronic configurations, and modify their charge-transfer path, leading to an enrichment in the catalytic performances toward the OER (Shi, Fu et al. 2019). In addition, the low mass penetrability caused by the partial meso- as well as macro-porosity of bulk MOFs and the obstruction of metal sites by organic ligands is furthermore an inevitable constraint for their OER activity (Du, Li et al. 2021). Such shortages can be overcome by converting bulk MOFs to an ultrathin MOF layer with few nanometer thicknesses. This approach benefits from rapid mass transport as well as electron transfer numbers during catalysis, enlarged surface area, and coordinatively unsaturated metal centers resulting in an increased OER performance.

Two-dimensional (2D) Co-MOF nanosheets of approximately 2 nm thickness were prepared in a single step by hydrothermal method with polyvinylpyrrolidone (PVP) as a surfactant (Xu, Li et al. 2018). The prepared nanosheets displayed low overpotential and high intrinsic activity under standard OER conditions than the bulk and micro–nano Co-MOFs. The observed high OER activity was attributed to the improved electrical conductivity, the existence of coordinatively unsaturated Co sites, and high ease of access to active metal centers. However, it was difficult to remove the organic surfactant completely, and as a result, the organic surfactant was shown to bind on the MOF surface, which partially blocks the active sites. To overcome this, ultrathin iron-cobalt MOF-74 nanosheets were synthesized using amorphous iron-cobalt-oxide nanosheets (FeCo-ONS) under hydrothermal conditions wherein FeCo-ONS served both as the metal source and template (Figure 9.3). The as-prepared FeCo-ONS nanosheet exhibited a thickness of 2.6 nm and showed a 22-mV low overpotential than the solvothermal synthesized sample. Density functional theory (DFT) calculations further showed that the OH* adsorption energy barrier on Co sites might be modified by the interaction of Co and Fe, that is, via partial electron transfer from Co to Fe atoms.

Zhao et al. reported NiCo bimetal–organic framework nanosheets (NiCo-UMOFNs), which exhibited high electrocatalytic activity for OER (Zhao, Wang et al. 2016). NiCo-ultrathin metal–organic framework nanosheets (UMOFNs) were loaded on Cu foam electrode, and the modified electrode displayed a low onset potential of 1.39 V and an overpotential of 189 mV at 10 mA cm^{-2} under standard OER conditions. X-ray

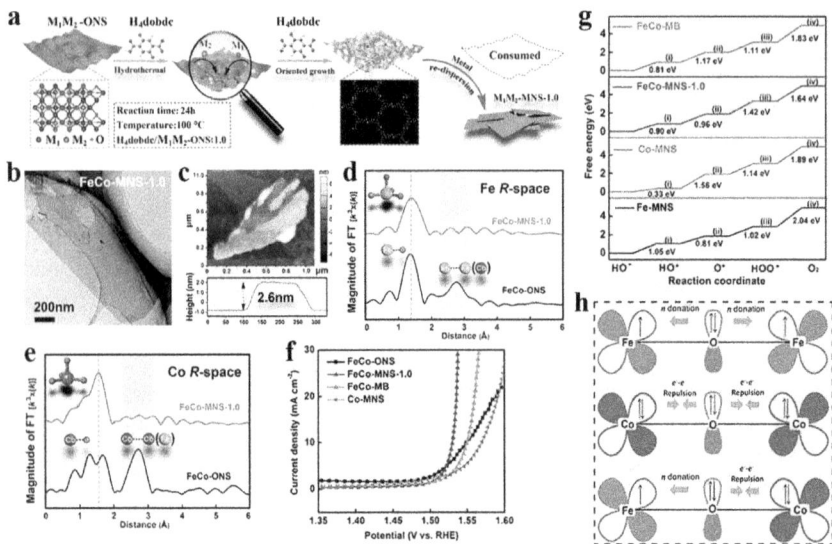

FIGURE 9.3 (a) Fabrication process, (b) transmission electron microscopy (TEM) image, and (c) atomic force microscopy (AFM) image of FeCo-MNS. The k^3-weighted Fourier transform extended X-ray absorption fine structure (EXAFS) spectra of (d) Fe R-space and (e) Co R-space. (f) OER polarization curves measured in 0.1 M KOH. (g) Energy profiles for the OER. (h) Schematic representation of the electronic coupling between Fe and Co.

Source: Reproduced with permission from Du, Li et al. (2021) Copyright 2021, RSC.

absorption spectroscopy (XAS) shows that the presence of coordinatively unsaturated metal atoms in the nanosheets is accountable for high OER activity. Furthermore, DFT calculations showed that fewer unfilled 3d e_g states of the Ni atoms in NiCo-UMOFNs acted as coordinatively unsaturated active metal centers for OER catalysis rather than Co. Furthermore, the addition of Co persuaded the coupling effect and enhanced the partial charge transfer between Ni^{2+} to Co^{2+} resulting in an improved performance.

Besides ultrathin MOF materials, well-regulated modification of the structure of bulk MOFs and specific modulation of their properties also benefit OER catalysis. A bimetallic MOF comprising terephthalic acid (A), 2-aminoterephthalic acid (B) ($A_{2.7}B$-MOF-$FeCo_{1.6}$) with structural heterogeneity was synthesized by a simple solvothermal process (Xue, Li et al. 2018). The electronic structure of MOF was modulated by altering the metal centers and the proportion of two organic linkers. The improved bimetallic MOF exhibited rapid water oxidation kinetics with a low Tafel slope and delivered an overpotential of 288 mV at a current density of 10 mA cm^{-2}. The $A_{2.7}B$-MOF-$FeCo_{1.6}$ MOF showed superior performance than its counterparts such as A-MOF-Fe, A-MOF-Co, and A-MOF-$FeCo_{1.6}$ as well as B-MOF-$FeCo_{1.6}$. Further insights from X-ray photoelectron spectroscopy (XPS) indicated charge transfer from oxygen to Fe and Co. Table 9.1 summarizes the OER performance of different MOF-based catalysts. Electrochemical conditions/parameters such as substrate, electrolyte, overpotential, Tafel slope, and more are given in detail for easy consideration.

TABLE 9.1

Electrocatalytic OER Activities of MOF-Based Catalysts. Reproduced with permission from Du, Li et al. (2021).

MOF	Substrate	Electrolyte	Catalyst Loading (mg cm^{-2})	Overpotential at 10 mA cm^{-2} (mV)	Tafel Slope (mV dec^{-1})	Stability	Ref.
NiCo-UMOFNs	GC/CF	1M KOH	0.2/N	250/189	42/N	200 h at 1.48 V	(Zhao, Wang et al. 2016)
NiFe-UMNs	GC	1M KOH	0.4	260	30	10,000 s at 1.51 V	(Hai, Jia et al. 2018)
Ultrathin Co-MOF	GC	1M KOH	0.25	263	74	12,000 s at 10 mA cm^{-2}	(Xu, Li et al. 2018)
FeCo-MNS	GC	0.1M KOH	0.36	298	21.6	10,000 s at 1.53 V	(Zhuang, Ge et al. 2019)
CoFe 2D MOFs	GC	1M KOH	0.35	274	46.7	10 h at 10 mA cm^{-2}	(Cai, Liu et al. 2020)
NiFe-MOF/OM-NFH	GC	1M KOH	0.4	270	123	40,000 s at 1.5 V	(Li, Ma et al. 2019)
Fe:2D-Co-NS	NF	1M KOH	N	211	46	96 h at 10 mA cm^{-2}	(Huang, Li et al. 2018)
A$_{2.7}$B-MOF-FeCo$_{1.6}$	GC	1M KOH	0.35	288	39	38,000 s at 10 mA cm^{-2}	(Xue, Li et al. 2018)
Fe$_2$Ni-BDC	GC	1M KOH	0.255	260	35	10 h at 10 mA cm^{-2}	(Li, Huang et al. 2019)
HHTP@ZIF-67	GC	1M KOH	N	238	104	12,000 s at 1.494 V	(Zhu, Ding et al. 2018)
Ni-Fe-MOF NSs	GC	1M KOH	N	221	56	20 h at 10 mA cm^{-2}	(Li, Wang et al. 2019)
CTGU-10c2	GC	0.1M KOH	0.0706	240	58	50 h at 10 mA cm^{-2}	(Zhou, Huang et al. 2019)
(Ni$_2$Co)$_{1-x}$Fe$_x$-MOF-NF	C-C	1M KOH	0.54	257	41.3	35 h at 1.488 V	(Qian, Li et al. 2019)
Co$_x$Fe$_{1-x}$-MOF-74	RDE	1M KOH	N	280	56	12 h at 1.51 V	(Zhao, Pattengale et al. 2018)

(Continued)

TABLE 9.1 (CONTINUED)

MOF	Substrate	Electrolyte	Catalyst Loading (mg cm^{-2})	Overpotential at 10 mA cm^{-2} (mV)	Tafel Slope (mV dec^{-1})	Stability	Ref.
NNU-23	CC	0.1M KOH	1.0	365	77.2	15 h at 10 mA cm^{-2}	(Wang, Dong, Qiao et al. 2018)
NiFe-PBA	GC	1M KOH	0.25	258	46	100 h at 10 mA cm^{-2}	(Su, Wang et al. 2018)
NiFe-MOF	NF	0.1M KOH	0.3	240	34	20,000 s at 1.42 V	(Nam, Shekhah et al. 2020)
NFN-MOF	NF	1M KOH	0.6	240	58.8	30 h at 500 mA cm^{-2}	(Senthil Raja, Chuah et al. 2018)
NiFeCP	NF	1M KOH	N	188	29	17 h at 50 mA cm^{-2}	(Li, Li et al. 2019)
NiFe	NFF	1M KOH	N	227	38.9	15 h at 20 mA cm^{-2}	(Cao, Ma et al. 2019)
LDH-MOF	NF	1M KOH	N	275.38	47	10 h at 10 mA cm^{-2}	(Chen, Zhuang et al. 2019)

9.3 OXYGEN REDUCTION REACTION

Oxygen reduction reaction (ORR) refers to the half-reaction where oxygen is reduced to water or hydrogen peroxide. Typically in fuel cells, the reduction of oxygen to water is preferred since the current is higher. However, the slow kinetics of ORR is accountable for major voltage loss in low-temperature H_2/air fuel cells even with the benchmark Pt-based catalysts. Classically, in ORR, electrochemically molecular oxygen is reduced to H_2O (acidic media) or OH- (alkaline media) intermediates in a 4-e⁻ pathway or to H_2O_2 (acidic media) or HO_2^- intermediates in a 2-e⁻ pathway (Figure 9.4; Xia, Mahmood et al. 2016, Kulkarni, Siahrostami et al. 2018) As a result of the higher reaction kinetics as well as efficacy, generally, the 4-e⁻ pathway is preferred. In practical applications, Pt-based electrocatalysts are best for the 4-e⁻ pathway; however, the high cost, paucity, and lower durability confine its well-known application in renewable energy technologies. As an alternate, earth-abundant transition metal–based catalysts are developed, and in this aspect, the development of MOFs based on transition metals with high activity and durability becomes indispensable. As already known, pristine MOFs lack conductivity, and hence the integration of MOFs with nanomaterials such as nano carbons, carbon nanotubes, or inorganic nanoparticles becomes an efficient method to achieve significant performance in ORR catalysis (Xia, Mahmood et al. 2016). Following Xu and co-workers' report on the derivatization of MOF using MOF-5 as the starting material, numerous MOF-derived materials were reported comprising porous carbons, metal/metal oxide, nitride, phosphide-doped carbons, heteroatom-doped carbon, and others using MOF as precursors (Li, He et al. 2019). Table 9.2 summarizes the performance of reported MOFs and their composites toward ORR electrocatalysis.

Numerous examinations have established that the local electronic structure, as well as physicochemical properties, of catalysts are influential factors for ORR activity. In addition to this, heteroatom doping and the introduction of M-N-C sites/

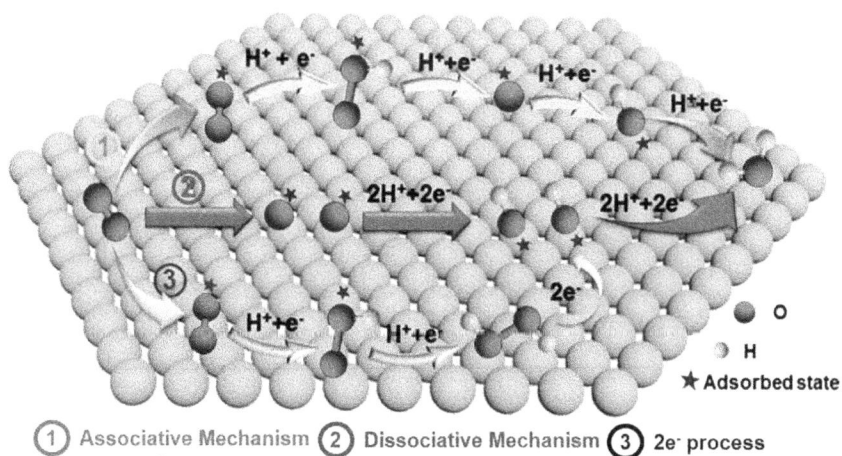

FIGURE 9.4 Illustration of the reaction pathway of the ORR and the resulting intermediates on the catalyst surface through different mechanisms.

Source: Reproduced with permission from RSC, Li, He et al. (2019)

TABLE 9.2

A List of MOF-Derived Nonnoble ORR Catalysts and Their Performance. Adapted with permission from RSC, Li, He et al. (2019).

MOF-derived Nonnoble Electrocatalysts	ORR Performance			Testing Conditions	Performance (peak power density, mW cm⁻²)	Ref.
	Onset Potential (V vs. RHE)	Half-Wave Potential (V vs. RHE)				
Co(im)₂	0.83	0.77		0.1 M HClO₄	—	(Ma, Goenaga et al. 2011)
ZIF-67–750*	0.86	0.71		0.1 M HClO₄	—	(Xia, Zhu et al. 2014)
ZIF-67–800*	0.92	0.82		0.1 M KOH	—	(Chen, Ma et al. 2017)
ZIF-67+Ni²⁺-800*	0.070**	−0.049**		0.1 M KOH	Methanol fuel cell: 4335.6	(Tang, Cai et al. 2016)
MIL-88B-NH₂_900*	1.03	0.92		0.1 M KOH	Alkaline direct fuel cell: 22.7	(Zhao, Yin et al. 2014)
ZIF-8 + Fe²⁺ + Phen + NH₃_1050*	—	—		Acid	H₂/O₂ PEMFC: 0.75 @ 0.64 V	(Lefèvre, Proietti et al. 2009)
Fe-ZIF-8–1000*	0.95	0.82		0.1 M HClO₄	—	(Wang, Zhang et al. 2016)
Zn/Co-ZIF-8–900*	0.982	0.881		0.1 M KOH	—	(Yin, Yao et al. 2016)
ZIF-8 + Fe(acac)₃–900*	0.986	0.900		0.1 M KOH	—	(Chen, Ji et al. 2017)
ZIF-8 + Fe(acac)₃ + Co(acac)₃–900*	0.995	0.920		0.1 M KOH	—	(Zhang, Chen et al. 2018)
ZIF-8 + FePc + DMSO-1000*	0.893	0.804		0.5 M H₂SO₄	—	(Liang, Chen et al. 2017)
Fe-ZIF-8–1000*	0.902	0.805		0.1 M HClO₄	H₂/O₂ PEMFC: 1141	(Liu, Liu et al. 2018)
Fe-ZIF-8–1100*	0.95	0.85		0.5 M H₂SO₄	—	(Zhang, Hwang et al. 2017)
Fe20-PCN-222–800*	0.93	0.776		0.1 M HClO₄	—	(Jiao, Wan et al. 2018)
Cu(BTC)(H₂O) + dicyandiamide-800*	0.94	0.869		0.1 M KOH	—	(Li, Han et al. 2018)

Mn-ZIF-8-Cyanimide-1100 & 900* (two step carbonization)	0.96	0.80	0.5 M H_2SO_4	H_2/O_2 PEMFC: 460	(Li, Chen et al. 2018)
Zn/Co-ZIF-8 + $FeCl_3$_900*	1.06	0.863	0.1 M $HClO_4$	H_2/Air PEMFC: 505	(Wang, Huang et al. 2017)
ZIF-67 + H_2S—1000*	−0.04***	−0.14***	0.1 M KOH	–	(Chen, Li et al. 2015)
ZIF-67 + Se—750*	0.935	0.806	0.1 M KOH	–	(Wu, Xue et al. 2016)
ZIF-67 + Thioacetamide—700*	0.85	0.78	0.1 M $HClO_4$	–	(Zhang, An et al. 2016)

* Carbonization temperature
** The potential is referenced with a standard calomel electrode.
*** The potential is referenced with an Ag/AgCl electrode.

defects were shown to enhance the active sites for ORR. Besides, volcano plots done on numerous materials suggested that the integration of heteroatom or metal dopants can refrain the electronic structure of catalysts by merging the spin density, charge effects, and ligand effect, generating a low overpotential closer to benchmark Pt. Hence, the identification of an active site is a crucial step. Furthermore, increasing porosity creates a large number of defects, as well as edge sites, which are considered active sites for ORR. In addition, hierarchical porous structures with both micro- and mesopores favor mass transfer ability. Combining all these requirements, such as electron transfer, mass transfer, interface compatibility, and so on, MOF deriva- tives have gained considerable interest as they serve as a promising platform for the construction of nonnoble ORR catalysts.

9.4 HYDROGEN EVOLUTION REACTION

A sustainable way of hydrogen production can be done by electrochemical water splitting, chemo-catalytic water splitting or by photocatalytic pathway to obtain pure hydrogen. Hydrogen generation is important as hydrogen can be used as an alternate energy source and can fuel power vehicles based on the hydrogen fuel cell (Zaman, Noor et al. 2021). Among these, the electrochemical hydrogen generation is limited by three factors, such as the short lifetime of the electrode material, the use of noble metal catalysts, and thermal efficiency less than the water-splitting thermodynamic limits, that is, 1.23 V. Typically HER reactions are carried out under acidic condi- tions in which most of the metal catalysts show insufficient stability except noble metal catalysts (Scheme 9.2). However, electrolyzers established under alkaline con- ditions are technically well built and commonly exist at the commercial scale. In addition, HER in alkaline media does not pose similar stability risks as the one tackled in acidic conditions. In alkaline media, HER is controlled by three subtle and important descriptors, such as (1) hydrogen adsorption on the surface of the catalyst, (2) restriction of the adsorption of the hydroxyl group on the surface of the catalyst to prevent catalyst poisoning, and (3) energy required for the dissociation of a water molecule (Scheme 9.2). However, the major setback in alkaline HER is the decreased reaction kinetics. HER reactions in acidic media are 2–3 magnitude higher than the rate of the reaction in alkaline media. In general, HER requires high overpotential, and in this case, Pt, Ru, and Pd show better efficiency. Principally, 20% Pt/C is used as HER benchmark catalyst, and henceforth, most of the research activities are geared toward finding an alternate electrocatalyst with similar or improved stability and performance of 20% Pt/C.

The combined effect of porous properties and framework structure of MOF in this regard provides a range of applications, especially in the field of electrocatalysis. By proper engineering of the framework and by incorporation of various ligands, metal components improved catalytic activity can be achieved. Furthermore, the composite preparation with a variety of conducting materials and/or carbonization approach generates new MOF-derived materials for electrocatalytic applications, especially towards HER. Such a design is expected to lower the HER overpotential, improve the kinetics of the reaction, and reduce the cost compared to expensive noble metal catalysts.

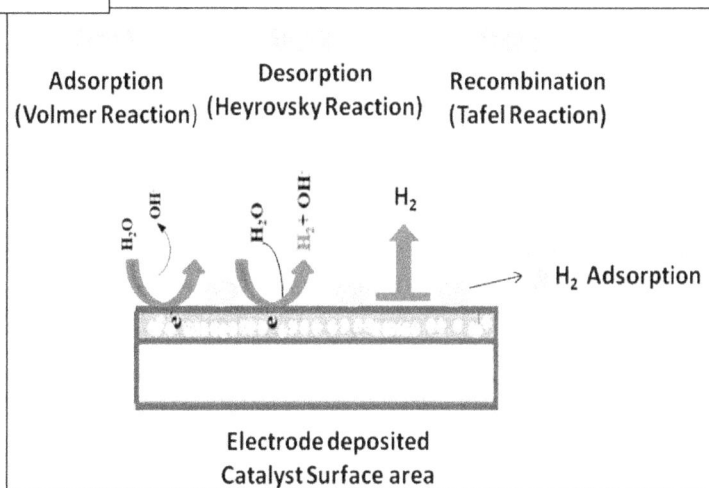

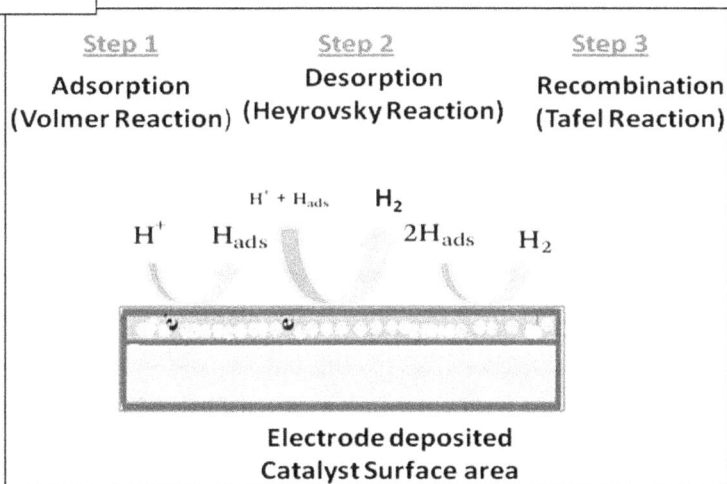

SCHEME 9.2 Schematic diagram of the Volmer–Heyrovsky and Volmer-Tafel processes on a catalyst surface in acidic and alkaline media. Reproduced with permission from Zaman, Noor et al. (2021).

It is recognized that 3D MOFs show better stability (structural and morphological) with high porosity. However, 1D/2D MOFs display improved usage of the active sites, with low stability and poor porosity. Thus, mingling the advantages of both 3D and 1D/2D MOFs is extremely desired but remains a great challenge. Also, the electrocatalysts that are efficient in acidic media are not effective under alkaline conditions requiring high overpotentials to initiate the catalysis. As discussed in other sections pristine MOFs show poor conductivity and poor stability under alkaline and acidic conditions; hence, their use as electrocatalysts requires special treatments such as composite formation and others. Table 9.3 summarizes the performance of some MOF-derived catalysts toward HER.

9.5 CO$_2$ REDUCTION

Electrochemical reduction of CO$_2$ to value-added products affords a pathway for the decrease of atmospheric CO$_2$ concentration and for the generation of valuable carbonaceous fuels/chemicals under ambient conditions (Zhang, Cai et al. 2021). The extreme chemical stability of CO$_2$ molecules poses a challenge in the effective conversion of CO$_2$ to value-added products due to the large kinetic and thermodynamic energy barrier. Electrochemical CO$_2$ reduction is a complex reaction that involves multiple proton and electron transfer steps as shown in Table 9.1. As can be seen from the equation, the process involves the formation of many complex intermediates followed by 2-, 4-, 6-, 8-, or 12-electron reduction in aqueous media, and the product selectivity is determined by their reaction pathway (Qiao, Liu et al. 2014). The product selectivity also depends on the mode of interaction and nature of binding with the metal centers, and this mode of interaction is different for different metals (Kortlever, Shen et al. 2015, Zhang, Hu et al. 2018). Initially, a one-electron reduction generates CO$_2^-$ radical, which is then absorbed on the surface of the catalyst; upon protonation, *COOH intermediate is formed. Successive proton-coupled-electron transfer reactions with *COOH intermediate generates CO and H$_2$O; then CO desorbs from the catalyst surface. Generation of C2, C3, C4 carbon products entails strong adsorption ability of CO intermediate on the catalyst surface which promotes further dimerization, multiple proton-coupled-electron-transfer reactions then affords multi-carbon products. Thus, it is evident that the catalytic activity and selectivity of the products rely on the number of surface active sites, coordination geometry, electronic structure, and chemical composition of the electrocatalyst (Arán-Ais, Gao et al. 2018). Hence, the development of high-performance electrocatalysts turns out to be essential for the efficient conversion of CO$_2$ to useful chemicals such as CO, CH$_4$, formic acid, methanol, ethane, ethylene, ethanol, and others (Shao, Yi et al. 2020). In this trend, MOFs with active metal ion sites, the high specific surface area turns out to be an ideal platform for the electrochemical reduction of CO$_2$, however, the poor electronic conductivity of pristine MOFs hampers their direct role towards electrocatalysis. To improve the electronic conductivity, MOFs were often composited with conductive materials like carbon nanotubes, graphene, conducting polymers, and others. The robust contact between the MOF and these substrates leads to an increase in the number of active sites, which further accelerates the charge transfer between the electrode and the catalysts. Besides, these MOFs were also carbonized under an inert atmosphere to prepare MOF-derived porous nanomaterial that shows improved electronic conductivity as well as enhanced electrocatalytic behavior (Shao, Yi et al. 2020). Carbonization leads to the formation of homogeneously dispersed metal/metal-oxide nanoparticles or single metal atom/metal oxide all over the defective graphitic carbon matrix. This approach facilitates the contact between dissolved CO$_2$ and active sites.

TABLE 9.3

A List of MOF-Derived Nonnoble HER Catalysts and Their Performance. Adapted with permission from RSC, Zaman, Noor et al. (2021).

Electrocatalyst	Testing Conditions	η10 (mV)	Tafel Slope (mV dec−1)	Stability	Ref.
3D Nibpyfcd Hp	0.5 M H2SO4	350	60	30 h	(Khrizanforova, Shekurov et al. 2020)
		400	65	2000 cycles	
3D Cobpyfcd Hp		400			
1D Zn(fcdHp)	0.5 M H2SO4	340	110	1000 cycles	(Shekurov, Khrizanforova et al. 2019)
2D MOF H3[NiIII3(tht)2]	0.5 M H2SO4	333	80.5	–	(Dong, Pfeffermann et al. 2015)
THTA—Co H3[Co3(tht)(thₑ)]	0.5 M H2SO4	283	71	300 cycles	(Dong, Zheng et al. 2017)
G/THTA—Co G/H3[Co3(thₑ)(tha)]		230	70	400 cycles, 4 h	
H3[Ni3(tht)(tha)]		315	76	–	
Au/Co Fe-MOFNs	0.1 M KOH		115	–	(Wang, Jiao et al. 2019)
Cu-MOF:					
HKUST-1 ED	0.5 M H2SO4	590	183.6	–	(Li, Lu et al. 2019)
HKUST-1HT		660	222.4	–	
2D Cu MOF, Cu6(C8H4O4)5(H2O)6·H3[P(W3O10)4]	0.5 M H2SO4	660	100	–	(Hu, Pang et al. 2019)
Hf12-CoDBP/(CNTs)	0.026 M TFA	650	178	7 h	(Micheroni, Lan et al. 2018)
CTGU-5	0.5 M H2SO4	388	125	–	(Wu, Zhou et al. 2017)
CTGU-6		425	176	–	
Acetylene Black&CTGU-5		44	45	96 h	

(Continued)

TABLE 9.3 (CONTINUED)

Electrocatalyst	Testing Conditions	η10 (mV)	Tafel Slope (mV dec−1)	Stability	Ref.
UiO-66-NH2-Mo-5	0.5 M H2SO4	200	59	7 h	(Dai, Liu et al. 2016)
NENU-500	0.5 M H2SO4	237	96	2000 cycles	(Qin, Du et al. 2015)
NENU-501		392	137		
HUST-201	0.5 M H2SO4	192	79	2000 cycles	(Zhang, Li et al. 2018)
HUST-200		131	51		
(GO 8 wt%) Cu-MOF composite	0.5 M H2SO4	209 at 30 mA cm−2	84		(Jahan, Liu et al. 2013)
Zn0.30Co2.70S4 polyhedra	0.5 M H2SO4	80	47.5	60 h	(Huang, Song et al. 2016)
CoP CPHs	0.5 M H2SO4	133	51	12 h	(Xu, Han et al. 2015)
CoP@BCN-1	1.0 M KOH	215	52	2000 cycles	(Tabassum, Guo et al. 2017)
CuCo@NC	0.5 M H2SO4	145	79	30000 s	(Kuang, Wang et al. 2017)
Cu3P@NPPC-650	0.5 M H2SO4	89	76	3000 cycles	(Wang, Dong, Du et al. 2018)

From Table 9.4, and the literature data, it is clear that the assembly of MOFs with a suitable metal center is of extreme importance. For example, for the electrocatalytic conversion of CO_2 to CO, Co, Fe, Zn, and noble metal–based MOFs were used, whereas for the synthesis of C2 and C3 hydrocarbons, Cu-based MOFs were employed. Table 9.2 summarizes the electrochemical performance of different MOF-based electrocatalysts toward CO_2 reduction. Fe and Co exist in several oxidation/reduction states, and hence, they offer a better catalytic activity for the

TABLE 9.4
Electrochemical Potentials of Possible CO_2 Reduction Products in Aqueous Solutions

Possible Half-Reactions of Electrochemical CO_2 Reduction	Electrode Potentials (V vs. SHE) at pH 7	Product Name
$CO_2(g) + e^- \rightarrow {}^*COO^-$	−1.90	CO_2 anion radical
$CO_2(g) + 2H^+ + 2e^- \rightarrow HCOOH(l)$	−0.61	Formic acid
$CO_2(g) + H_2O(l) + 2e^- \rightarrow HCOO^-(aq) + OH^-$	−0.43	Formate
$CO_2(g) + 2H^+ + 2e^- \rightarrow CO(g) + H_2O(l)$	−0.53	Carbon monoxide
$CO_2(g) + H_2O(l) + 2e^- \rightarrow CO(g) + 2OH^-$	−0.52	Carbon monoxide
$CO_2(g) + 4H^+ + 2e^- \rightarrow HCHO(l) + H_2O(l)$	−0.48	Formaldehyde
$CO_2(g) + 3H_2O(l) + 4e^- \rightarrow HCHO(l) + 4OH^-$	−0.89	Formaldehyde
$CO_2(g) + 6H^+(l) + 6e^- \rightarrow CH_3OH(l) + H_2O(l)$	−0.38	Methanol
$CO_2(g) + 5H_2O(l) + 6e^- \rightarrow CH_3OH(l) + 6OH^-$	−0.81	Methanol
$CO_2(g) + 8H^+ + 8e^- \rightarrow CH_4(g) + 2H_2O(l)$	−0.24	Methane
$CO_2(g) + 6H_2O(l) + 8e^- \rightarrow CH_4(g) + 8OH^-$	−0.25	Methane
$2CO_2(g) + 12H^+ + 12e^- \rightarrow C_2H_4(g) + 4H_2O(l)$	0.06	Ethylene
$2CO_2(g) + 8H_2O(l) + 12e^- \rightarrow C_2H_4(g) + 12OH^-$	−0.34	Ethylene
$2CO_2(g) + 12H^+ + 12e^- \rightarrow CH_3CH_2OH(l) + 3H_2O(l)$	0.08	Ethanol
$2CO_2(g) + 9H_2O(l) + 12e^- \rightarrow CH_3CH_2OH(l) + 12OH^-(l)$	−0.33	Ethanol

electroreduction of CO_2. Hod and co-workers (Hod, Sampson et al. 2015) electrodeposited a thin film of Fe-MOF-525 on fluorine-doped tin oxide (FTO) glass. Fe-MOF catalytically reduced CO_2 to syngas CO and H_2 with 100% Faradaic efficiency. Similarly, Dong and co-workers (Dong, Qian et al. 2018) prepared a composite catalyst with carbon black and Fe-porphyrin-based MOF and used it as an electrocatalyst for the electroreduction of CO_2 to CO in a CO_2-saturated 0.5 M $KHCO_3$ solution with 91% Faradaic efficiency. Co displays oxidation state-dependent coordination geometries, such as square planar, tetrahedral, pyramidal, and octahedral. Lan and co-workers reported the hydrothermal synthesis of polyoxometalate (POM)-metalloporphyrin organic frameworks (M-PMOF) and tested their electrocatalytic reduction activity in CO_2-saturated 0.5M $KHCO_3$ solution (Wang, Huang et al. 2018). The synergistic effects of POM and Co-porphyrin contributed to the excellent electrocatalytic performance with a remarkable Faradaic efficiency of 99% for CO production.

Cu metal has been well studied for its electrocatalytic behavior towards CO_2 reduction. However, the first study on the use of copper-based MOF as an electrocatalyst was reported by Hinogami and co-workers (Hinogami, Yotsuhashi et al. 2012), wherein a copper rubeanate MOF was synthesized by mixing rubeanic acid and aqueous $CuSO_4$ solution. Cu-rubeanate MOF on carbon paper showed a highly selective reduction to formic acid (98% selectivity) in 0.5M $KHCO_3$ with a current efficiency of 30%. The performance of the MOF electrocatalyst was compared to that of the Cu metal electrode under the same conditions. Results showed that the MOF electrode displayed 0.2V more positive onset potential than the Cu metal electrode and produced a 13-fold greater amount of formic acid than the Cu metal electrode. The high selectivity of the Cu-MOF electrode was ascribed to the low electron density on metallic sites and weak adsorption of CO_2 on the MOF surface compared with the Cu metal electrode. This weak adsorption is responsible for the selective production of formic acid. Besides the structure of MOF, the solvent also plays an important role in the catalytic activity and product selectivity. Kumar et al. (Senthil Kumar, Senthil Kumar et al. 2012) reported the electrochemical preparation of $Cu_3(BTC)_2$ MOF and its electrocatalytic activity towards the preparation of oxalic acid with 51% Faradaic efficiency and 90% purity (Figure 9.5). Zinc-based MOFs with 4–6 coordination numbers were synthesized and their electrocatalytic activities towards CO_2 reduction were investigated. Wang and co-workers reported the synthesis of Zn-based zeolitic imidazolate framework (ZIF-8). ZIF-8 synthesized using $ZnSO_4$ showed the highest electrocatalytic activity for the conversion of CO_2 to CO with a Faradaic efficiency of 65% (Wang, Hou et al. 2017). Apart from these transition metal-based MOFs, noble metal–based electrocatalysts were also synthesized and their activity toward the electrocatalytic reduction of CO_2 was investigated. Re-SURMOF-based electrocatalyst showed excellent conversion of CO_2 to CO with 93% Faradaic efficiency (Ye, Liu et al. 2016). The superior performance of Re-SURMOF is attributed to the charge transport among the electrode interface and Re-MOF redox catalyst.

Apart from pristine MOF materials, MOF-derived materials were also employed as electrocatalysts (Singh, Mukhopadhyay et al. 2021). It is well known that metal nodes in MOFs act as active centers for catalysis. The high-temperature treatment of

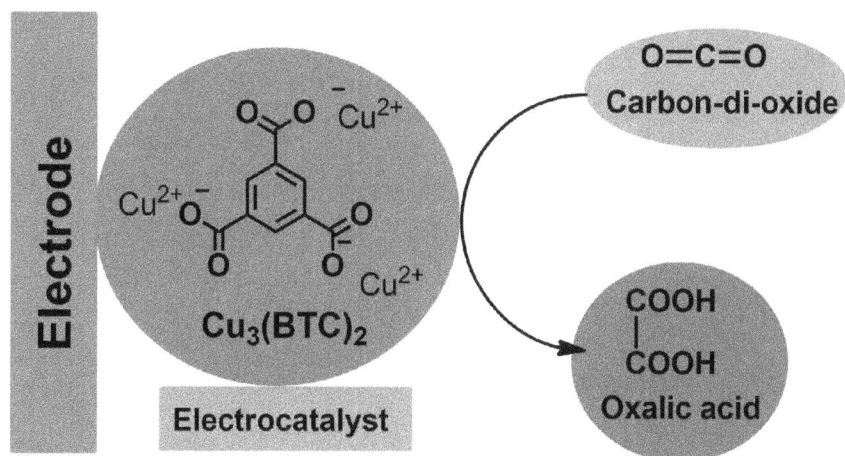

FIGURE 9.5 Electrochemical reduction of CO_2 to oxalic acid on $Cu_3(BTC)_2$ MOF.

MOFs yields metal nanoparticles or single-metal atom-embedded porous carbon materials that show distinct advantages for electrocatalytic CO_2 reduction with improved electronic conductivity, specific surface area, and the like. Table 9.5 summarizes the performance of MOF-based electrocatalysts toward electrochemical CO_2 reduction.

9.6 CONCLUSION AND FUTURE PROSPECTS

The development of effectual nonnoble electrocatalysts for oxygen evolution, oxygen reduction (ORRs), hydrogen reduction, and CO_2 reduction (CO_2RRs) reactions have attracted widespread attention owing to their prominence in energy conversion applications. Therefore, the exploitation of the basis of catalytic activity of electrocatalysts has become an everlasting topic for both theoretical and experimental researchers. The research on the modulation of isolated components to improve the catalytic activity does not solve the obstinate challenges. While designing these catalysts, structure-level thinking is required rather than restraining the catalyst from one dimension. MOFs, with their hierarchical structure, intrinsic porosity, high surface area, and structural and morphological tunability, show promising aspects as electrocatalysts; however, there are other issues that have to be considered for the successful deployment of these materials. As far as electrocatalysis is concerned, MOFs were used as electrode materials both on the anodic side for the oxidation and on the cathodic side for the reduction of specific substrates (water, CO_2, etc.); however, the poor conducting nature of these materials hinders such activity. MOFs are frequently functionalized/composited with conducting materials such as CNT, graphene, conducting polymers, and the like or carbonized to render graphitized metal-oxide materials as a step to improve their conductivity. Typically, such composites show improved conductivity with active metal sites, and hence, they are ideal electrocatalysts compared to pristine MOF materials. Water splitting and ORRs, as

TABLE 9.5

MOF-Based Electrocatalysts and Their Performance Toward Electrochemical CO_2 Reduction

Electrode	Potential	Products	Faradaic Efficiency (%)	Ref.
Zn-BTC MOF/CP	−2.2 V vs. Ag/AgCl	CH_4	80.1 ± 6.6	(Kang, Zhu et al. 2016)
		CO	7.9 ± 2.6	
ZIF-8	−1.8 V vs. SCE	CO	65	(Wang, Hou et al. 2017)
Re-SURMOF/FTO	−1.6 V vs. NHE	CO	93 ± 5	(Ye, Liu et al. 2016)
Fe-MOF-525/FTO	−1.3 V vs. NHE	CO	54 ± 2	(Hod, Sampson et al. 2015)
		H_2	45 ± 1	
ZIF-8-derived Fe-N-C	−0.6 V vs. RHE	CO	91	(Huan, Ranjbar et al. 2017)
ZIF-8-derived Ni SAs/N-C	−1.0 V vs. RHE	CO	71.9	(Zhao, Dai et al. 2017)
Ag_2O/layered ZIF	−1.2 V vs. RHE	CO	80.5	(Jiang, Wu et al. 2017)
ZIF-Fe-CNT-FA-p	−0.86 V vs. RHE	CO	100	(Guo, Yang et al. 2017)
C-AFC/ZIF-8	−0.43 V vs. RHE	CO	93	(Ye, Cai et al. 2017)
$Al_2(OH)_2$TCPP-Co	−0.7 V vs. RHE	CO	76	(Kornienko, Zhao et al. 2015)
M-PMOF	−0.8 V vs. RHE	CO	98.7	(Wang, Huang et al. 2018)
Co/Zn ZIFs derived Co-N_2	Overpotential 520 mV	CO	94	(Wang, Chen et al. 2018)
Copper rubeanate MOF	−1.2 V vs. SHE	HCOOH	65	(Hinogami, Yotsuhashi et al. 2012)
$Cu_3(BTC)_2$	−1.12 V vs. Ag/Ag⁺	Oxalic acid	51	(Senthil Kumar, Senthil Kumar et al. 2012)
HKUST-1 mediate Cu	−1.07 V vs. RHE	C_2H_4	45	(Nam, Bushuyev et al. 2018)
GDE-Cu-MOF	−1.8 V vs. SCE	C_2H_4	16	(Qiu, Zhong et al. 2018)
Cu-based HKUST-1	10 mA.cm⁻²	CH_3OH	5.6	(Albo, Vallejo et al. 2017)
		C_2H_5OH	10.3	
OD Cu/C-1000 (Cu-based HKUST-1)	−0.1 V vs. RHE	CH_3OH	45.2	(Zhao, Liu et al. 2017)
		C_2H_5OH		
$Cu_2(L)$-e/Cu	−1.4 V vs. Ag/Ag⁺	HCOOH	90.5	(Kang, Li et al. 2020)

well as CO_2RRs, are frequently studied electrocatalytic reactions that are well established. In all these electrocatalytic reactions, noble metal catalysts, such as Pt or IrO_2/RuO_2, serve as the benchmark catalysts and their scarcity, as well as high cost, demand the search for cheaper alternatives with excellent electrocatalytic performance. As alternatives, inorganic transition metals, metal oxides, metal-free carbon composites, and others were widely investigated for their electrocatalytic behavior. Among these, MOF-derived catalysts have secured a special place as they contain both the metal sites as well as conductive carbon matrices with high surface area, intrinsic porosity, and other properties well suited for electrocatalysis. From previous studies, it is evident that an ideal catalyst should contain active metal sites, carbon support, and high surface area. MOF-derived metal/metal oxide–based carbon materials satisfy these criteria. In addition, the chances to design MOF composites are plenty and offer unique physicochemical properties, stability under alkaline/acidic conditions, and more. Theoretical studies supported by experimental tuning at the molecular level would offer plenty of chances for the generation of possible catalysts with efficient electrocatalytic activity.

REFERENCES

Albo, J., D. Vallejo, G. Beobide, O. Castillo, P. Castaño and A. Irabien (2017). "Copper-Based Metal—Organic Porous Materials for CO_2 Electrocatalytic Reduction to Alcohols." *ChemSusChem* **10**(6): 1100–1109.

Arán-Ais, R. M., D. Gao and B. Roldan Cuenya (2018). "Structure- and Electrolyte-Sensitivity in CO_2 Electroreduction." *Accounts of Chemical Research* **51**(11): 2906–2917.

Cai, M., Q. Liu, Z. Xue, Y. Li, Y. Fan, A. Huang, M.-R. Li, M. Croft, T. A. Tyson, Z. Ke and G. Li (2020). "Constructing 2D MOFs from 2D LDHs: A Highly Efficient and Durable Electrocatalyst for Water Oxidation." *Journal of Materials Chemistry A* **8**(1): 190–195.

Cao, C., D.-D. Ma, Q. Xu, X.-T. Wu and Q.-L. Zhu (2019). "Semisacrificial Template Growth of Self-Supporting MOF Nanocomposite Electrode for Efficient Electrocatalytic Water Oxidation." *Advanced Functional Materials* **29**(6): 1807418.

Carrasco, S. (2018). "Metal-Organic Frameworks for the Development of Biosensors: A Current Overview." *Biosensors* **8**(4).

Chen, B., R. Li, G. Ma, X. Gou, Y. Zhu and Y. Xia (2015). "Cobalt sulfide/N,S Codoped Porous Carbon Core—Shell Nanocomposites as Superior Bifunctional Electrocatalysts for Oxygen Reduction and Evolution Reactions." *Nanoscale* **7**(48): 20674–20684.

Chen, B., G. Ma, Y. Zhu and Y. Xia (2017). "Metal-Organic-Frameworks Derived Cobalt Embedded in Various Carbon Structures as Bifunctional Electrocatalysts for Oxygen Reduction and Evolution Reactions." *Scientific Reports* **7**(1): 5266.

Chen, J., P. Zhuang, Y. Ge, H. Chu, L. Yao, Y. Cao, Z. Wang, M. O. L. Chee, P. Dong, J. Shen, M. Ye and P. M. Ajayan (2019). "Sublimation-Vapor Phase Pseudomorphic Transformation of Template-Directed MOFs for Efficient Oxygen Evolution Reaction." *Advanced Functional Materials* **29**(37): 1903875.

Chen, Y., S. Ji, Y. Wang, J. Dong, W. Chen, Z. Li, R. Shen, L. Zheng, Z. Zhuang, D. Wang and Y. Li (2017). "Isolated Single Iron Atoms Anchored on N-Doped Porous Carbon as an Efficient Electrocatalyst for the Oxygen Reduction Reaction." *Angewandte Chemie International Edition* **56**(24): 6937–6941.

Dai, X., M. Liu, Z. Li, A. Jin, Y. Ma, X. Huang, H. Sun, H. Wang and X. Zhang (2016). "Molybdenum Polysulfide Anchored on Porous Zr-Metal Organic Framework to Enhance the Performance of Hydrogen Evolution Reaction." *Journal of Physical Chemistry C* **120**(23): 12539–12548.

Dang, S., Q.-L. Zhu and Q. Xu (2017). "Nanomaterials Derived from Metal—Organic Frameworks." *Nature Reviews Materials* **3**(1): 17075.

Dong, B.-X., S.-L. Qian, F.-Y. Bu, Y.-C. Wu, L.-G. Feng, Y.-L. Teng, W.-L. Liu and Z.-W. Li (2018). "Electrochemical Reduction of CO_2 to CO by a Heterogeneous Catalyst of Fe—Porphyrin-Based Metal—Organic Framework." *ACS Applied Energy Materials* **1**(9): 4662–4669.

Dong, R., M. Pfeffermann, H. Liang, Z. Zheng, X. Zhu, J. Zhang and X. Feng (2015). "Large-Area, Free-Standing, Two-Dimensional Supramolecular Polymer Single-Layer Sheets for Highly Efficient Electrocatalytic Hydrogen Evolution." *Angewandte Chemie International Edition* **54**(41): 12058–12063.

Dong, R., Z. Zheng, D. C. Tranca, J. Zhang, N. Chandrasekhar, S. Liu, X. Zhuang, G. Seifert and X. Feng (2017). "Immobilizing Molecular Metal Dithiolene—Diamine Complexes on 2D Metal—Organic Frameworks for Electrocatalytic H_2 Production." *Chemistry—A European Journal* **23**(10): 2255–2260.

Du, J., F. Li and L. Sun (2021). "Metal—Organic Frameworks and Their Derivatives as Electrocatalysts for the Oxygen Evolution Reaction." *Chemical Society Reviews* **50**(4): 2663–2695.

Guo, Y., H. Yang, X. Zhou, K. Liu, C. Zhang, Z. Zhou, C. Wang and W. Lin (2017). "Electrocatalytic Reduction of CO_2 to CO with 100% Faradaic Efficiency by Using Pyrolyzed Zeolitic Imidazolate Frameworks Supported on Carbon Nanotube Networks." *Journal of Materials Chemistry A* **5**(47): 24867–24873.

Hai, G., X. Jia, K. Zhang, X. Liu, Z. Wu and G. Wang (2018). "High-Performance Oxygen Evolution Catalyst using Two-Dimensional Ultrathin Metal-Organic Frameworks Nanosheets." *Nano Energy* **44**: 345–352.

Hinogami, R., S. Yotsuhashi, M. Deguchi, Y. Zenitani, H. Hashiba and Y. Yamada (2012). "Electrochemical Reduction of Carbon Dioxide using a Copper Rubeanate Metal Organic Framework." *ECS Electrochemistry Letters* **1**(4): H17–H19.

Hod, I., M. D. Sampson, P. Deria, C. P. Kubiak, O. K. Farha and J. T. Hupp (2015). "Fe-Porphyrin-Based Metal—Organic Framework Films as High-Surface Concentration, Heterogeneous Catalysts for Electrochemical Reduction of CO_2." *ACS Catalysis* **5**(11): 6302–6309.

Hu, A., Q. Pang, C. Tang, J. Bao, H. Liu, K. Ba, S. Xie, J. Chen, J. Chen, Y. Yue, Y. Tang, Q. Li and Z. Sun (2019). "Epitaxial Growth and Integration of Insulating Metal—Organic Frameworks in Electrochemistry." *Journal of the American Chemical Society* **141**(28): 11322–11327.

Huan, T. N., N. Ranjbar, G. Rousse, M. Sougrati, A. Zitolo, V. Mougel, F. Jaouen and M. Fontecave (2017). "Electrochemical Reduction of CO_2 Catalyzed by Fe-N-C Materials: A Structure—Selectivity Study." *ACS Catalysis* **7**(3): 1520–1525.

Huang, J., Y. Li, R.-K. Huang, C.-T. He, L. Gong, Q. Hu, L. Wang, Y.-T. Xu, X.-Y. Tian, S.-Y. Liu, Z.-M. Ye, F. Wang, D.-D. Zhou, W.-X. Zhang and J.-P. Zhang (2018). "Electrochemical Exfoliation of Pillared-Layer Metal—Organic Framework to Boost the Oxygen Evolution Reaction." *Angewandte Chemie International Edition* **57**(17): 4632–4636.

Huang, Z.-F., J. Song, K. Li, M. Tahir, Y.-T. Wang, L. Pan, L. Wang, X. Zhang and J.-J. Zou (2016). "Hollow Cobalt-Based Bimetallic Sulfide Polyhedra for Efficient All-pH-Value Electrochemical and Photocatalytic Hydrogen Evolution." *Journal of the American Chemical Society* **138**(4): 1359–1365.

Jahan, M., Z. Liu and K. P. Loh (2013). "A Graphene Oxide and Copper-Centered Metal Organic Framework Composite as a Tri-Functional Catalyst for HER, OER, and ORR." *Advanced Functional Materials* **23**(43): 5363–5372.

Jiang, X., H. Wu, S. Chang, R. Si, S. Miao, W. Huang, Y. Li, G. Wang and X. Bao (2017). "Boosting CO_2 Electroreduction Over Layered Zeolitic Imidazolate Frameworks Decorated with Ag_2O Nanoparticles." *Journal of Materials Chemistry A* **5**(36): 19371–19377.

Jiao, L., G. Wan, R. Zhang, H. Zhou, S.-H. Yu and H.-L. Jiang (2018). "From Metal—Organic Frameworks to Single-Atom Fe Implanted N-doped Porous Carbons: Efficient Oxygen Reduction in Both Alkaline and Acidic Media." *Angewandte Chemie International Edition* **57**(28): 8525–8529.

Kang, X., L. Li, A. Sheveleva, X. Han, J. Li, L. Liu, F. Tuna, E. J. L. McInnes, B. Han, S. Yang and M. Schröder (2020). "Electro-Reduction of Carbon Dioxide at Low Over-Potential at a Metal—Organic Framework Decorated Cathode." *Nature Communications* **11**(1): 5464.

Kang, X., Q. Zhu, X. Sun, J. Hu, J. Zhang, Z. Liu and B. Han (2016). "Highly Efficient Electrochemical Reduction of CO_2 to CH_4 in an Ionic Liquid using a Metal—Organic Framework Cathode." *Chemical Science* **7**(1): 266–273.

Khrizanforova, V., R. Shekurov, V. Miluykov, M. Khrizanforov, V. Bon, S. Kaskel, A. Gubaidullin, O. Sinyashin and Y. Budnikova (2020). "3D Ni and Co Redox-Active Metal—Organic Frameworks based on Ferrocenyl Diphosphinate and 4,4′-Bipyridine Ligands as Efficient Electrocatalysts for the Hydrogen Evolution Reaction." *Dalton Transactions* **49**(9): 2794–2802.

Kitagawa, S., R. Kitaura and S.-I. Noro (2004). "Functional Porous Coordination Polymers." *Angewandte Chemie International Edition* **43**(18): 2334–2375.

Kornienko, N., Y. Zhao, C. S. Kley, C. Zhu, D. Kim, S. Lin, C. J. Chang, O. M. Yaghi and P. Yang (2015). "Metal—Organic Frameworks for Electrocatalytic Reduction of Carbon Dioxide." *Journal of the American Chemical Society* **137**(44): 14129–14135.

Kortlever, R., J. Shen, K. J. P. Schouten, F. Calle-Vallejo and M. T. M. Koper (2015). "Catalysts and Reaction Pathways for the Electrochemical Reduction of Carbon Dioxide." *Journal of Physical Chemistry Letters* **6**(20): 4073–4082.

Kuang, M., Q. Wang, P. Han and G. Zheng (2017). "Cu, Co-Embedded N-Enriched Mesoporous Carbon for Efficient Oxygen Reduction and Hydrogen Evolution Reactions." *Advanced Energy Materials* **7**(17): 1700193.

Kulkarni, A., S. Siahrostami, A. Patel and J. K. Nørskov (2018). "Understanding Catalytic Activity Trends in the Oxygen Reduction Reaction." *Chemical Reviews* **118**(5): 2302–2312.

Lefèvre, M., E. Proietti, F. Jaouen and J.-P. Dodelet (2009). "Iron-Based Catalysts with Improved Oxygen Reduction Activity in Polymer Electrolyte Fuel Cells." *Science* **324**(5923): 71.

Li, F., G.-F. Han, H.-J. Noh, S.-J. Kim, Y. Lu, H. Y. Jeong, Z. Fu and J.-B. Baek (2018). "Boosting Oxygen Reduction Catalysis with Abundant Copper Single Atom Active Sites." *Energy & Environmental Science* **11**(8): 2263–2269.

Li, F.-L., P. Wang, X. Huang, D. J. Young, H.-F. Wang, P. Braunstein and J.-P. Lang (2019). "Large-Scale, Bottom-Up Synthesis of Binary Metal—Organic Framework Nanosheets for Efficient Water Oxidation." *Angewandte Chemie International Edition* **58**(21): 7051–7056.

Li, H., M. Eddaoudi, M. O'Keeffe and O. M. Yaghi (1999). "Design and Synthesis of an Exceptionally Stable and Highly Porous Metal-Organic Framework." *Nature* **402**(6759): 276–279.

Li, J., M. Chen, D. A. Cullen, S. Hwang, M. Wang, B. Li, K. Liu, S. Karakalos, M. Lucero, H. Zhang, C. Lei, H. Xu, G. E. Sterbinsky, Z. Feng, D. Su, K. L. More, G. Wang, Z. Wang and G. Wu (2018). "Atomically Dispersed Manganese Catalysts for Oxygen Reduction in Proton-Exchange Membrane Fuel Cells." *Nature Catalysis* **1**(12): 935–945.

Li, J., W. Huang, M. Wang, S. Xi, J. Meng, K. Zhao, J. Jin, W. Xu, Z. Wang, X. Liu, Q. Chen, L. Xu, X. Liao, Y. Jiang, K. A. Owusu, B. Jiang, C. Chen, D. Fan, L. Zhou and L. Mai (2019). "Low-Crystalline Bimetallic Metal—Organic Framework Electrocatalysts with Rich Active Sites for Oxygen Evolution." *ACS Energy Letters* **4**(1): 285–292.

Li, L., J. He, Y. Wang, X. Lv, X. Gu, P. Dai, D. Liu and X. Zhao (2019). "Metal—Organic Frameworks: A Promising Platform for Constructing Non-Noble Electrocatalysts for the Oxygen-Reduction Reaction." *Journal of Materials Chemistry A* **7**(5): 1964–1988.

Li, W., F. Li, H. Yang, X. Wu, P. Zhang, Y. Shan and L. Sun (2019). "A Bio-Inspired Coordination Polymer as Outstanding Water Oxidation Catalyst Via Second Coordination Sphere Engineering." *Nature Communications* **10**(1): 5074.

Li, X., D.-D. Ma, C. Cao, R. Zou, Q. Xu, X.-T. Wu and Q.-L. Zhu (2019). "Inlaying Ultrathin Bimetallic MOF Nanosheets into 3D Ordered Macroporous Hydroxide for Superior Electrocatalytic Oxygen Evolution." *Small* **15**(35): 1902218.

Li, X.-F., M.-Y. Lu, H.-Y. Yu, T.-H. Zhang, J. Liu, J.-H. Tian and R. Yang (2019). "Copper-Metal Organic Frameworks Electrodeposited on Carbon Paper as an Enhanced Cathode for the Hydrogen Evolution Reaction." *ChemElectroChem* **6**(17): 4507–4510.

Liang, S., R. Chen, P. Yu, M. Ni, Q. Zhang, X. Zhang and W. Yang (2017). "Ionically Dispersed Fe(ii)—N and Zn(ii)—N in Porous Carbon for Acidic Oxygen Reduction Reactions." *Chemical Communications* **53**(83): 11453–11456.

Liu, Q., X. Liu, L. Zheng and J. Shui (2018). "The Solid-Phase Synthesis of an Fe-N-C Electrocatalyst for High-Power Proton-Exchange Membrane Fuel Cells." *Angewandte Chemie International Edition* **57**(5): 1204–1208.

Ma, S., G. A. Goenaga, A. V. Call and D.-J. Liu (2011). "Cobalt Imidazolate Framework as Precursor for Oxygen Reduction Reaction Electrocatalysts." *Chemistry—A European Journal* **17**(7): 2063–2067.

Micheroni, D., G. Lan and W. Lin (2018). "Efficient Electrocatalytic Proton Reduction with Carbon Nanotube-Supported Metal—Organic Frameworks." *Journal of the American Chemical Society* **140**(46): 15591–15595.

Moosavi, S. M., A. Nandy, K. M. Jablonka, D. Ongari, J. P. Janet, P. G. Boyd, Y. Lee, B. Smit and H. J. Kulik (2020). "Understanding the Diversity of the Metal-Organic Framework Ecosystem." *Nature Communications* **11**(1): 4068.

Nam, D.-H., O. S. Bushuyev, J. Li, P. De Luna, A. Seifitokaldani, C.-T. Dinh, F. P. García de Arquer, Y. Wang, Z. Liang, A. H. Proppe, C. S. Tan, P. Todorović, O. Shekhah, C. M. Gabardo, J. W. Jo, J. Choi, M.-J. Choi, S.-W. Baek, J. Kim, D. Sinton, S. O. Kelley, M. Eddaoudi and E. H. Sargent (2018). "Metal—Organic Frameworks Mediate Cu Coordination for Selective CO_2 Electroreduction." *Journal of the American Chemical Society* **140**(36): 11378–11386.

Nam, D.-H., O. Shekhah, G. Lee, A. Mallick, H. Jiang, F. Li, B. Chen, J. Wicks, M. Eddaoudi and E. H. Sargent (2020). "Intermediate Binding Control Using Metal—Organic Frameworks Enhances Electrochemical CO_2 Reduction." *Journal of the American Chemical Society* **142**(51): 21513–21521.

Qian, Q., Y. Li, Y. Liu, L. Yu and G. Zhang (2019). "Ambient Fast Synthesis and Active Sites Deciphering of Hierarchical Foam-Like Trimetal—Organic Framework Nanostructures as a Platform for Highly Efficient Oxygen Evolution Electrocatalysis." *Advanced Materials* **31**(23): 1901139.

Qiao, J., Y. Liu, F. Hong and J. Zhang (2014). "A Review of Catalysts for the Electroreduction of Carbon Dioxide to Produce Low-Carbon Fuels." *Chemical Society Reviews* **43**(2): 631–675.

Qin, J.-S., D.-Y. Du, W. Guan, X.-J. Bo, Y.-F. Li, L.-P. Guo, Z.-M. Su, Y.-Y. Wang, Y.-Q. Lan and H.-C. Zhou (2015). "Ultrastable Polymolybdate-Based Metal—Organic Frameworks as Highly Active Electrocatalysts for Hydrogen Generation from Water." *Journal of the American Chemical Society* **137**(22): 7169–7177.

Qiu, Y.-L., H.-X. Zhong, T.-T. Zhang, W.-B. Xu, P.-P. Su, X.-F. Li and H.-M. Zhang (2018). "Selective Electrochemical Reduction of Carbon Dioxide Using Cu Based Metal Organic Framework for CO_2 Capture." *ACS Applied Materials & Interfaces* **10**(3): 2480–2489.

Senthil Kumar, R., S. Senthil Kumar and M. Anbu Kulandainathan (2012). "Highly Selective Electrochemical Reduction of Carbon Dioxide using Cu Based Metal Organic Framework as an Electrocatalyst." *Electrochemistry Communications* **25**: 70–73.

Senthil Raja, D., X.-F. Chuah and S.-Y. Lu (2018). "In Situ Grown Bimetallic MOF-Based Composite as Highly Efficient Bifunctional Electrocatalyst for Overall Water Splitting with Ultrastability at High Current Densities." *Advanced Energy Materials* **8**(23): 1801065.

Shao, P., L. C. Yi, S. M. Chen, T. H. Zhou and J. Zhang (2020). "Metal-Organic Frameworks for Electrochemical Reduction of Carbon Dioxide: The Role of Metal Centers." *Journal of Energy Chemistry* **40**: 156–170.

Shekurov, R., V. Khrizanforova, L. Gilmanova, M. Khrizanforov, V. Miluykov, O. Kataeva, Z. Yamaleeva, T. Burganov, T. Gerasimova, A. Khamatgalimov, S. Katsyuba, V. Kovalenko, Y. Krupskaya, V. Kataev, B. Büchner, V. Bon, I. Senkovska, S. Kaskel, A. Gubaidullin, O. Sinyashin and Y. Budnikova (2019). "Zn and Co Redox Active Coordination Polymers as Efficient Electrocatalysts." *Dalton Transactions* **48**(11): 3601–3609.

Shi, Q., S. Fu, C. Zhu, J. Song, D. Du and Y. Lin (2019). "Metal—Organic Frameworks-Based Catalysts for Electrochemical Oxygen Evolution." *Materials Horizons* **6**(4): 684–702.

Singh, C., S. Mukhopadhyay and I. Hod (2021). "Metal—Organic Framework Derived Nanomaterials for Electrocatalysis: Recent Developments for CO_2 and N_2 Reduction." *Nano Convergence* **8**(1): 1.

Su, X., Y. Wang, J. Zhou, S. Gu, J. Li and S. Zhang (2018). "Operando Spectroscopic Identification of Active Sites in NiFe Prussian Blue Analogues as Electrocatalysts: Activation of Oxygen Atoms for Oxygen Evolution Reaction." *Journal of the American Chemical Society* **140**(36): 11286–11292.

Tabassum, H., W. Guo, W. Meng, A. Mahmood, R. Zhao, Q. Wang and R. Zou (2017). "Metal—Organic Frameworks Derived Cobalt Phosphide Architecture Encapsulated into B/N Co-Doped Graphene Nanotubes for All pH Value Electrochemical Hydrogen Evolution." *Advanced Energy Materials* **7**(9): 1601671.

Tang, H., S. Cai, S. Xie, Z. Wang, Y. Tong, M. Pan and X. Lu (2016). "Metal—Organic-Framework-Derived Dual Metal- and Nitrogen-Doped Carbon as Efficient and Robust Oxygen Reduction Reaction Catalysts for Microbial Fuel Cells." *Advanced Science* **3**(2): 1500265.

Wang, H.-F., L. Chen, H. Pang, S. Kaskel and Q. Xu (2020). "MOF-Derived Electrocatalysts for Oxygen Reduction, Oxygen Evolution and Hydrogen Evolution Reactions." *Chemical Society Reviews* **49**(5): 1414–1448.

Wang, J., Z. Huang, W. Liu, C. Chang, H. Tang, Z. Li, W. Chen, C. Jia, T. Yao, S. Wei, Y. Wu and Y. Li (2017). "Design of N-Coordinated Dual-Metal Sites: A Stable and Active Pt-Free Catalyst for Acidic Oxygen Reduction Reaction." *Journal of the American Chemical Society* **139**(48): 17281–17284.

Wang, Q. and D. Astruc (2020). "State of the Art and Prospects in Metal—Organic Framework (MOF)-Based and MOF-Derived Nanocatalysis." *Chemical Reviews* **120**(2): 1438–1511.

Wang, R., X.-Y. Dong, J. Du, J.-Y. Zhao and S.-Q. Zang (2018). "MOF-Derived Bifunctional Cu3P Nanoparticles Coated by a N,P-Codoped Carbon Shell for Hydrogen Evolution and Oxygen Reduction." *Advanced Materials* **30**(6): 1703711.

Wang, S.-S., L. Jiao, Y. Qian, W.-C. Hu, G.-Y. Xu, C. Wang and H.-L. Jiang (2019). "Boosting Electrocatalytic Hydrogen Evolution over Metal—Organic Frameworks by Plasmon-Induced Hot-Electron Injection." *Angewandte Chemie International Edition* **58**(31): 10713–10717.

Wang, X., Z. Chen, X. Zhao, T. Yao, W. Chen, R. You, C. Zhao, G. Wu, J. Wang, W. Huang, J. Yang, X. Hong, S. Wei, Y. Wu and Y. Li (2018). "Regulation of Coordination Number over Single Co Sites: Triggering the Efficient Electroreduction of CO_2." *Angewandte Chemie International Edition* **57**(7): 1944–1948.

Wang, X., H. Zhang, H. Lin, S. Gupta, C. Wang, Z. Tao, H. Fu, T. Wang, J. Zheng, G. Wu and X. Li (2016). "Directly Converting Fe-Doped Metal—Organic Frameworks into Highly Active and Stable Fe-N-C Catalysts for Oxygen Reduction in Acid." *Nano Energy* **25**: 110–119.

Wang, X.-L., L.-Z. Dong, M. Qiao, Y.-J. Tang, J. Liu, Y. Li, S.-L. Li, J.-X. Su and Y.-Q. Lan (2018). "Exploring the Performance Improvement of the Oxygen Evolution Reaction in a Stable Bimetal—Organic Framework System." *Angewandte Chemie International Edition* **57**(31): 9660–9664.

Wang, Y., P. Hou, Z. Wang and P. Kang (2017). "Zinc Imidazolate Metal—Organic Frameworks (ZIF-8) for Electrochemical Reduction of CO_2 to CO." *ChemPhysChem* **18**(22): 3142–3147.

Wang, Y.-R., Q. Huang, C.-T. He, Y. Chen, J. Liu, F.-C. Shen and Y.-Q. Lan (2018). "Oriented Electron Transmission in Polyoxometalate-Metalloporphyrin Organic Framework for Highly Selective Electroreduction of CO_2." *Nature Communications* **9**(1): 4466.

Wu, R., Y. Xue, B. Liu, K. Zhou, J. Wei and S. H. Chan (2016). "Cobalt Diselenide Nanoparticles Embedded within Porous Carbon Polyhedra as Advanced Electrocatalyst for Oxygen Reduction Reaction." *Journal of Power Sources* **330**: 132–139.

Wu, Y.-P., W. Zhou, J. Zhao, W.-W. Dong, Y.-Q. Lan, D.-S. Li, C. Sun and X. Bu (2017). "Surfactant-Assisted Phase-Selective Synthesis of New Cobalt MOFs and Their Efficient Electrocatalytic Hydrogen Evolution Reaction." *Angewandte Chemie International Edition* **56**(42): 13001–13005.

Xia, W., A. Mahmood, Z. Liang, R. Zou and S. Guo (2016). "Earth-Abundant Nanomaterials for Oxygen Reduction." *Angewandte Chemie International Edition* **55**(8): 2650–2676.

Xia, W., J. Zhu, W. Guo, L. An, D. Xia and R. Zou (2014). "Well-Defined Carbon Polyhedrons Prepared from Nano Metal—Organic Frameworks for Oxygen Reduction." *Journal of Materials Chemistry A* **2**(30): 11606–11613.

Xu, M., L. Han, Y. Han, Y. Yu, J. Zhai and S. Dong (2015). "Porous CoP Concave Polyhedron Electrocatalysts Synthesized from Metal—Organic Frameworks with Enhanced Electrochemical Properties for Hydrogen Evolution." *Journal of Materials Chemistry A* **3**(43): 21471–21477.

Xu, Y., B. Li, S. Zheng, P. Wu, J. Zhan, H. Xue, Q. Xu and H. Pang (2018). "Ultrathin Two-Dimensional Cobalt—Organic Framework Nanosheets for High-Performance Electrocatalytic Oxygen Evolution." *Journal of Materials Chemistry A* **6**(44): 22070–22076.

Xue, Z., Y. Li, Y. Zhang, W. Geng, B. Jia, J. Tang, S. Bao, H.-P. Wang, Y. Fan, Z.-w. Wei, Z. Zhang, Z. Ke, G. Li and C.-Y. Su (2018). "Modulating Electronic Structure of Metal-Organic Framework for Efficient Electrocatalytic Oxygen Evolution." *Advanced Energy Materials* **8**(29): 1801564.

Yaghi, O. M., M. O'Keeffe, N. W. Ockwig, H. K. Chae, M. Eddaoudi and J. Kim (2003). "Reticular Synthesis and the Design of New Materials." *Nature* **423**(6941): 705–714.

Ye, L., J. Liu, Y. Gao, C. Gong, M. Addicoat, T. Heine, C. Wöll and L. Sun (2016). "Highly Oriented MOF Thin Film-Based Electrocatalytic Device for the Reduction of CO_2 to CO Exhibiting High Faradaic Efficiency." *Journal of Materials Chemistry A* **4**(40): 15320–15326.

Ye, Y., F. Cai, H. Li, H. Wu, G. Wang, Y. Li, S. Miao, S. Xie, R. Si, J. Wang and X. Bao (2017). "Surface Functionalization of ZIF-8 with Ammonium Ferric Citrate Toward High Exposure of Fe-N Active Sites for Efficient Oxygen and Carbon Dioxide Electroreduction." *Nano Energy* **38**: 281–289.

Yin, P., T. Yao, Y. Wu, L. Zheng, Y. Lin, W. Liu, H. Ju, J. Zhu, X. Hong, Z. Deng, G. Zhou, S. Wei and Y. Li (2016). "Single Cobalt Atoms with Precise N-Coordination as Superior Oxygen Reduction Reaction Catalysts." *Angewandte Chemie International Edition* **55**(36): 10800–10805.

Zaman, N., T. Noor and N. Iqbal (2021). "Recent Advances in the Metal—Organic Framework-Based Electrocatalysts for the Hydrogen Evolution Reaction in Water Splitting: A Review." *RSC Advances* **11**(36): 21904–21925.

Zhang, C., B. An, L. Yang, B. Wu, W. Shi, Y.-C. Wang, L.-S. Long, C. Wang and W. Lin (2016). "Sulfur-Doping Achieves Efficient Oxygen Reduction in Pyrolyzed Zeolitic Imidazolate Frameworks." *Journal of Materials Chemistry A* **4**(12): 4457–4463.

Zhang, D., W. Chen, Z. Li, Y. Chen, L. Zheng, Y. Gong, Q. Li, R. Shen, Y. Han, W.-C. Cheong, L. Gu and Y. Li (2018). "Isolated Fe and Co Dual Active Sites on Nitrogen-Doped Carbon for a Highly Efficient Oxygen Reduction Reaction." *Chemical Communications* **54**(34): 4274–4277.

Zhang, H., S. Hwang, M. Wang, Z. Feng, S. Karakalos, L. Luo, Z. Qiao, X. Xie, C. Wang, D. Su, Y. Shao and G. Wu (2017). "Single Atomic Iron Catalysts for Oxygen Reduction in Acidic Media: Particle Size Control and Thermal Activation." *Journal of the American Chemical Society* **139**(40): 14143–14149.

Zhang, J., W. Cai, F. X. Hu, H. Yang and B. Liu (2021). "Recent Advances in Single Atom Catalysts for the Electrochemical Carbon Dioxide Reduction Reaction." *Chemical Science* **12**(20): 6800–6819.

Zhang, L., S. Li, C. J. Gómez-García, H. Ma, C. Zhang, H. Pang and B. Li (2018). "Two Novel Polyoxometalate-Encapsulated Metal—Organic Nanotube Frameworks as Stable and Highly Efficient Electrocatalysts for Hydrogen Evolution Reaction." *ACS Applied Materials & Interfaces* **10**(37): 31498–31504.

Zhang, W., Y. Hu, L. Ma, G. Zhu, Y. Wang, X. Xue, R. Chen, S. Yang and Z. Jin (2018). "Progress and Perspective of Electrocatalytic CO_2 Reduction for Renewable Carbonaceous Fuels and Chemicals." *Advanced Science* **5**(1): 1700275.

Zhao, C., X. Dai, T. Yao, W. Chen, X. Wang, J. Wang, J. Yang, S. Wei, Y. Wu and Y. Li (2017). "Ionic Exchange of Metal—Organic Frameworks to Access Single Nickel Sites for Efficient Electroreduction of CO_2." *Journal of the American Chemical Society* **139**(24): 8078–8081.

Zhao, K., Y. Liu, X. Quan, S. Chen and H. Yu (2017). "CO2 Electroreduction at Low Overpotential on Oxide-Derived Cu/Carbons Fabricated from Metal Organic Framework." *ACS Applied Materials & Interfaces* **9**(6): 5302–5311.

Zhao, S., H. Yin, L. Du, L. He, K. Zhao, L. Chang, G. Yin, H. Zhao, S. Liu and Z. Tang (2014). "Carbonized Nanoscale Metal—Organic Frameworks as High Performance Electrocatalyst for Oxygen Reduction Reaction." *ACS Nano* **8**(12): 12660–12668.

Zhao, S., Y. Wang, J. Dong, C.-T. He, H. Yin, P. An, K. Zhao, X. Zhang, C. Gao, L. Zhang, J. Lv, J. Wang, J. Zhang, A. M. Khattak, N. A. Khan, Z. Wei, J. Zhang, S. Liu, H. Zhao and Z. Tang (2016). "Ultrathin Metal—Organic Framework Nanosheets for Electrocatalytic Oxygen Evolution." *Nature Energy* **1**(12): 16184.

Zhao, X., B. Pattengale, D. Fan, Z. Zou, Y. Zhao, J. Du, J. Huang and C. Xu (2018). "Mixed-Node Metal—Organic Frameworks as Efficient Electrocatalysts for Oxygen Evolution Reaction." *ACS Energy Letters* **3**(10): 2520–2526.

Zhou, W., D.-D. Huang, Y.-P. Wu, J. Zhao, T. Wu, J. Zhang, D.-S. Li, C. Sun, P. Feng and X. Bu (2019). "Stable Hierarchical Bimetal—Organic Nanostructures as High Performance Electrocatalysts for the Oxygen Evolution Reaction." *Angewandte Chemie International Edition* **58**(13): 4227–4231.

Zhu, R., J. Ding, Y. Xu, J. Yang, Q. Xu and H. Pang (2018). "π-Conjugated Molecule Boosts Metal—Organic Frameworks as Efficient Oxygen Evolution Reaction Catalysts." *Small* **14**(50): 1803576.

Zhuang, L., L. Ge, H. Liu, Z. Jiang, Y. Jia, Z. Li, D. Yang, R. K. Hocking, M. Li, L. Zhang, X. Wang, X. Yao and Z. Zhu (2019). "A Surfactant-Free and Scalable General Strategy for Synthesizing Ultrathin Two-Dimensional Metal—Organic Framework Nanosheets for the Oxygen Evolution Reaction." *Angewandte Chemie International Edition* **58**(38): 13565–13572.

Index

Note: Page numbers in *italics* refer to figures, those in **bold** refer to tables.

For Product Safety Concerns and Information please contact our EU
representative GPSR@taylorandfrancis.com
Taylor & Francis Verlag GmbH, Kaufingerstraße 24, 80331 München, Germany